INTELLIGENT SYSTEMS

Technology and Applications

VOLUME III

Signal, Image, and Speech Processing

Edited by
Cornelius T. Leondes

INTELLIGENT SYSTEMS

Technology and Applications

VOLUME III

Signal, Image, and Speech Processing

CRC PRESS

Boca Raton   London   New York   Washington, D.C.

Library of Congress Cataloging-in-Publication Data

Intelligent systems : technology and applications / edited by Cornelius T. Leondes.
    p. cm.
    Includes bibliographical references and index.
    Contents: v. 1. Implementation techniques -- v. 2. Fuzzy systems, neural networks, and expert systems -- v. 3. Signal, image, and speech processing -- v. 4. Database and learning systems -- v. 5. Manufacturing, industrial, and management systems -- v. 6. Control and electric power systems.
    ISBN 0-8493-1121-7 (alk. paper)
    1. Intelligent control systems. I. Leondes, Cornelius T.

TJ217.5 .I5448 2002
629.8--dc21
                                                        2002017473

Visit the CRC Press Web site at www.crcpress.com

Foreword

Intelligent Systems: Technology and Applications is a significant contribution to the artificial intelligence (AI) field. Edited by Professor Cornelius Leondes, a leading contributor to intelligent systems, this set of six well-integrated volumes on the subject of intelligent systems techniques and applications provides a valuable reference for researchers and practitioners. This landmark work features contributions from more than 60 of the world's foremost AI authorities in industry, government, and academia.

Perhaps the most valuable feature of this work is the breadth of material covered. Volume I looks at the steps in implementing intelligent systems. Here the reader learns from some of the leading individuals in the field how to develop an intelligent system. Volume II covers the most important technologies in the field, including fuzzy systems, neural networks, and expert systems. In this volume the reader sees the steps taken to effectively develop each type of system, and also sees how these technologies have been successfully applied to practical real-world problems, such as intelligent signal processing, robotic control, and the operation of telecommunications systems. The final four volumes provide insight into developing and deploying intelligent systems in a wide range of application areas. For instance, Volume III discusses applications of signal, image, and speech processing; Volume IV looks at intelligent database management and learning systems; Volume V covers manufacturing, industrial, and business applications; and Volume VI considers applications in control and power systems. Collectively this material provides a tremendous resource for developing an intelligent system across a wide range of technologies and application areas.

Let us consider this work in the context of the history of artificial intelligence. AI has come a long way in a relatively short time. The early days were spent in somewhat of a probing fashion, where researchers looked for ways to develop a machine that captured human intelligence. After considerable struggle, they fortunately met with success. Armed with an understanding of how to design an intelligent system, they went on to develop useful applications to solve real-world problems. At this point AI took on a very meaningful role in the area of information technology.

Along the way there were a few individuals who saw the importance of publishing the accomplishments of AI and providing guidance to advance the field. Among this small group I believe that Dr. Leondes has made the largest contribution to this effort. He has edited numerous books on intelligent systems that provide a wealth of information to individuals in the field. I believe his latest work discussed here is his most valuable contribution to date and should be in the possession of all individuals involved in the field of intelligent systems.

Jack Durkin

Preface

For most of our history the wealth of a nation was limited by the size and stamina of the work force. Today, national wealth is measured in intellectual capital. Nations possessing skillful people in such diverse areas as science, medicine, business, and engineering, produce innovations that drive the nations to a higher quality of life. To better utilize these valuable resources, intelligent systems technology has evolved at a rapid and significantly expanding rate to accomplish this purpose. Intelligent systems technology can be utilized by nations to improve their medical care, advance their engineering technology, and increase their manufacturing productivity, as well as play a significant role in a very wide variety of other areas of activity of substantive significance.

Intelligent systems technology almost defines itself as the replication to some effective degree of human intelligence by the utilization of computers, sensor systems, effective algorithms, software technology, and other technologies in the performance of useful or significant tasks. Widely publicized earlier examples include the defeat of Garry Kasparov, arguably the greatest chess champion in history, by IBM's intelligent system known as "Big Blue." Separately, the greatest stock market crash in history, which took place on Monday, October 19, 1987, occurred because of a poorly designed intelligent system known as computerized program trading. As was reported, the Wall Street stockbrokers watched in a state of shock as the computerized program trading system took complete control of the events of the day. Alternatively, a significant example where no intelligent system was in place and which could have, indeed no doubt would have, prevented a disaster is the Chernobyl disaster which occurred at 1:15 A.M. on April 26, 1987. In this case the system operators were no doubt in a rather tired state and an effectively designed class of intelligent system known as "backward chaining" EXPERT System would, in all likelihood, have averted this disaster.

The techniques which are utilized to implement Intelligent Systems Technology include, among others:

Knowledge-Based Systems Techniques
EXPERT Systems Techniques
Fuzzy Theory Systems
Neural Network Systems
Case-Based Reasoning Methods
Induction Methods
Frame-Based Techniques
Cognition System Techniques

These techniques and others may be utilized individually or in combination with others.

The breadth of the major application areas of intelligent systems technology is remarkable and very impressive. These include:

Agriculture
Business
Chemistry
Communications
Computer Systems
Education
Electronics
Engineering
Environment
Geology
Image Processing
Information Management

Law
Manufacturing
Mathematics
Medicine
Meteorology
Military
Mining
Power Systems
Science
Space Technology
Transportation

It is difficult now to find an area that has not been touched by Intelligent Systems Technology. Indeed, a perusal of the tables of contents of these six volumes, *Intelligent Systems: Technology and Applications*, reveals that there are substantively significant treatments of applications in many of these areas.

Needless to say, the great breadth and expanding significance of this field on the international scene requires a multi-volume set for an adequately substantive treatment of the subject of intelligent systems technology. This set of volumes consists of six distinctly titled and well-integrated volumes. It is appropriate to mention that each of the six volumes can be utilized individually. In any event, the six volume titles are:

1. Implementation Techniques
2. Fuzzy Systems, Neural Networks, and Expert Systems
3. Signal, Image, and Speech Processing
4. Database and Learning Systems
5. Manufacturing, Industrial, and Management Systems
6. Control and Electric Power Systems

The contributors to these volumes clearly reveal the effectiveness and great significance of the techniques available and, with further development, the essential role that they will play in the future. I hope that practitioners, research workers, students, computer scientists, and others on the international scene will find this set of volumes to be a unique and significant reference source for years to come.

Cornelius T. Leondes
Editor

About the Editor

Cornelius T. Leondes, B.S., M.S., Ph.D., Emeritus Professor, School of Engineering and Applied Science, University of California, Los Angeles, has served as a member or consultant on numerous national technical and scientific advisory boards. Dr. Leondes served as a consultant for numerous Fortune 500 companies and international corporations. He has published over 200 technical journal articles and has edited and/or co-authored more than 120 books. Dr. Leondes is a Guggenheim Fellow, Fulbright Research Scholar, and IEEE Fellow, as well as a recipient of the IEEE Baker Prize award and the Barry Carlton Award of the IEEE.

Contributors

L. Andreadis
Democritus University of Thrace
Xanthi, Greece

Kenneth E. Barner
University of Delaware
Newark, Delaware

Douglas R Campbell
University of Paisley
Scotland, UK

Henrik Flyvbjerg
Risø National Laboratory
Roskilde, Denmark

Gian Luca Foresti
University of Udine
Udine, Italy

Colin Fyfe
University of Paisley
Scotland, UK

Mark Girolami
University of Paisley
Scotland, UK

Leontios J. Hadjileontiadis
Aristotle University of Thessaloniki
Thessaloniki, Greece

John G. Harris
University of Florida
Gainesville, Florida

Qianhui Liang
University of Florida
Gainesville, Florida

G. Louverdis
Democritus University of Thrace
Xanthi, Greece

Hirobumi Nishida
Ricoh Software Research Center
Tokyo, Japan

Stavros M. Panas
Aristotle University of Thessaloniki
Thessaloniki, Greece

Björn O. Peters
John von Neumann Institute for
  Computing
Jülich, Germany

Gert Pfurtscheller
Institute of Biomedical Engineering
Graz, Austria

Chai Quek
Nanyang Technological University
Singapore

Yannis A. Tolias
Aristotle University of Thessaloniki
Thessaloniki, Greece

Ph. Tsalides
Democritus University of Thrace
Xanthi, Greece

M.I. Vardavoulia
Democritus University of Thrace
Xanthi, Greece

Abdul Wahab
Nanyang Technological University
Singapore

Contents

Volume III: Signal, Image, and Speech Processing

Contents

Volume I: Implementation Techniques

Contents

Volume II: Fuzzy Systems, Neural Networks, and Expert Systems

Contents

Volume IV: Database and Learning Systems

Contents

Volume V: Manufacturing, Industrial, and Management Systems

Contents

1

Artificial Intelligence Systems Techniques and Applications in Speech Processing

Douglas R. Campbell
University of Paisley

Colin Fyfe
University of Paisley

Mark Girolami
University of Paisley

1.1 Introduction

Speech is a complex signal from which we extract many types of information, from the message content and meaning, to the nature of the transmission medium, to the identity and condition of the speaker. How well we perform these tasks depends on the quality of the speech signal, the efficacy of our hearing, the nature of the listening environment, and our accumulated auditory experience.

Because of its importance in human communications, and since we do not have to consciously direct our attention (or even be awake) to detect it, sound and in particular, speech, provides a channel for communication which combines a degree of immediacy with the release of sight and touch for other tasks.

The commercial incentive to develop spoken language interfaces has stimulated substantial worldwide research and development activity, e.g., automatic speech recognition (ASR) for voice command and automatic dictation (speech-to-text), synthetic speech generation for voice response and automatic reading (text-to-speech), automatic spoken language translation (speech-to-text-to-text-to-speech), and speaker recognition for security purposes.

Current active areas of research include:

- The development of reliable speech recognition, mobile and hands-free telecommunications terminals and hearing aids, robust to the presence of background noise and reverberation.
- The infusion of synthetic speech with cues to interpretation, by automatically generating the culturally appropriate inflections, and the infusion of synthetic speech with cues to emotional state, by automatically producing appropriate pitch, rate, rhythm, and inflections. These are necessary to the production of high quality synthetic speech, e.g., for text-to-speech systems which will not sound "robotic," or for the regeneration of stored speech or speech which has undergone high levels of data compression, such as are desired in modern communication systems.
- The development of speaker recognition to the point where it provides reliable support to other means of verifying identity, i.e., PIN, fingerprint, iris print, or facial measures.
- Speech recognition for pronunciation training.
- Accurate and reliable automated voice activity detection (VAD), speech segmentation, speech quality assessment, and pronunciation quality assessment.
- The separation of simultaneous speakers.

1.2   Aims

Two aims were at the forefront of the authors' minds in writing this chapter:

1. Speech processing research is necessarily a cross-disciplinary activity, usually carried out by persons trained within traditional disciplinary boundaries. Workers, especially new post-graduate students in this field, should have some degree of familiarity with the terminology, knowledge base, and some of the more esoteric results from the main contributory fields.
2. As information on the complexity of the problems has accrued, solutions have grown in sophistication, and relatively simplistic linear deterministic and adaptive approaches have been expanded to embrace stochastic and nonlinear schemes of which artificial intelligence (AI) approaches evolving from, or incorporating, artificial neural networks (ANNs) are particularly important, and encouragement should be given to further such work.

1.3   Outline

This chapter starts with an introduction to, and discussion of, the complex features of speech signals. It continues with a presentation and discussion of some important aspects of human hearing that may stimulate and inform engineering ideas for approaching the human facility in dealing with speech signals. Following this material, a condensed survey of speech enhancement strategies is presented which aims to provide background to the AI approaches later discussed. An introduction to artificial neural networks is then presented, with the features of the main schemes and application areas identified. This is followed by a review of some currently relevant applications of ANN-based or -derived paradigms, with the particular examination of an application to the speaker separation problem.

1.4 Caveat

In a relatively short chapter it would never be possible to do proper justice to the breadth of this field. We have therefore taken the opportunity to be unashamedly partial in selecting material in which we have particular interest, hoping it may stimulate the readers and encourage them to access the reference material.

1.5 Speech Signals

1.5.1 Characteristics of Speech

1.5.1.1 Generalities

Human speech is produced by a learned, coordinated process involving: drawing air through the larynx, varying the tension on the vocal cords, positioning the articulators of the vocal tract, and performing these actions under the conscious control of learned language skills. The character and frequency content of the basic sound source are altered by voluntary muscular control of the tension on the vocal cords. Thus are generated voiced sounds (e.g., "oo") having a periodic structure, or unvoiced sounds (e.g., "sh") having an aperiodic structure. This is the basis of sounds that are then amplified by the resonant features of the vocal tract and shaped by articulation. In engineering terms, this generates a spectral shaping which is characterized by identifiable peaks (formants). Articulation is the process of interrupting and shaping the sound signal from the larynx to form the basic sounds of speech (phonemes) and combining them to form words. The articulators used are the jaw, lips, soft palate, teeth, and tongue.

Although speech in different languages may use different formant structures and may exhibit different probability of occurrence of the various phonemes, several studies have shown that for long-term measures of conversational speech there is a great deal of similarity across many languages. The differences between individuals are usually much greater than the differences between languages. Byrne et al.[1] performed an international comparison of long-term average speech spectra (LTASS) over 13 languages (Figure III.1.1): Arabic, Cantonese, Danish, English (several dialects), French (Canadian), German, Japanese, Mandarin, Russian, Singhalese, Swedish, Vietnamese, and Welsh. To quote from their findings, "there appears to be no systematic separation between the English versus the non-English languages, or between the non-tonal versus the tonal languages"; thus, "it is reasonable to propose a universal LTASS which should be satisfactory for many purposes ... ".

The amplitude of vibration of the vocal cords and the frequency-dependent sustaining effect of acoustic resonances modifies the intensity of the voice. On average, the amplitude of conversational speech peaks at about 63 dB sound pressure level (SPL) in the frequency range 300 to 600 Hz. Figure III.1.2 shows the SPL in dB for speech at four levels of vocal effort plotted against frequency. It also shows the percentage importance of each frequency band to the understanding of speech averaged across several speakers and languages. Note that the importance is peaking around 3 kHz where the speech energy is relatively low. The variation of speech amplitude with time can be considered as a modulation process, and the modulation frequencies of speech lie in the range 0.25 to 25 Hz (Houtgast and Steeneken[2]), tending to peak around 3 to 6 Hz.

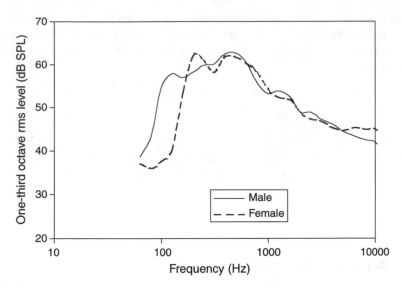

FIGURE III.1.1   Long-term conversational speech spectrum averaged over 12 languages. (Plotted from the tabulated data of Byrne, D. et al., *J. Acoust. Soc. Am.*, 96(4), 2104–2120, 1994.)

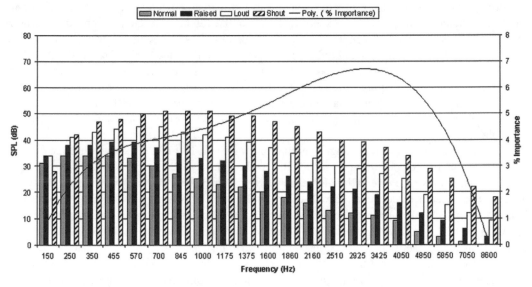

FIGURE III.1.2   Speech levels for four different vocal efforts, and importance to understanding, by frequency band. (Plotted from the tabulated data of Gatehouse and Robinson in Martin, M. (Ed.), *Speech Audiometry*, 2nd ed., Whorr Publ. Ltd., London, U.K., 1997.)

This type of information is of value in the design of the broad features of speech systems, e.g., the required dynamic range and bandwidth of a telephone system. However, many present-day problems in speech processing require more information on the fine structure and variability of speech signals and models of its production and perception.

1.5.1.2 Individuality and Variability

The frequency range of an individual voice is limited by the length and mass of the vocal cords. At puberty structural changes during growth result in most male voices deepening markedly relative to

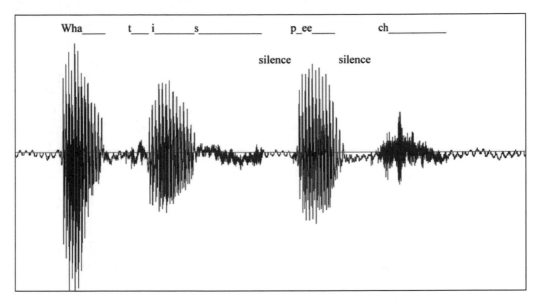

FIGURE III.1.3   Time history of phrase "What is speech" — duration 1.6 sec.

child and female voices. Additionally, the acoustic resonances of the air volumes of the chest, throat, oral and nasal cavities, and the sinuses of the skull, change as an individual grows, suffers infections, and ages, determining the quality of the individual voice at any time. Thus, speech processing systems have to accept a wide range of inter- and intra-talker variability in the speech signal.

As can be seen in Figure III.1.3, speech exhibits bursts of localized and differentiated activity; i.e., it is statistically nonstationary. However, examining a section more closely, as in Figure III.1.4, it can be considered to be relatively stationary in the short term, e.g., within a 25-msec time frame. It can also be seen that there are periods of no voice activity. There are two such silence periods, both about 90 msec, in the word "speech" shown in Figure III.1.3. In a telecommunication system, quiet periods are wasteful of system capacity and if they can be detected, then they can be utilized to release system resources for greater efficiency. This requires a reliable voice activity detector (VAD). A further step in refinement is required for ASR where end-point detection of words and isolation of phonemes is required. Since a word may have a silence period within it, more complex strategies are required in addition to VAD. In both these cases, overall system performance and acceptability to the user are dependent on the reliability of the detector.

The manner in which connected speech is delivered is under the control of the speaker. Typically the rate, rhythm, intensity, and precision of articulation will vary, depending on the speaker's emotional state, state of health, and environmental circumstances. When speaking in high noise levels (i.e., >80 dB SPL), speakers show reflex changes in pitch, formant frequency, duration, and intensity distribution across frequency, known as the Lombard effect (Junqua[3]).

Individuals vary one from another in the details of their vocal structures and the control they are capable of exercising, and each individual varies daily in emotional state and in state of health. Such is our familiarity with speech that normal-hearing individuals routinely use it to infer properties of the speaker from their speech; e.g., their emotional state, a practice capitalized on often to very convincing effect by actors. Vocal controls allow inflections that provide culturally specific cues to the required interpretation of spoken statements; e.g., the same words spoken with different inflection may indicate a statement or a question. In everyday communication we commonly use remembered voices as strong identifiers of individuals and compensate for the normal range of variations any

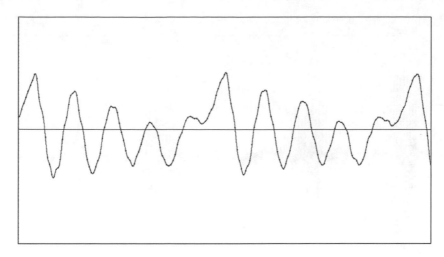

FIGURE III.1.4   Section of "*i*" in phrase "What is speech" — duration 25 msec.

individual voice may evidence. However, given the variability of the individual and the existence of good mimics, voice alone is not sufficient to provide the degree of security required for some identification purposes. The variability also impacts on the acceptability of ASR systems.

In addition to variability inherent in the speech production process, speech signals suffer environmental modifications in transmission to the listener/receiver. Background noise may seriously erode the signal-to-noise ratio (SNR) at a receiver, and this noise corruption may be additive (e.g., electrical noise on a telephone line) or multiplicative (e.g., modulation of the speech by interaction with vibrating structures such as a vehicle body). The process of transmitting the speech signal may introduce a convolutive distortion, e.g., the limited bandwidth of some telephone systems or security screen microphones. Reverberation effects are common in enclosed spaces (most people find an anechoic environment unpleasant), so speech processing systems will often be dealing with signals that have suffered reverberation, typically with T_{60} times in the range of 0.2 to 2 sec (Mackenzie[4]). Reverberation causes corruption of the clean speech signal with modified time-delayed echoes of itself and other sources. This is particularly difficult to deal with in the common situation where the speaker may be free to move within the reverberant environment, such as with hands-free telephones and hearing aids. Since the speaker is varying his/her distance from the reflecting surfaces, and corrupting sources such as competing background speakers may also be doing so, the acoustic transfer functions linking sound sources to receivers may be frequently changing.

Thus, an ASR system for an auto-teller may have to operate in situations such as an echoic shopping mall, a noisy street with traffic, or a quiet bank lobby. An airline ticket reservation system using ASR will have to be able to reliably process calls from mobile phones which may be used in acoustic environments as dissimilar as an automobile, a domestic living room, or a public park. Particularly for hands-free telephones, the behavior of the ASR with signals exhibiting reverberation components is an important issue.

1.5.2   Aspects of Hearing

The human auditory nervous system is considered by some to be the most complex sensory processing system in the body, and much of its functional mechanism is not yet understood. In spite, or perhaps because, of this, consideration of our ability to separate desired from undesired signals, and extract meaning from them, can stimulate engineering ideas for speech processing at both mechanism and system modeling levels. However, two conflicting viewpoints should be balanced:

1. "In Nature's infinite book of secrecy a little I can read" (Shakespeare; *Antony and Cleopatra*). Natural processes of animal development can be viewed as a huge uncontrolled experiment in evolutionary strategies, in which the fitness function is often hidden from, and changes in a manner unpredictable to, the observer, never mind the participants. In such a system, signs of convergence on a common solution to a problem faced by individuals across different genus, Family, Order, or Class of Animalia should be considered an important pointer at least to possible solution procedures or system structures, if not their engineering implementation.

2. "Accuse not Nature, she hath done her part, do thou but thine" (Milton; *Paradise Lost*). Without denying the place of detailed engineering models as vehicles for aiding understanding, e.g., in cochlear mechanics or neurological functioning, we should keep in mind when seeking engineering solutions that the biological solution to a problem may have no great validity when biological cost factors are removed. Objective equivalence is seldom essential for the required subjective (perceptual) equivalence.

Selected features of this complex signal processing system are now discussed.

1.5.2.1   Outer Ear to Auditory Nerve

In engineering terms, the peripheral processing functions from the outer ear to the auditory nerve leaving the cochlea are accepted as transduction, filtering, amplification, nonlinear compression, impedance matching, and spectral analysis. While recent work has clarified the functioning and discriminatory power of the cochlea, it is still possible to agree with Allen[5] that, although an adaptive system, the cochlea fundamentally performs the function of a filter bank over the frequency range 20 Hz to 20 kHz.

This has encouraged efforts to afford aspects of human performance to speech input devices by incorporating engineering models of peripheral human hearing (Lyon,[6,7] Ghitza,[8,9] Yang, Wang, and Shama,[10] Cooke,[11] Giguere and Woodland[12]). The expectations are that if the models were sufficiently representative, any benefits in processing experienced by humans would be transplanted to the machines. Early attempts delivered equivocal results when employed as a front end for ASR systems (Gagnon,[13] Bourlard and Dupont[14]), but more recent sub-band processing attempts have reported encouraging results.

1.5.2.2   Processing Operations of the Auditory Brainstem

In the following discussions, "channel" refers to a signal path from the left or right ear, and "sub-band" refers to a contiguous frequency group within one channel.

The processing centers of the auditory brainstem receive from the inner ears a two-channel set of time-domain signals in contiguous nonlinearly spaced frequency bands. This frequency segregation appears to be maintained by the neural pathways all the way to the auditory cortex. Hermansky[15] makes the important point that the main reason that human auditory perception does a spectral analysis of the incoming acoustic signal may not be to provide a spectral basis for direct phonetic classification, but rather to provide freedom to ignore or de-emphasize low signal-to-noise spectral regions prior to applying temporal analysis to the high SNR sub-bands of the signal, as a mechanism for supporting a more graceful degradation of performance.

Figure III.1.5 presents a simplified representation of afferent (feedforward) signal characteristics and the neural processing functions of the auditory brainstem. The flow of information is not one-way up to the higher centers; various efferent (feedback) paths also exist but are not shown. The processing nuclei are labeled with their common acronyms and can be referenced in Pickles.[16] A key is not necessary here, since the main information to take from this figure includes the complexity of

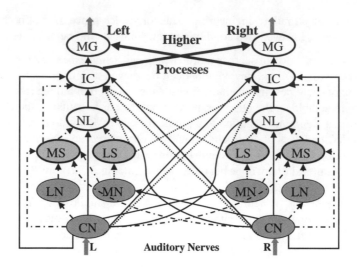

FIGURE III.1.5   Simplified representation of afferent (feedforward) signal characteristics and the neural processing centers of the auditory brainstem.

even this simplified and incomplete picture, the hierarchical structure, the existence of parallel paths, and paths that cross over between the two channels at various levels in the hierarchy. The detailed function performed by the feedback paths is not yet clear; however, it seems reasonable to assume that their function is to support selective "attending" by providing, in engineering terms, a constraint or weight modification pathway which fine-tunes the brainstem processing.

Current interpretations of experimental evidence (Palmer in Moore[17]) identify the separation of left- from right-ear signals, low- from high-frequency signals, timing from intensity information, and their re-integration at various processing centers and levels in the hierarchy. Cross-fed signals from the contralateral side are often inhibitory to the ipsilateral centers, i.e., act as cancellation inputs. In particular, both superior olivary complexes combine auditory information from the ipsilateral ear with inhibitory signals from the contralateral ear along frequency-specific neural pathways; thus, at some centers information appears to be processed in sub-bands.

The organization and functional units in the mammalian auditory brainstem are very similar across members of that class; indeed, birds also have very similar structures and neural pathways (Konishi[18]). Thus, it seems reasonable to assume that processing functions of the auditory brainstem are nonspecific to speech, and are the more general operations of acoustic signal detection, location, extraction, and enhancement, with the auditory cortex operating and controlling functions associated with selective attention, interpretation, and ascribing meaning.

1.5.2.3 Phase Sensitivity

It is often stated in publications on audio engineering that the ear is insensitive to phase. This is obviously not meant to apply to the binaural listening situation since interaural phase difference at low-audio frequencies is used to spatially locate sound sources.

Not surprisingly in monaural listening, human subjects have not been shown to be capable of discriminating between pure tones (single frequency sinusoid) simply on the basis of their starting phase. In the case of mixtures of pure tones, the situation is less clear-cut. Changing only the phase relationships between the tonal components shown in Figure III.1.6 can markedly alter the temporal shape of an audio waveform without altering the perceived sound. This has important implications for schemes that depend on time-domain error measures to assess the quality of a speech processing technique. However, if the ear were completely insensitive to phase, e.g., it responded only to the

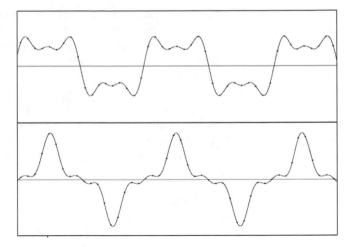

FIGURE III.1.6   Each of the two time histories shown are formed by the sum of three pure tones at 400, 1200, and 2000 Hz, the components differing only in phase.

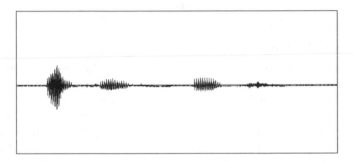

FIGURE III.1.7a   Narrow band-pass (800–1.3 kHz) filtered speech phrase "What is speech."

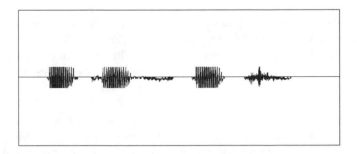

FIGURE III.1.7b   Center and outer clipped speech phrase "What is speech."

magnitude spectrum, then the wave shape would be irrelevant to the perceived sound. Experiments that examine the perception of pitch have shown that phase changes are more detectable in waveforms consisting of a few harmonics than for those with many harmonics (Moore[19]).

Experiments examining sensitivity to modulation effects have found that relative phase changes produced by different forms of modulation were not detectable in waveforms that consist of tonal components at widely spaced frequencies, but could be distinguished in waveforms that consist of a small number of tones having a small frequency difference (Moore[17]). Sensitivity decreased with increasing frequency, and the critical modulation frequency above which the effect was not detectable was around 60 Hz. The range of modulation frequencies seen in conversational speech is about 0.25 to 25 Hz (Houtgast and Steeneken[2]), peaking in the range 3 to 6 Hz. Intelligibility experiments applying band-limiting to speech modulation spectra indicate that from the reception viewpoint, the most important region is from 2 to 8 Hz (Hermansky[15]).

The implication is that the human auditory system is capable of exploiting both the spectral and temporal structure of a speech signal. Drastic amplitude limiting or filtering (Stickney and Assmann[20]) of speech will usually influence quality long before it degrades intelligibility for normal-hearing listeners. Figures III.1.7a and b show the time histories of narrow band-pass filtered and amplitude clipped versions of Figure III.1.3 that are still intelligible, although obviously badly distorted.

1.5.2.4   Simultaneous, Post- and Pre-Masking

When two sound sources are presented simultaneously to the ear, one may mask our perception of the other, partially or totally. In controlled experiments, the degree of masking of one pure tone by another has been shown to be dependent on their relative amplitudes and frequencies. Perceptual models incorporating this effect are core elements of the audio processing section of present-day MPEG coding schemes, e.g., MP3 music files. Masking also operates in nonsimultaneous scenarios where the masker may be presented before the tone to be detected (post-masking) or after it (pre-masking). The latter is a relatively weak effect.

1.5.2.5   Intermittent Masking

Many of the natural sounds we experience are intermittent. To achieve a specific intelligibility score, a speech signal has to be about 3 dB higher if the background (masking) noise is continuous, with a spectrum shaped to match the speech long-term spectrum, than if it is competing connected speech (Plomp[21]). Dermody[22] discusses the importance of this intermittent masking in yielding cues independent of language processing, a commentary supported by others (Gustafsson and Arlinger;[23] Ma and O'Shaughnessy;[24] Van der Heijden and Kohlrausch[25]). It appears that the "windows" in either the desired or the competing signal can be used to view and estimate the other, allowing interference reduction.

1.5.2.6   Channel Modeling

It is well known that our visual system adapts to changes in the color balance of environmental illumination and thus provides us with perceptual stability. Our acoustic environment modifies much of what we hear, and familiar signals may provide us with information that allows our hearing system to adapt to the environmental modifications. Experimental evidence from Summerfield and Assman,[26] Watkins,[27] and Darwin, McKeown, and Kirby[28] suggests that the auditory system is able to perform a communication channel modeling operation. Thus it is possible that enhancement and source location may be aided by estimation of the acoustic transfer function path difference, from the sources of interference and desired signal to the two ears. A channel modeling operation can be much less computationally intensive than signal modeling, since the transfer function is governed by the acoustic properties of the human head, and the characteristics of the reverberant environment, which are relatively slowly changing compared to the acoustic spectra. If channel modeling is effected through binaural hearing, then "binaural sluggishness" (Kollmeier and Gilkey[29]) may be desirable in that, to be useful for signal enhancement, the identified transfer function should exhibit a degree of perceptual stability, in spite of the temporal variations of the spectra of everyday acoustic signals. Averaging mechanisms that impart a degree of "sluggishness" appear in many adaptive signal processing schemes.

1.5.2.7 Binaural Processing: Lateralization and Unmasking

Having two sensors (ears) on either side of an acoustic absorber (about 20 dB for the human head) supports the possibility of using similarity and difference operations to aid the extraction of signals of interest from the acoustic environment (e.g., by coherent addition as with the simple two-element beamformer).

Evidence from mammals and birds indicates that binaural processing occurs in the auditory brainstem nuclei before operations peculiar to conceptual processing are applied in the higher auditory cortex (Moore[17]). The ability to separate sound sources and to reduce the effects of reverberation depends heavily but not exclusively on binaural hearing (Butler,[30] Koenig et al.,[31] Butler and Flannery,[32] Oldfield and Parker,[33] Wickesberg and Oertel[34]). The ability to extract location information is essential to the concept of computational auditory scene analysis (CASA) (Brown and Cooke[35]).

In addition to locating sources in space, and using coherent addition, binaural information can be used to de-emphasize an undesired signal dependent on its binaural correlation properties, as demonstrated by "binaural unmasking" (Durlach[36]), which can lower the hearing threshold (Carhart, Tillman, and Johnson[37,38]). Binaural unmasking or binaural masking level difference refers to the initially surprising result that a target signal at one ear is less effectively masked if a correlated version of the masking noise is available to the other ear.

In engineering terms, this can be interpreted as an adaptive noise cancellation operation, presumably performed in the biological system by contralateral inhibition, specialized centers for which are present both in humans and other mammals (Hine, Martin and Moore),[39] apparently located in auditory brainstem nuclei of the superior olivary complex (Pickles,[16] Feng[40]). There is evidence from narrow-band studies that suggests it may operate in frequency sub-bands (Jeffres,[41] Bernstein and Trahiotis[42]). This is supported by the fact that binaural unmasking is most effective with wide-band noise signals that are in anti-phase at the two ears, a situation that would not occur in nature. However, if binaural unmasking had evolved to operate in sub-bands, then it would result in a system that would best "cancel" wide-band signals when they were in anti-phase, whether or not such signals were practically important. Support for this can be constructed from consideration (Allen[43]) of the work of Fletcher[44] in identifying sub-band processing of speech signals.

Binaural unmasking applies to audio signals, including those where context cannot be inferred, and in humans appears to be established as a selective process by the age of 12 months (Schneider, Dale, and Trehub[45]). The inference is that the process may be fundamental to the separation and extraction of competing signals, an important operation both for our survival and to the learning of vocal communication.

Moncur and Dirks[46] and Plomp[21] found binaural hearing superior to monaural hearing at maintaining the intelligibility of speech in the presence of reverberation, continuous speech-shaped noise, or competing connected speech. To quote Moncur and Dirks, "Binaural hearing was superior to monaural hearing in all conditions of reverberation, excepting only the anechoic-quiet condition . . . even when speech intelligibility was very low (or even zero) for the far ear, this ear continued to make a substantial contribution to the binaural system . . . mere duplication of a monaural signal at both ears does not materially improve intelligibility". This last comment, supported by the results of Carhart, Tillman, and Johnson,[38] emphasizes the advantage of binaural noise-cancellation over simple coherent addition of the signals at the two ears.

Durlach et al.[47] and Bernstein and Trahiotis[42] confirmed the involvement of interaural correlation in binaural unmasking. One means by which interaural correlations could aid desired signal extraction would be by grouping sounds with similar sub-band interaural correlation, categorizing them as emanating from a particular source, and locating that source in an auditory scene model using interaural time difference (ITD) and interaural level difference (ILD) information.

However, lateralization is not necessary for effective binaural unmasking. Carhart, Tillman, and Johnson[38] and Culling and Summerfield[48] found that masking release was largely independent of the pattern of interaural correlations across frequency, and did not support masking release by source segregation through grouping frequency components with common ITD. In anechoic conditions, ILD, which increases with frequency, exceeds ITD in contributing to lowering speech reception thresholds (Bronkhurst and Plomp[49]), while ITD has been reported to be the dominant lateralization process when low-frequency (i.e., <1500 Hz) information is available (Wightman and Kistler[50]). Thus, grouping by ITD is more likely to be implicated in lateralization than in binaural unmasking.

In addition to providing redundancy for enhanced survival, binaural hearing appears necessary for effective source separation and signal enhancement in common SNRs. In this it may be unique and unlike the visual system, e.g., where binocular vision is important over only a very restricted portion of the visual field, and binocular unmasking appears to be a much smaller effect than its auditory equivalent (Schneider, Moraglia, and Jepson[51]).

Binaural unmasking appears to be a fundamental processing operation allowing selective enhancement of binaural signals by a cancellation mechanism. The performance of noise cancellation schemes improves with the accuracy of the noise estimate, and thus tends to yield greater improvements at low SNR (Widrow and Stearns[52]). Speech signal energy is low in the high-frequency range (e.g., unvoiced speech) and a noise cancellation approach would be a reasonable processing choice, or perhaps evolutionary result, given two sensor channels and neural building blocks exhibiting excitatory/inhibitory processing operations.

Engineering possibilities for utilizing binaural processing to enhance a desired signal include:

- Source separation through lateralization, e.g., by correlation (Bodden and Blauert[53]), and characterization, supporting construction of an internal map of the auditory scene apparently being performed in the auditory cortex where functional columns have been identified organized by binaural dominance (Pickles[16]), and where other sense information can be integrated. The map could be parameterized by sound source locations (azimuth and elevation) and characteristics (properties or names), e.g., frequency content, timbre, Violin, John, etc. (see Banks[54] for an attempt at an engineering implementation). While this appears a reasonable engineering operation to perform, it seems clear, as discussed above, that human hearing utilizes this approach for map formation as distinct from signal enhancement, a conclusion that should be of interest to CASA researchers.
- Coherent addition as in the delay-sum beamformer (Hoffman et al.[55]).
- Adaptive noise cancellation (ANC) (Widrow and Stearns[52]), an operation at least superficially analogous to binaural unmasking. The ANC scheme has structural similarities to the sub-band equalization and cancellation (EC) model which Durlach[36] proposed to explain binaural unmasking.

An engineering implementation of sub-band noise cancellation offers the possibility of performing "binaural unmasking" outside the body. If such a mechanism of binaural processing is central to human hearing performance, it may be a useful component of conventional hearing aids, cochlear implant processors, ASR, and hands-free telephone systems that seek equivalent or better performance. It is likely that the ability to exploit adaption, intermittency, channel modeling, and lateral inhibition will be necessary features.

1.5.3   Speech Processing

The following material presents a condensed survey of non-AI speech processing strategies to provide a background to the AI approaches later discussed. The emphasis in this section is on the speech enhancement and ASR areas. The books by Deller, Proakis, and Hansen[56] and Gold and Morgan,[57] supplemented by the surveys of Engebretson,[58] Gong,[59] and Cole et al.,[60] provide useful discussions of many concerns that are still current, and comprehensive bibliographies on speech enhancement and speech recognition techniques.

In aiming to improve the noise robustness of ASR systems, hands-free telephones, hearing aids, etc., one strategy is to develop speech enhancement by focusing not on the detailed modeling of physiological mechanisms, but on the processing system functions and their hierarchy of application. The advantage is that "by concentrating on the auditory function, rather than on the characteristics of speech, the resultant signal processing will be effective independent of the auditory stimulus" (Kates[61]), and independent of the nature of the later process, machine or human.

The review of the previous section suggests that particularly promising methods are likely to take advantage of binaural approaches, channel identification, signal intermittency, correlation, and sub-band processing.

1.5.3.1 A Signal Model

Ignoring modulation effects, a useful model of the spatially distributed noisy speech signal $x_i(t)$ consists of the "clean" speech signal $s(t)$ convolved (*) with the acoustic transfer function $a_i(t)$ from the speech source to the ith sensor, the result exhibiting correlation over space and time, a convolved noise signal from the jth noise source to the ith sensor $a_{ij}(t)*n_j(t)$ also exhibiting such correlation, and measurement noise $v_i(t)$ uncorrelated with $s(t)$ or $n_j(t)$:

$$x_i(t) = a_i(t)^*s_i(t) + \sum_j a_{ij}(t)^*n_j(t) + v_i(t)$$

Estimating the speech requires separating the correlated from uncorrelated components, the remaining correlated noise from speech, then deconvolution of the speech. Intermittency of $s(t)$ and $n_j(t)$ can allow separation of $a_i(t)*s(t)$ from $\sum_j a_{ij}(t)*n_j(t)$ when their spectra overlap. Estimates of a_i and a_{ij} can be obtained if space is scanned, e.g., by a sensor (microphone) array. Methods of enhancement attempt to reduce the effects of v and n, the transfer functions a_i and a_{ij}, the acoustic noise field $\sum_j a_{ij}(t)^*n_j(t) + v_i(t)$, and the processed estimate of $\sum_j a_{ij}(t)^*n_j(t) + v_i(t)$.

1.5.3.2 Fixed Filtering

Conventional fixed filters cannot discriminate between overlapping noise and speech spectra and thus attenuate both; neither can they track time-varying properties. Fixed deconvolution processing to reduce the effects of reverberation allows no source motion or time-varying transfer functions.

1.5.3.3 Wiener Filtering

This provides an optimum estimate of the desired signal $s(t)$ when dealing with statistically stationary signals and uncorrelated additive noise $n(t)$, $v(t)$. The design process requires the noisy signal auto-spectrum, the signal/noise cross-spectrum, and the extraction of a causal transfer function. Approximations can be made for the case of short-time stationary signals, taking advantage of the intermittent nature of speech to obtain spectral estimates during "noise only" and "noisy speech" periods (Lim and Oppenheim,[62] Moir, Campbell, and Dabis[63]) to allow the separation of $a_i(t)^*s(t)$ and $\sum_j a_{ij}(t)^*n_j(t) + v_i(t)$. The Wiener filter is optimal only in the least mean square error (MSE) sense, but the averaging nature of this criterion is not well matched to perceptual criteria. High coefficient update rates generate "musical noise" artifacts, while low rates result in perceptible convolution distortion of the speech.

1.5.3.4 Spectral Subtraction

Related to Wiener filtering, spectral subtraction (Lim and Oppenheim[62]) involves transformation of the signal from time to the frequency domain, estimating the magnitude, power, or some higher moment of a "noise-only" spectral record, and subtracting it from the spectrum of a "noisy speech" record. The result is recombined with the phase of the original "noisy speech" and inverse transformed to yield the processed record. Negative values occurring after subtraction due to changes in the noise spectrum may be set to some small value. The resulting spectrum contains randomly appearing spectral lines that generate short tone bursts resulting in the distracting artifact known as "musical noise." Overestimation of the noise and application of a non-zero threshold allows some manipulation of this effect, but amounts to adding noise of a more acceptable character.

Various extensions of the basic technique, e.g., log spectrum nonlinear methods, have been examined for use in speech recognizers (Van Compernolle, Ma, and Van Diest,[64] Mokbel, Barbier, and Chollet[65]). However, these methods are generally significantly more complex, and the attempt to minimize distortion in the frequency domain using short-time transform approximations to a generally nonstationary noise signal is conceptually unsatisfying. In practice, spectral subtraction tends to be unsuccessful for speech recognizers at low SNR (e.g., <12 dB), although Hirsch[66] found it benefited cochlear implant wearers down to 0 dB SNR. This may have been due to the artificiality of the speech/noise mixing process, or perhaps that the very restricted frequency response of implant wearers means they do not perceive the high levels of "musical noise" generated at such SNRs.

Spectral subtraction is a form of sub-band processing and may bear some relationship to human auditory processing, e.g., the self-normalization discussed by Wang and Shamma.[67]

1.5.3.5 Sub-Band Processing

Filterbank or transform methods derive a set of sub-bands within which continuous time signals can be processed, allowing the operation of distortion reduction in the time domain rather than the frequency domain. The flexibility supported by such processing has been found to help alleviate reverberation effects (Allen, Berkley, and Blauert,[68] Gilloire,[69] Goulding and Bird[70]). Sub-band processing has been investigated, particularly for hearing aid applications. The approaches have concentrated mainly on spectral feature enhancement and multichannel compression (MCC) systems (Walker, Byrne, and Dillon,[71] White,[72] Plomp,[73] Kollmeier,[74] Stone and Moore,[75] Baer, Moore, and Gatehouse[76]). Although mentioned together here, spectral feature enhancement and MCC approaches are essentially antagonists, since MCC tends to reduce spectral contrast. While some improvement of intelligibility in noise has been reported as the number of channels increases (Yund and Buckles[77]), the overall results are equivocal, suffering in some cases from distortion of the speech envelope or SNR reduction.

Recently there has been substantial interest in sub-band strategies in order to improve the robustness of ASR to noise. Critically, the new work (Ming and Jack Smith,[78] Morris et al.[79]) focuses on schemes for maintaining joint spectral information, such as the magnitude spectrum profile, that is lost by the independent sub-band ASR attempts which resulted in reduced recognition performance with clean speech. Although these approaches might seem to reflect the sub-band and hierarchical organization of human hearing, they are too specifically based on phoneme recognition to claim this.

1.5.3.6 Multi-Sensor Methods

Yund and Buckles[80] comment that "the distinction between the SNR within a channel and the overall SNR is important. . . . When the SNR is sufficiently negative within a narrow frequency channel, virtually nothing can be done to recover the signal information there." This is true for single sensor applications of the above methods that cannot deal effectively with spectral overlap of the desired and interfering signals at low SNR. Multi-sensor methods can take advantage of the distribution of the signals over space and time to aid separation (Toner and Campbell,[81] Weinstein, Feder, and Oppenheim[82]).

Classical beamforming is essentially a delay and sum operation that aligns signals from the desired direction to allow coherent addition. Silverman,[83] Silverman and Kirtman,[84] and Grenier[85] have investigated acoustic beamformers for speaker location and signal enhancement. Peterson et al.,[86]

Zurek, Greenberg, and Peterson,[87] Greenberg and Zurek,[88] Soede, Bilsen, and Berkhout,[89] Soede, Berkhout, and Bilsen,[90] Stadler and Rabinowitz,[91] Hoffman et al.,[55] and Van Hoesel and Clark[92] have investigated fixed and adaptive beamformers acting as directional microphones for hearing aid use with improvements of around 7 dB in speech reception threshold in noisy anechoic conditions being recorded. In general, a deterioration is seen in moving to reverberant surroundings although, compared to single sensor approaches, multi-sensor methods have been shown to compensate to a degree for multiple noise sources or reverberation effects (Bloom and Cain,[93] Yamada, Wang, and Itakura[94]).

1.5.3.7 Adaptive Noise Cancellation

Classical ANC is predicated on access to a good estimate of the noise, which can then be adjusted and subtracted from the noisy signal. The adjustment is accomplished using a transfer function model of the differential path between a noise sensor and a "speech + noise" sensor. The operation essentially separates correlated and uncorrelated signals from the two sensors, the error output estimating the uncorrelated components while the model output estimates the correlated components (Ferrara and Widrow[95]).

Houtgast and Steeneken[2] reported a high correlation between the acoustic transfer function of the environment and speech intelligibility. The ability of the auditory system to perform a communication channel modeling operation supports such an approach for speech enhancement. Adaption is essential to compensate for statistical nonstationarity of the sources, source or sensor motion, or time-varying acoustic paths. Griffith and Jim[96] described a combination of adaptive noise cancellation with beamforming that has been investigated for speech enhancement (Farrel, Mammone, and Flanagan,[97] Oh, Viswanathan, and Papamichalis[98]) and hearing aid applications (Chazan, Medan, and Shvadron,[99] Van Compernolle,[100] Peterson et al.[101]).

The adaptive filter is usually moving-average (MA) to avoid possible model instability. The model is identified by minimizing a performance criterion in which the gradient of the ideally quadratic mean square error performance surface is usually approximated by the instantaneous error gradient. In a realistic echo cancellation experiment, Gudvangen and Flockton[102] were unable to show any advantage of recursive over MA adaptive models in reducing the number of coefficients and therefore the computational complexity of the algorithm.

Various algorithms have been proposed for adapting the model, the most common implementation being the computationally simple least mean squares (LMS) algorithm (Widrow and Stearns[52]). A trade-off is implied between LMS algorithm stability and convergence rate in the choice of adaption step size. When applied in its classical wide-band form, high-order filters are required to model realistic room acoustic transfer functions, which can generate troublesome misadjustment noise. Rabiner, Crochiere, and Allen[103] have verified that the LMS algorithm is robust and useful for band-limited as well as wide-band inputs. Cowan[104] showed that in the presence of significant noise levels, the LMS method exhibited more robust behavior than the recursive least squares (RLS) algorithm. In attempting to account for the robustness observed in three decades of practical use, Hassibi, Sayed, and Kailath[105] reported in 1996 that the LMS algorithm is optimal not in the least mean square sense but in the H^∞ sense. Eweda[106] has described several practical situations in which the LMS algorithm performs as well as, or better than, RLS.

Such noise cancellation approaches to identification of differential acoustic transfer functions can produce impressive results in anechoic environments with localized sound radiators (Strube,[107] Dabis, Moir, and Campbell[108]); however, performance deteriorates in reverberant environments (Brey et al.,[109] Weiss,[110] Greenberg and Zurek,[88] Lu and Clarkson,[111] Van Hoesel and Clark[92]). Wallace and Goubran[112] applied the LMS noise canceller to a recording of automobile noise, and found that it reduced the low-frequency noise, which was significantly correlated between channels, but increased noise at high frequencies where both correlation and noise energy were low. Various researchers have utilized cross-channel correlation to control processing functions; e.g., Allen, Berkley, and Blauert[68] adjusted a sub-band gain dependent on cross-channel correlation. Le Bouquin-Jeannes et al.[113] have applied a similar approach in an automobile hands-free telecommunication application. Their results suggest that the approach has advantages when operating in environments with statistically nonstationary noise. Greenberg and Zurek[88] incorporated a band-pass cross-channel correlation operation over a single rather wide frequency band to inhibit the adaption process.

Environmental reflections of acoustic energy will be dependent on the geometry and materials of the reflecting structures and the relative locations of the sound sources and receivers. The reflection coefficients of many structural materials are frequency dependent, while the geometry and contents of an enclosure exhibit natural resonant modes yielding preferential contribution of some frequencies. Guy and Abdou[114] demonstrated large directional variations of sound energy ratio at frequencies below 1 kHz in a small test room. Correlation between signals from two microphones within an average echoic room will also, therefore, vary with frequency.

These variations suggest that an approach involving diverse processing dependent on cross-channel correlation within multiple sub-bands should be more successful.

1.5.3.8   Multi-Microphone Sub-Band Adaptive Processing

Kollmeier, Peissig, and Hohmann[115] reported a binaural frequency domain processing technique for a hearing aid application. The method is related to spectral subtraction and beamforming approaches. It selects frequency bands with phase and amplitude corresponding to a desired target direction, and can be modified to suppress reverberation. However, the subjective assessment by hearing-impaired listeners did not demonstrate an unequivocal improvement in speech intelligibility in reverberant surroundings. As with related spectral subtraction methods, processing artifacts can become troublesome.

Speech enhancement combining multi-microphone methods with intermittent adaptive processing and diversity of processing within sub-bands has been suggested (Toner and Campbell[81]). Two or more "closely spaced" microphones are used in an adaptive noise cancellation scheme involving the identification of a differential acoustic path transfer function during a "noise-only" period in intermittent speech. It allows noise features within sub-bands, such as the noise power, the correlation or coherence between signals from multiple sensors, and the behavior of the adaptive algorithm, to influence the subsequent processing during the "noisy speech" period. The same structure is used to process the signals in both the correlated and uncorrelated noise cases. In the former, the filter is adapted during a known "noise only" period and then frozen, performing fixed filtering during the "speech + noise" period, with the speech estimate taken from the error output. In the latter, the filter is constantly adapting and the speech estimate is taken from the filter output. Both continue until a new noise correlation estimate can be made during the next detected "noise-only" interval.

1.5.3.9   Discussion

Gagnon and McGee,[116] Hansen and Clements,[117] and Bateman, Bye, and Hunt[118] have compared various single sensor algorithms including Wiener and spectral subtraction variants in a speech recognition application. Grossly summarizing their findings, the enhancement algorithms were unhelpful below 20 dB SNR, yielding recognition error rates between 40 and 90%. Gong[59] tabulated 19 reports of speech recognition approaches operating with an SNR of 10 dB in Gaussian noise and yielding recognition accuracies of between 30 and 97%, with the highest relating to very restricted vocabularies. Lippman[119] compared the performance of humans and ASR over a range of six speech corpora with noisy speech data and vocabularies from 10 to 65,000 words, finding that "human word error rates are often more than an order of magnitude lower than those of current recognizers in both quiet and degraded environments".

Multiple sensor methods (Van Compernolle, Ma, and Van Diest,[64] Campbell, Moir, and Dabis,[120] Oh, Viswanathan, and Papamichalis[98]) have achieved SNR improvements over single sensor methods of around 20 dB in anechoic and 12 dB in semi-realistic test conditions with SNRs below 5 dB. In real life this improvement may drop to below 6 dB (Shields and Campbell[121]), but many factors remain to be optimized. Other typical results reported by Hoffman et al.[55] and Van Hoesel and Clark[92] for hearing aid applications achieve noise suppression of around 20 dB in anechoic conditions dropping to 3 to 6 dB in reverberant surroundings.

Altered spectra (e.g., spectral tilt, coloration) and processing artifacts (e.g., musical tones, misadjustment noise) are inherent limitations of many of the methods discussed. Often, more effort seems expended on controlling the artifacts introduced by complex enhancement processing than is spent on treating the original corruption.

Many of the methods require a voice activity detector (VAD), which attempts to take advantage of the intermittency of speech signals by allowing either spectral estimation or system adaption during the "noise-only" period, to an extent simulating the intermittent masking capability of human hearing. However, in realistic situations, the VAD may be required to operate at low SNRs with noise of unknown and varying structure. Approaches using a coherence or correlation VAD as reported by Le Bouquin-Jeannes and Faucon[122] and Agaiby and Moir[123] are of particular interest for their ability to operate at low SNRs.

The multi-microphone approach using closely spaced sensors with sub-band processing has shown promise as a means of improving intelligibility in reverberant surroundings (Shields and Campbell[121]). "Close" spacing of the sensors reduces the required order (complexity) of the adaptive filter and thus the sub-band computational load both for adaptive and intermittent fixed processing. It also reduces the misadjustment noise when using continuously adapting schemes. Sub-band operation gives faster adaption through the freedom to use different adaptive step-sizes in each band (Toner and Campbell,[124] Mahalanobis et al.[125]). The individual adaptive filters converge to narrow-band functions, and adaption speed could be increased by starting at a suitable narrow-band set of coefficients, i.e., operating within environmentally reasonable constraints. Separate decisions can be made on the appropriate form of processing for each sub-band, and cross-band effects such as lateral inhibition could be introduced. The inherent parallelism of the approach allows implementation by parallel processors spreading the computational load which may be further reduced by applying sub-sampling techniques if care is taken with the filter-bank design to avoid aliasing problems (Gilloire and Vetterli[126]).

Conceptual similarities with binaural unmasking and the EC model suggest that ANC systems and the LMS adaptive scheme, which has a very close association with engineering neural networks, could provide a modeling scheme for binaural hearing; however, Knecht, Schenkel, and Moschytz[127] have reported rather disappointing performance from an ANN speech enhancement scheme. Work by Van Compernolle and Van Gerven[129] and Van Gerven et al.[130] developed a symmetric structure related to the ANC approach, interpreted as a blind separation system, utilizing a wide-band cross-channel correlation measure which also has possibilities as a binaural processing model.

1.6 Artificial Intelligence and Neural Networks

Traditional AI is based on high-level symbol processing, i.e., logic programming, expert systems, semantic nets, etc. All rely on there existing some high-level representation of knowledge which can be manipulated by using some type of formal syntax manipulation scheme; the rules of grammar. Such approaches have proved to be very successful in emulating human prowess in a number of fields:

- Playing chess at the Grand Master level
- Matching professional expertise in medicine or the law using expert systems
- Creating mathematical proofs for solving complex mathematical problems

Yet there are still major areas of human expertise which we are unable to mimic using software; e.g., our machines have difficulty reliably reading human handwriting, recognizing human faces, or exhibiting "common sense." Notice how low-level the last list seems compared to the list of achievements. It has been said that the difficult things have proved easy to program, whereas the easy things have proved difficult.

1.6.1 Artificial Neural Networks (ANNs)

Tasks such as those discussed above seemingly require no great human expertise; young children are adept at many of these tasks. The underlying presumption of creating artificial neural networks is that such human expertise is due to the nature of the "hardware" from which the brain is built. Therefore, if we are to emulate biological proficiencies in these areas, we should base our AI machines on hardware structures (or simulations of such hardware) that are modeled on that found within our heads. Artificial neural networks (Haykin[130]) is one of the new information processing methods which are intended to emulate those found in biological systems.

The attractive properties of human neural information processing may be described as:

- Biological information processing is robust and fault-tolerant: early in life, we have our greatest number of neurons; yet though we daily lose many thousands of neurons, we continue to function for many years without an associated deterioration in our capabilities.
- Biological information processors are flexible: we do not require to be reprogrammed when we go into a new environment; we adapt to the new environment, i.e., we learn.
- We can handle fuzzy, probabilistic, noisy, and inconsistent data in a way that, with computer programs, is only possible with a great deal of sophisticated programming and then only when the context of such data has been analyzed in detail. Contrast this with our innate ability to handle uncertainty.
- The machine that is performing these functions is highly parallel, small, compact, and dissipates little power. Discussion of these properties usually begins with a look at the biological machine we are attempting to emulate.

1.6.2 Biological and Silicon Neurons

In a biological neuron, information is received by the neuron at synapses on its dendrites. Each synapse represents the junction of an incoming axon from another neuron with a dendrite of the neuron represented in Figure III.1.8; an electrochemical transmission occurs at the synapse that allows information to be transmitted from one neuron to the next. The information is then transmitted along the dendrites until it reaches the cell body, where a summation of the electrical impulses reaching the body takes place and some function of this summation is performed. If this function is greater than a particular threshold, the neuron will fire: this means that it will send a signal (in the form of a wave of ionization) along its axon in order to communicate with other neurons. In this way, information is passed from one part of the network of neurons to another. Synapses are thought to have different efficiencies, and these efficiencies appear to change during the neuron's lifetime. We will return to this feature when we discuss learning.

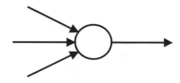

FIGURE III.1.8   A simplified neuron. Information is summed at the cell body and then transmitted onward via the axon.

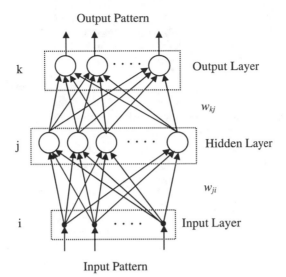

FIGURE III.1.9   An artificial neural network with three layers of neurons. This is often called a two-layer network since there are two layers of weights.

Mathematically, the inputs are represented by the vector **x** and the synapses' efficiencies are modeled by a weight vector **w**. Therefore, the single output value of this neuron is given by:

$$y = f\left(\sum_i w_i x_i\right) = f(\mathbf{w}{\cdot}\mathbf{x}).$$

Sometimes $f(\ )$ will be the identity function. Notice that if the weight between two neurons is positive, the input neuron's effect may be described as excitatory; if the weight between two neurons is negative, the input neuron's effect may be described as inhibitory.

Thus, the single neuron is an extremely simple processing unit. The power of neural networks arises from the combination and interaction of many of these simple processing units; i.e., we throw lots of simple and robust computational power at a problem. Often the engineering neurons are arranged to act in concert in layers as shown in Figure III.1.9. Again, we may be thought to be emulating nature, as the typical human has several hundred billion neurons and there are six major layers recognizable in the human cortex.

In Figure III.1.9, the set of inputs (the input vector, **x**) enter the network from the bottom and are propagated through the network via the weights until the activation reaches the output layer. The middle layer is known as the hidden layer as it is invisible (unobservable) from outside the net: we may not affect its activation directly.

1.6.3   Learning in Artificial Neural Networks

There are two operational modes in standard ANNs:

1. Activation transfer mode: when activation is transmitted throughout the network
2. Learning mode: when the network organizes usually on the basis of the most recent activation transfer

The latter is a desired feature of neural networks so that they need not be re-programmed when they encounter novel environments. Their behavior should change in order to adapt to the new environment, and such behavioral changes appear as changes in the weights in the network, which we call learning. It is believed that human learning is due to changes in the efficiency with which synapses pass information between neurons, and the changes in weights in an artificial neural network are intended to model the changing synaptic efficiencies in biological neural networks.

There are two main types of learning in an ANN:

1. *Supervised learning:* with this type of learning, we provide the network with input data and the correct answer; i.e., what output we wish to receive given that input data. The input data is propagated forward through the network until activation reaches the output neurons. We can then compare the answer that the network has calculated with that which we wished to get. If the error is not sufficiently small, we adjust the weights to ensure that the network is more likely to give the correct answer in the future if it is again presented with the same (or similar) input data. This weight adjustment scheme is known as supervised learning or learning with a teacher.

2. *Unsupervised learning:* with this type of learning, we only provide the network with the input data. The network is required to self-organize (i.e., to teach itself), depending on some structure in the input data. Typically, this structure may be some form of redundancy in the input data or clusters in the data.

The interesting facet of learning in many problems is not just that the input patterns may be learned/classified/identified precisely, but that this learning has the capacity to generalize. That is, while learning will take place on a set of *training patterns*, an important property is that the network can generalize its results over a set of *test patterns* that it has not seen during learning. One of the important consequences here is that there is a danger of over-learning a set of training patterns so that new patterns that are not part of the training set are not properly classified.

1.6.4   Supervised Learning

The most common supervised learning network is the multilayer perceptron (MLP), which is a feedforward, fully connected hierarchical network consisting of an input layer, one or more middle or hidden layers, and an output layer.

MLP networks are capable of performing arbitrary mappings:

$$g : R^n \rightarrow R^m$$

where $g(x) = z$.

Such mappings are possible if a sufficient number of hidden units are provided and if the network can be trained; i.e., if a set of weights that performs the desired mapping can be found.

Activity in the network is propagated forward via weight vectors from the input layer to the hidden layer where some function of the net activation is calculated. Then the activity is propagated via more weight vectors to the output neurons. Now two sets of weight vectors should be updated: those between the hidden and output layers, and those between the input and hidden layers. The error due to the first set of weight vectors is calculable by the Least Mean Square Error rule. It is required to propagate backward that part of the error due to the errors that exist in the second set of weights, and assign the error proportionately to the weights that cause it. This is solved by the back-propagation method.

The network operation is as follows:

Feedforward:
    Hidden layer:

$$h_i = \sum_{j=1}^{H} w_{ij} x_j, \forall_i$$

Output layer:

$$y_k = \sum_{j=1}^{N} v_{kj} h_j, \forall k$$

Change of weights:
  Error on the output layer:

$$e_k = t_k - y_k$$

where t_k is the target.
  Error on the hidden layer:

$$e_i = \sum_{k=1}^{0} v_{ki} e_k, \forall i$$

Weight vectors between the input layer and the hidden layer: $w_{ij} = w_{ij} + \eta e_i x_j$. Weight vectors between the hidden and the output layer: $v_{ki} = v_{ki} + \eta e_k h_i$, where η is the learning rate.

1.6.5   Issues in Back-Propagation

1.6.5.1   Batch vs. On-line Learning

The back-propagation algorithm is only theoretically guaranteed to converge if used in batch mode; i.e., if all patterns in turn are presented to the network, the total error calculated, and the weights updated in a separate stage at the end of each training epoch. However, it is more common to use the on-line version where the weights are updated after the presentation of each individual pattern. It has been found empirically that this leads to faster convergence, though there is the theoretical possibility of entering a cycle of repeated changes. In an attempt to avoid this, in on-line mode we usually ensure that the patterns are presented to the network in a random and changing order.

1.6.5.2   Activation Functions

The most popular activation functions are the logistic function $1/(1 + \exp(-bx))$ and the tanh(bx) function. Both of these functions satisfy the basic criterion that they are differentiable. In addition, they are both monotonic and have the important property that their rate of change is greatest at an intermediate value and least at extreme values. This makes it possible to saturate a neuron's output at one or another of their extreme values. The final point worth noting is the ease with which their derivatives can be calculated.

1.6.5.3   Momentum and Speed of Convergence

The basic back-propagation method described above is not known for its fast speed of convergence, and though we could simply increase the learning rate η, this tends to introduce instability into the learning rule, causing wild oscillations in the learned weights. It is possible to speed up the basic method in a number of ways, the simplest being to add a "momentum" term to the change of weights. The basic idea is to make the new change of weights large if it is in the direction of the previous changes of weights, and make it smaller if it is in a different direction.
  Thus we use:

$$\Delta w_{ij}(t+1) = \eta.e_j.o_i + \alpha \Delta w_{ij}(t),$$

where the α determines the influence of the momentum. The second term is sometimes known as the "flat spot avoidance" term. Clearly, the momentum parameter α must be between 0 and 1. Momentum has the additional property that it helps to slide the learning rule over local minima.

1.6.5.4 Local Minima

Error descent is bedeviled by local minima. Popular opinion has it that local minima are not much problem to ANNs, in that a network's weights will typically converge to solutions that, even if they are not globally optimal, are good enough. There is as yet little analytical evidence for this belief. A heuristic often quoted is to ensure that the initial (random) weights are such that the average input to each neuron is approximately, or just below, unity. This suggests randomizing the initial weights of neuron j around the value $\frac{1}{\sqrt{N}}$, where N is the number of weights into the jth neuron. A second heuristic is to introduce into the network a little random noise, which acts like an annealing schedule.

1.6.5.5 Weight Decay and Generalization

While we wish to see as good a performance as possible on the training set, we are even more interested in the network's performance on the test set, since this is a measure of how well the network generalizes. Generalization is important not only because we wish a network to perform on new data which it has not seen during learning, but also because we often have to operate with data which is noisy, distorted or incomplete.

If a neural network has a large number of weights (i.e., degrees of freedom), we may be in danger of over-fitting the network to the training data, and this can lead to poor performance on the test data. One approach is to remove connections either explicitly or by giving each weight a tendency to decay toward zero. The simplest method is to calculate:

$$w_{ij}^{new} = (1 - \varepsilon)w_{ij}^{old}$$

after each update of the weights. This does have the disadvantage that it discourages the use of large weights, since a single large weight may be decayed by more than a lot of small weights. More complex decay routines can be applied, which will encourage small weights to disappear.

1.6.5.6 Adaptive Parameters

Since it is not easy to choose the parameter values *a priori*, one approach is to change them dynamically; e.g., if we use too small a learning rate, we find that the error E is decreasing consistently, but by too small an amount each time. If our learning rate is too large, we find that the error is decreasing and increasing haphazardly. This suggests adapting the learning rate according to a schedule such as:

$$\Delta\eta = +a, \text{ if } \Delta E < 0$$

consistently, and:

$$\Delta\eta = -b\eta, \text{ if } \Delta E > 0.$$

This schedule may be thought of as increasing the learning rate if it seems that we are consistently going in the correct direction, but decreasing the learning rate if the direction sometimes changes.

1.6.5.7 The Number of Hidden Neurons

The number of hidden nodes has a particularly large effect on the generalization capability of the network. A network with too many weights (degrees of freedom) will tend to memorize the data, while a network with too few will be unable to perform the task allocated to it.

1.6.6 Support Vector Machines

Multilayer artificial neural networks can be viewed within the framework of statistical learning theory, and a large amount of research investigation has recently been devoted to the study of machines for classification and regression based on empirical risk minimization (ERM) called Support Vector Machines (SVMs). The reader is directed to Vapnik[131] for a detailed treatment of this particular area of research; but briefly, SVMs are powerful discriminative classifiers that have been demonstrated to provide state-of-the-art results for many standard benchmark classification problems (Muller et al.[132]).

The nonlinear discriminative power of SVM classifiers comes from employing a nonlinear kernel that returns a distance between two input feature vectors within an implicitly defined high-dimensional nonlinear feature space (Scholkopf et al.[133]). As the kernel implicitly defines a metric in the associated feature space, it has been proposed that a probabilistic generative model, such as a Hidden Markov Model (HMM), can be used to provide a kernel which would return the distance between points in the parameter space of the particular generative model chosen (Jaakkola and Haussler[134]). A useful online resource considering SVMs posed as regression-based function approximators is available at the URL http://www.kernel-machines.org.

1.6.7 Unsupervised Learning

Supervised networks require a training set to exist, on which we already have the answers to the questions that we are going to pose to the network. Yet humans appear to be able to learn (indeed some would say can only learn) without explicit supervision. The aim of unsupervised learning is to mimic this aspect of human capabilities, and as such, this type of learning tends to use more biologically plausible methods than those using the error descent methods above. The network must self-organize; to do so, it must react to some aspect of the input data, typically either redundancy in the input data or clustering in the data (i.e., there must be some structure in the data to which it can respond). There are two major paradigms used:

- Hebbian learning
- Competitive learning

1.6.7.1 Hebbian Learning

Hebbian learning is named after Donald Hebb who conjectured (Hebb[135]):

> *"When an axon of a cell A is near enough to excite a cell B and repeatedly or persistently takes part in firing it, some growth process or metabolic change takes place in one or both cells such that A's efficiency as one of the cells firing B, is increased."*

Consider the simplest feedforward neural network which has a set of input neurons with associated input vector **x** and a set of output neurons with associated output vector **y**. Then we have, as before:

$$y_i = \sum_j w_{ij} x_j,$$

where now the (Hebbian) learning rule is defined by:

$$\Delta w_{ij} = \eta x_j y_i.$$

That is, the weight w_{ij} between each input and output neuron is increased proportional to the magnitude of the simultaneous firing of these neurons. Now we can substitute the value of y calculated by feeding forward the activation into the learning rule to obtain:

$$\Delta w_{ij} = \eta x_j \sum_k w_{ik} x_k = \eta \sum_k w_{ik} x_k x_j.$$

Writing the learning rule in this way emphasizes that it depends on the correlation between different parts of the input data's vector components.

Because of this last feature, many artificial neural networks that use Hebbian learning, usually in combination with a weight decay term, can be shown to perform an approximate Principal Component Analysis (PCA) of a data set. Since standard statistics provide more efficient solutions, this was merely interesting until nonlinearity was included either in the learning or the projection. These types of methods have found favor in Independent Component Analysis (ICA) (Comon[136]).

It is sometimes of interest, particularly in the context of speech processing, to consider so-called anti-Hebbian learning. Consider two neurons connected by a single weight v, which learns by $\Delta v = -y_i y_j$, where $y_i = y_i + v y_i$ and $y_j = y_j + v y_i$. Then the lateral connection (v) will grow proportional to the correlation between the two neurons, and the net effect is thus to decorrelate the outputs of these neurons. Unlike Hebbian learning, there is no need to limit the weight changes, since the process is inherently self-limiting.

1.6.7.2 Competitive Learning

In competitive learning, after the activation has been sent forward, only that output neuron which wins a competition is allowed to fire. Such output neurons are often called winner-take-all units. The aim of competitive learning is to categorize the data by clustering it. However, as with the Hebbian learning networks, we provide no correct answer (i.e., no labeling information) to the network. It must self-organize on the basis of the structure of the input data. The basic method is to ensure that the similarities of instances within a class is as great as possible, while making the differences between instances of different classes as great as possible.

Hertz et al.[137] point out that simple competitive learning leads to the creation of grandmother cells, the proverbial neuron which would fire if and only if your grandmother hovered in sight. The major difficulty with such neurons is their lack of robustness: if you lose your grandmother cell, you will never again recognize her. In addition, we should note that if we have N grandmother cells, we could only recognize N categories, whereas if we were using a binary code, we could distinguish between 2^N categories.

The basic mechanism of simple competitive learning is to find a winning unit and update its weights to make it more likely to win in the future should a similar input be given to the network; i.e.:

$$\Delta w_{ij} = \eta(x_j - w_{ij})$$

for the winning neuron i.

Note that the change in weights is a function of the *difference* between the weights and the input. This rule will move the weights of the winning neuron directly toward the input. If used over a distribution, the weights will tend to the mean value of the distribution since:

$$\Delta w_{ij} \rightarrow 0 \Leftrightarrow w_{ij} \rightarrow \langle x_j \rangle,$$

where the angled brackets indicate the ensemble average. A geometric analogy is often given for understanding simple competitive learning. Considering Figure III.1.10, we have two groups of points lying on the surface of the sphere, and the two radii represent the weights of the network. The weights have converged to the mean of each group and will be used to classify any future input to one or another group.

A potential problem with this type of learning is that some neurons can come to dominate the process; i.e., the same neuron continues to win all the time, while other neurons (*dead neurons*) never win. While this can be desirable if we wish to preserve some neurons for possible new sets of input patterns, it can be undesirable if we wish to develop the most efficient neural network. In this situation, it pays to ensure that all weights are normalized at all times (and so already on the surface of the sphere), so that one neuron is not winning just because it happens to be larger than the others. Another possible approach is to employ leaky learning, where the winning neuron is updated by a greater amount than all other neurons. The amount of the "leak" can be varied during

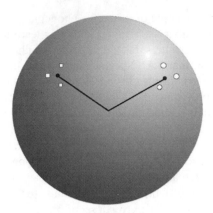

FIGURE III.1.10   The input vectors are represented by points (circles and squares) on the surface of a sphere and the lines represent the directions along which the weights have converged. Each is terminating on the mean of the group of input vectors surrounding it.

the course of a simulation. Another possibility is to have a variable threshold so that neurons that have won often in the past have a higher threshold than others. This is sometimes known as learning with a conscience. Finally, noise in the input vectors can help in the initial approximate weight learning process.

Simple competitive learning has been augmented by the work of Kohonen,[138] who has added a topology preservation property to it; nearby neurons respond to similar data, and data close in the original space are responded to by adjacent neurons. This provides a graceful degradation in the quantization.

Carpenter and Grossberg[139] have developed a growing network in an attempt to solve the stability-plasticity problem; they want the network to remember what it has already learned, but still be capable of learning new things.

1.6.8   Typical Application Areas

The number of application areas in which ANNs are used is growing daily, and some representative types of problems to which neural networks have been applied are surveyed here.

Pattern completion. ANNs can be trained on sets of visual patterns represented by pixel values. If, subsequently, a part of an individual pattern (or a noisy pattern) is presented to the network, we can allow the network's activation to propagate through the network until it converges to the original (memorized) visual pattern. The network is acting like a content-addressable memory. Typically, such networks have a recurrent (feedback as opposed to feedforward) aspect to their activation passing.

Classification. An early example of this type of network was trained to differentiate between male and female faces. It is actually rather difficult to create an algorithm to do this, yet an ANN has been shown to have near-human capacity to do so.

Optimization. It is notoriously difficult to find algorithms for solving optimization problems. A famous optimization problem is the Traveling Salesman Problem in which a salesman must travel to a number of cities, visiting each one once and only once in an optimal (i.e., least distance or least cost) route. There are several types of neural networks which have been shown to converge to "good-enough" solutions to this problem; i.e., solutions which may not be globally optimal, but can be shown to be close to the global optimum for any given set of parameters.

Feature detection. An early example of this is the phoneme-producing feature map of Kohonen. The network is provided with a set of inputs and must learn to pronounce the words; in doing so, it must identify a set of features that are important in phoneme production.

Data compression. There are many ANNs which have been shown to be capable of representing input data in a compressed format, losing as little of the information as possible; e.g., in image compression we may show that a great deal of the information given by a pixel-to-pixel representation of the data is redundant, and a more compact representation of the image can be found by ANNs.

Approximation. Given examples of an input to output mapping, a neural network can be trained to approximate the mapping so that a future input will give approximately the correct relationship.

Association. We may associate a particular input with a particular output so that given the same (or similar) input again, the network will give the same (or a similar) output again.

Prediction. This task may be stated as: given a set of previous examples from a time series, such as a set of closing prices for the FTSE, to predict the next (future) sample.

Control. For example, to control the movement of a robot arm (or truck, or any nonlinear process) to learn what inputs (actions) will have the correct outputs (results).

In the context of this chapter, ANNs have been used for signal processing applications in system identification, target detection, blind beam forming, adaptive equalization, and time series prediction.

1.7 Artificial Neural Network Applications

This section considers some applications to speech processing of signal processing methods that are based on the artificial neural network (ANN) paradigm. The distinction between supervised and unsupervised ANN models has been made previously and this distinction is further considered.

Perhaps the most well-known application of ANN methods to speech processing has been in the domain of speech recognition. The core technology now employed in commonly available speech recognition software is the Hidden Markov Model (HMM). The HMM is utilized to probabilistically model the acoustic behavior of basic speech units such as phonemes or words. The emission densities of the HMM are modeled by employing finite mixture models such as a mixture of Gaussians; however, hybrid HMM and ANN approaches to speech recognition employ an ANN to directly estimate the HMM state posterior probability (Bourland and Morgan[140]). Multi-layer Perceptrons (MLP) are employed for this task, due to their ability to "learn" nonlinear classification boundaries and their outputs being interpretable as class posterior probabilities. However, it should be noted that the MLP has more in common with nonlinear statistical pattern recognition methods than biologically inspired models of neural activity.

Another application of ANNs to speech processing is the removal of noise from corrupted speech. ANN-based filtering methods for noise-corrupted speech employ available samples of clean speech as a desired signal for training. Such ANN approaches tend to generalize poorly for actual measured speech with varying power and noise levels, and in addition take no account of the nonstationary nature of natural speech. To overcome these shortcomings, the Dual Extended Kalman Filter (EKF) was proposed by Wan and Nelson.[141] The Dual EKF takes only the noise-corrupted speech, and trains a sequence of ANNs which results in a nonstationary model that is then used to remove the noise from the speech. A nonlinear model of speech is devised based on MLP ANNs and an EKF is used in the state estimation of the model. The weights of the ANN are estimated by an additional EKF and a dual state and weight estimation process is devised. The results of experimental work by Wan and Nelson[141] indicated that this approach was promising for the enhancement of noise corrupted speech. The Dual EKF makes the assumption that the nonlinear autoregressive speech model lies in the class of feedforward ANN models. This of course can then be generalized to any nonlinear regression model, and as such is not restricted to the class of nonlinear feedforward ANN models.

The applications of ANN to speech processing considered so far have been motivated by the requirement for a method of modeling nonlinearity within a larger system, e.g., hybrid HMM and ANN speech recognition systems, or noise filtering of speech employing extended Kalman filters. Interestingly, of the ten papers appearing in the "Neural Networks for Speech Processing" session at the 1998 IEEE International Conference on Acoustics, Speech, and Signal Processing (ICASSP), all of them were focused on the problem of speech recognition where the ANNs were employed to provide state posterior probabilities.

1.7.1 Blind Source Separation of Convolutive Mixtures

An ANN may be specifically devised to model hypothesized forms of synaptic learning, and such models are thought to have the potential to be effective in the suppression of unwanted signals in speech. This section considers an extension of anti-Hebbian learning that includes temporal context, applied to the classical "cocktail party" problem, where wanted speech needs to be separated from unwanted competing speech or noise.

Following the signal model of Section 1.5.3.1, listening to, or recording a speech signal at the ith sensor in a noisy acoustic environment is described as:

$$x_i(t) = a_i(t)^* s(t) + \sum_j a_{ij}(t)^* n_j(t) \qquad \text{(III.1.1)}$$

in the deterministic case where there is no measurement noise at the receiver and the sensor does not distort the summation of the signals. This can be written compactly using matrix notation such as

$$\mathbf{x}(z) = \mathbf{A}(z)\mathbf{s}(z). \qquad \text{(III.1.2)}$$

We shall use z-domain notation throughout for convenience. The polynomial matrix $\mathbf{A}(z)$ will have individual entries of the form

$$A_{ij} = \sum_{k=0}^{\infty} a_{ij}(k)z^{-k},$$

indicating that the mixing components are now right-sided causal filters of infinite order. We impose stability constraints on each entry of $\mathbf{A}(z)$ such that bounded inputs produce bounded outputs (BIBO), i.e.,

$$\sum_{k=0}^{\infty} |a_{ij}(k)| < \infty.$$

In the case of acoustic environments, the impulse response of a typical room will have a finite duration, typically being of the order of 0.35 sec (Mackenzie[4]). Thus, for a sampling rate of 11.025 kHz, the finite causal mixing filter will have a length of typically 4000 tap delays. We can therefore approximate the causal filter by a finite impulse response of order L such that:

$$A_{ij} = \sum_{k=0}^{L} a_{ij}(k)z^{-k}.$$

As we are considering an inverse problem, we require that both the finite convolving filter and its inverse are stable and causal. This then requires that all poles and zeros of the mixing filters be within the unit circle (Oppenheim and Shafer[142]). This requirement is realistic in most simple acoustic situations where there are no excessive levels of reverberation or complicating multi-path signal arrivals. For illustrative purposes, if we consider an N source/N receiver situation we have, in the z-domain, the observed signals given as:

$$\mathbf{x}_1(z) = A_{11}(z)\,\mathbf{s}_1(z) + \cdots + A_{1N}(z)\,\mathbf{s}_N(z)$$

$$\vdots$$

$$\mathbf{x}_N(z) = A_{N1}(z)\,\mathbf{s}_1(z) + \cdots + A_{NN}(z)\,\mathbf{s}_N(z)$$

and equivalently in the time domain:

$$x_1(t) = \sum_{k=0}^{L} a_{11}(k)s_1(t-k) + \cdots + \sum_{k=0}^{L} a_{1N}(k)s_N(t-k)$$

$$\vdots$$

$$x_N(t) = \sum_{k=0}^{L} a_{N1}(k)s_1(t-k) + \cdots + \sum_{k=0}^{L} a_{NN}(k)s_N(t-k)$$

The notation $x(t)$ and $x(t-k)$ indicates the instantaneous sample value at time t and the lagged value at time $t-k$. As in the linear case, we seek a transformation that will approximately invert the effect of the mixing, and so in this case seek a filter $\mathbf{W}(z)$:

$$\mathbf{y}(z) = \mathbf{W}(z)\mathbf{x}(z) = \mathbf{W}(z)\mathbf{A}(z)\mathbf{s}(z) \qquad\qquad (\text{III.1.3})$$

As in the instantaneous case, we seek a polynomial matrix that will tend to satisfy:

$$\mathbf{W}(z) \to \mathbf{A}^{-1}(z)$$

where the inverse of the polynomial matrix is (Barnett,[143] Chan, Godshill, and Rayner[144]):

$$\mathbf{A}^{-1}(z) = \frac{1}{\det(\mathbf{A}(z))}[(-1)^{(i+j)}\det(\mathbf{A}_{ji}(z))]_{ij} \qquad\qquad (\text{III.1.4})$$

with the determinant given as the matrix Laplace expansion:

$$\det(\mathbf{A}(z)) = \sum_{j=1}^{N}(-1)^{(j-1)}A_{1j}(z)\det(\mathbf{A}_{1j}(z)),$$

and the term:

$$[(-1)^{(i+j)}\det(\mathbf{A}_{ij}(z))]_{ij}$$

being the matrix of cofactors of $\mathbf{A}(z)$ as in the standard linear case.

If we consider the simplified 2×2 case of (III.1.4), we then have:

$$\mathbf{W}(z) = \mathbf{A}^{-1}(z) = \frac{1}{(A_{11}(z)A_{22}(z) - A_{12}(z)A_{21}(z))} \begin{bmatrix} A_{22}(z) & -A_{12}(z) \\ -A_{21}(z) & A_{11}(z) \end{bmatrix}. \qquad (\text{III.1.5})$$

As in the general expression (III.1.3), we have the additional requirement that for a stable and causal separating solution $\mathbf{W}(z)$ to exist, the determinant of the polynomial mixing matrix must have all poles and zeros within the unit circle. In this section, we have considered the general convolutive

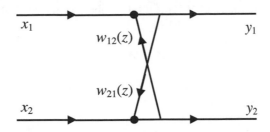

FIGURE III.1.11   Instantaneous lateral network model.

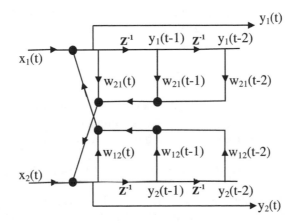

FIGURE III.1.12   Temporal extension of lateral network model.

mixing model, which consists of stable and causal finite length filters. The inverse system will be causal and stable if the mixing system is minimum phase (Oppenheim and Schafer[142]), i.e., all poles and zeros exist within the unit circle region of convergence. It is important to highlight and be aware of the shortcomings of this simplified model. The transfer functions A_{ij} describe the point-to-point acoustic path, but this will change if the sources or receivers are physically nonstationary. These functions are only approximations of the complex acoustic field. Van Gerven[145] gives a detailed discussion on the validity of the proposed acoustic mixing model. For our purposes, this model will allow investigation into the signal separating properties of temporal anti-Hebbian learning.

1.7.2 Temporal Linear Anti-Hebbian Model

We now consider a temporal extension of Foldiak's[146] initial decorrelation model of anti-Hebbian learning, Figure III.1.11, applying "memory based" synaptic lateral connections, we can represent this by Figure III.1.12, which shows the temporal model with the weights and the z^{-1} operator shown explicitly for two tap delays. The addition of memory to the synaptic weights reflects the notion that information may be conveyed and therefore extracted from the temporal relations within the data stream. The network output is now given as:

$$y_i(t) = x_i(t) + \sum_{j=1}^{N} \sum_{k=0}^{M} w_{ij}(t-k) y_j(t-k) \qquad \text{(III.1.6)}$$

where N is the number of neurons and M is the total number of delays. Written as a Z domain matrix, we have the compact form of Equation III.1.6:

$$y(z) = (\mathbf{I} - \mathbf{W}(z))^{-1} \, \mathbf{x}(z) \qquad (\text{III.1.7})$$

The anti-Hebbian rule has been shown to yield an output with diagonal covariance. For second-order independence, the following should be satisfied:

$$E\{y_i(t)y_j(t-k)\} = 0 \quad \forall i \neq j \wedge k = 1 \ldots M$$

and so we require an adaptive algorithm which adapts the dependant variable of the network output (i.e., the weights) to yield temporally decorrelated outputs in the expectation. Figure III.1.13 shows the adaptive filter model equivalent of the temporal anti-Hebbian model. By utilizing the temporally extended anti-Hebbian rule:

$$\Delta w_{ij}(k) = -\eta y_i(t)y_j(t-k) \quad \forall i \neq j \wedge k = 1 \ldots M \qquad (\text{III.1.8})$$

for each temporal output, we can then force the output cross-correlation terms to zero. So the cross weights will grow in an inhibitory fashion if there is correlation between the output $y_i(t)$ and each of the tapped outputs of the neuron $y_j(t-k)$ until the following holds:

$$E\{y_i(t)y_j(t-k)\} = 0 \quad \forall i \neq j \wedge k = 1 \ldots M$$

The weight connecting the current output at node i to the historical output at time $(t-k)$ of node j is denoted as $w_{ij}(k)$. Note that in the nontemporal model $\Delta w_{ij} = -\eta \, y_i y_j = -\eta \, y_j y_i = \Delta w_{ji}$ and so the anti-Hebbian learning will give a symmetric weight matrix. In the temporal case:

$$\Delta w_{ij}(k) \neq \Delta w_{ji}(k)$$

since in general:

$$-\eta \, y_i(t)y_j(t-k) \neq -\eta \, y_j(t)y_i(t-k)$$

and so an asymmetric polynomial weight matrix is generated. The linear anti-Hebbian learning decorrelates the delayed outputs of all neurons $i : (i \neq j \forall i)$ with respect to the instantaneous outputs of neuron j.

Consider that two signal sources (s_1 and s_2) the signals received at two points displaced in space (x_1 and x_2) from the sources will be given as the matrix multiplication of the transfer function matrix $\mathbf{A(z)}$ and the source vector:

$$\begin{bmatrix} x_1 \\ x_2 \end{bmatrix} = \begin{bmatrix} a_{11}(z) & a_{12}(z) \\ a_{21}(z) & a_{22}(z) \end{bmatrix} \begin{bmatrix} s_1 \\ s_2 \end{bmatrix} \qquad (\text{III.1.9})$$

$$\mathbf{x} = \mathbf{As} \qquad (\text{III.1.10})$$

For clarity we drop the use of (z) as it is now implied in the relevant equations.

To recover the original sources, $\mathbf{s} = \mathbf{A}^{-1}\mathbf{x}$ must be satisfied up to an arbitrary filter, using Equation III.1.7; for $\mathbf{y}$ to approximate $\mathbf{s}$ at convergence we can write:

$$\mathbf{y} = \hat{\mathbf{s}} = (\mathbf{I} - \mathbf{W})^{-1}\mathbf{x} = \mathbf{A}^{-1}\mathbf{x} \qquad (\text{III.1.11})$$

Some algebraic manipulation will yield, in the discrete z domain:

$$(1 - w_{12}w_{21})\, y_1 = (a_{11} + w_{12}a_{21})s_1 + (a_{12} + w_{12}a_{22})\, s_2 \qquad \text{(III.1.12)}$$

$$(1 - w_{12}w_{21})\, y_2 = (a_{22} + w_{21}a_{12})s_2 + (a_{21} + w_{21}a_{11})\, s_1$$

and so:

$$w_{12}(z) = a_{12}(z)a_{22}^{-1}(z)$$

and

$$w_{21}(z) = -a_{21}(z)a_{11}^{-1}(z) \qquad \text{(III.1.13)}$$

The values of the network output will then be a filtered representation of the original uncorrupted signal source as given in the more general case of Equation III.1.5:

$$\begin{bmatrix} \hat{s}_1 \\ \hat{s}_2 \end{bmatrix} = (\mathbf{I} - \mathbf{W})^{-1}\, \mathbf{As} \cong \begin{bmatrix} s_1 \\ s_2 \end{bmatrix} \qquad \text{(III.1.14)}$$

The weights of the network can then be forced to converge to the inverse filters given in Equation III.1.13 using a suitable criterion. The criterion that will be considered is initially the cross-correlation between the received signals and is a second-order independence criterion. We can use the temporal linear anti-Hebbian learning given in Equation III.1.8 to stochastically maximize the second-order independence criterion. Matsuoka, Masahiro, and Kawamoto[147] developed a blind separation algorithm for separation of an instantaneous symmetric mixture of time-varying Gaussian signals based on second-order statistics. As the signal statistics are nonstationary, they show that a normalized form of anti-Hebbian learning based on the feedback network (Figure III.1.11) can separate symmetric mixtures of Gaussian sources. The nonstationarity is exploited by on-line estimation of the signal variance that is time dependent. They then propose a simple algorithm:

$$\Delta w_{ij}(t) = -\frac{\eta\, y_i(t)y_j(t)}{E\{y_i^2(t)\}} \equiv -\frac{\eta\, y_i(t)y_j(t)}{\Phi(t)} \;\; \forall i \neq j$$

The variance is estimated using a moving average estimator:

$$\Phi_i(t + 1) = \alpha\Phi_i(t) + (1 - \alpha)y_i^2(t)$$

However, by using a temporal form of the Matsuoka, Masahiro, and Kawamoto[147] anti-Hebbian rules, statistical independence of the network output may also be achieved for convolutive mixing without resorting to higher-order statistics:

$$\Delta w_{ij}(k) = \frac{\eta\, y_i(t)y_j(t - k)}{\Phi_i(t)} \;\; \forall i \neq j$$

$$\Phi_i(t + 1) = \alpha\Phi_i(t) + (1 - \alpha)y_i^2(t) \qquad \text{(III.1.15)}$$

This is effectively a normalized version of the temporal anti-Hebbian update rule (Equation III.1.8).

In developing the Symmetric Adaptive Decorrelation (SAD) algorithm, Van Gerven[145] showed that nulling of the cross-spectrum of mixtures of convolved sources is a necessary condition for separation. The adaptation algorithm of Equation III.1.8 or the normalized version (Equation III.1.15) will then be potentially suitable for separation of mixtures of convolved sources.

1.7.3   Review of Existing Work on Adaptive Separation of Convolutive Mixtures

Van Gerven[145] takes the adaptive noise cancellation (ANC) configuration and extends it to a symmetric structure with an additional filter in the primary channel. The major problem with the direct structure is that the recovered signals can be severely distorted (Van Gerven,[145] Torkkola,[148]) and so requires additional post-processing for equalization. An alternative feedback structure is considered which does not require the additional equalization. This structure is exactly that given in Figure III.1.13 and can be considered as an adaptive filter representation of the temporal anti-Hebbian network. Van Gerven et al.[128] carried out a comparative study of second- and fourth-order adaptation update algorithms based on the feedback filter structure (Figure III.1.13) using the following:

$$\Delta w_{ij}(k) = -\eta \, y_i(t) y_j(t - k) \quad \forall i \neq j \wedge k = 1 \ldots M \qquad \text{(III.1.16)}$$

and

$$\Delta w_{ij}(k) = -\eta \, (y_i(t))^3 y_j(t - k) \quad \forall i \neq j \wedge k = 1 \ldots M \qquad \text{(III.1.17)}$$

They show through extensive simulation that in the case of strictly causal convolutive mixing, second-order adaptation provides statistically equivalent separating performance as fourth-order weight adaptation.

These results show that no benefit is gained from moving to higher-order statistics from simple second-order decorrelation. Platt and Faggin[149] extend the network structure proposed by Jutten and Herault[150] to incorporate time delays (Figure III.1.12). They consider explicit minimization of second- and fourth-order moments, and demonstrate the performance of the network with artificial mixtures of speech and music. Nguyen and Jutten[151] also extend the Jutten and Herault network structure to one possessing Finite Impulse Response (FIR) filter weights (Figure III.1.13). They develop algorithms for weight update, attempting to cancel out fourth-order cross-cumulants at all finite time delays within the filter structure. An alternative algorithm reported in the paper is the use of a decorrelation criterion for all delays, and cancellation of fourth-order cross-cumulants for zero delay.

Weinstein, Feder, and Oppenheim[82] and Yellin and Weinstein[152] consider decorrelation-based algorithms for multi-channel signal separation. Chan, Godsill, and Rayner[144] develop a batch-based algorithm based on constrained output decorrelation. They demonstrate the efficacy of their algorithm on recordings of mixtures of up to four speakers and excerpts of music. Recordings for the reported simulations are made in an anechoic chamber that exhibits little acoustic dispersion, and so the inverting filters will only be required to identify the cross-microphone delays. Principe, Wang, and Wu[153] utilize a "teacher forcing" temporal decorrelation network for separation of artificially

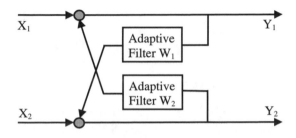

FIGURE III.1.13   Adaptive filter equivalent of temporal model.

convolved mixtures of speech. Empirical comparisons are made between linear and nonlinear FIR temporal learning as well as the gamma filter. As with the previous researchers considered, Principe, Wang, and Wu[153] find no benefit in moving from simple temporal decorrelation to nonlinear filtering.

A novel use of the network proposed in the paper is to calculate the cross-correlation matrix of a number of signals in an on-line adaptive manner. Algebraic subspace methods have been proposed by Gorokhov and Loubaton,[154] but these methods are slow due to the manipulation of large matrices. Lindgren, Sahlin, and Broman[155] consider source separation using a feedforward structure and a damped Newton algorithm based on a modified Hessian matrix. Their technique relies only on second-order statistics; but as the weight adaptation in most second-order methods is driven by the level of signal cross-correlation, the adaptation speed slows as the signals become less correlated; they proposed the use of the Newton-type update to overcome this problem.

Lambert[156] considers extensions of Bussgang-type algorithms for source separation. He develops an FIR algebra that allows computation in the frequency domain giving significant computational advantages, such as convolution in the time domain transforming to simple multiplication in the frequency domain. Lambert reports on separation of speakers in real-room environments using the Equivariant Bussgang-type algorithms (Haykin[130]).

Cichocki, Amari, and Cao[157] extend their feedforward and recurrent network algorithms to take into account delays and convolutions. A novel learning rate adaptation algorithm is proposed for signals that may be mixed in time-varying environments. The standard cubic nonlinearity is used within the reported simulations, which are run on artificial data. The injection of noise into the learning is empirically considered and shown to improve the convergence properties of the algorithm.

As has been discussed, Torkkola[148] extends the Infomax algorithm to consider convolved mixtures of sources. A feedback network similar to that proposed by Nguyen Thi and Jutten,[151] or Platt and Faggin[149] is considered. This removes the problem of the network output being temporally whitened due to temporal redundancies within the signals being removed by the direct filters. Results reported include separation of a mixture of two speakers recorded in a small conference room with cross-channel interference reduction of 12 dB reported. Lee, Bell, and Orglmeister[158] extend the work of Torkkola,[148] and perform source separation of speakers and music recorded in real-acoustic environments. Comparisons of separation performance are made based on recognition rates of ASR systems. For speech recorded with music, increases of 50% in recognition rates are reported after separation processing. Speech corrupted with competing speech yielded a 19% increase in recognition rate after processing. The novel use of filters with acausal extensions ensures that the separating filters will be stable even if the inverse mixing filters are nonminimum phase (Oppenheim and Shafer[142]). More recently, Charkani and Deville[159] and Deville and Charkani[160] have considered time domain convolutive source separation algorithms. By explicitly considering the minimization of the asymptotic error variance separating functions were developed which are related to the probability density functions of the underlying sources. In the following section we will observe that the maximum likelihood framework will naturally asymptotically minimize the parameter estimation error variance.

It can be concluded from this review that most researchers (from both the signal processing and artificial neural network communities) have found that for convolutive mixtures of signals (speech is the most often considered source signal), second-order statistics appear to suffice. No benefit has been found from explicit use of either higher-order statistics or sigmoidal nonlinearities. In the case of the Infomax temporal extensions, maximizing the information through a hyperbolic tangent is shown analytically to yield independent components; second-order temporal anti-Hebbian learning yields substantially similar separation performance.

1.7.4 Temporal Anti-Hebbian Learning Based on Maximum Likelihood Estimation

From Figure III.1.12, the output of each node at time t for an $N \times N$ network with memory-based synaptic weights of length L is given as Equation III.1.6:

$$y_i(t) = x_i(t) + \sum_{j \neq i} \sum_{k=0}^{L} w_{ij}(t-k)y_j(t-k) \tag{III.1.18}$$

where the subscripts denote spatial relations between the nodes within the network and the $(t-k)$ terms denote delays of k samples from time t. The maximum likelihood estimation (MLE) is given as:

$$\hat{\theta}_{ML} = \arg\ \max \left(\frac{1}{N} \sum_{i=1}^{N} \log p(\mathbf{x}_i; \hat{\theta}) \right) \tag{III.1.19}$$

whose value is given by the solution of:

$$\frac{\partial}{\partial \hat{\theta}} \left(\frac{1}{N} \sum_{i=1}^{N} \log p(\mathbf{x}_i; \hat{\theta}) \right) = 0. \tag{III.1.20}$$

We note that in the situation where the probability density function (PDF) satisfies regularity conditions (Cox and Hinckley[161]), we have an expectation and we may change the order of operation to:

$$E \left\{ \frac{\partial}{\partial \hat{\theta}} \log(p(\mathbf{x}_i; \hat{\theta})) \right\} = 0. \tag{III.1.21}$$

This can now be solved iteratively using the Robbins-Monro algorithm, such that parameter updates will be:

$$\hat{\theta}_{n+1} = \hat{\theta}_n + \mu_n \frac{\partial}{\partial \hat{\theta}} \log(p(\mathbf{x}_{n+1}; \hat{\theta}_n)) \tag{III.1.22}$$

The parametric form of the transformed observation (Equation III.1.6) has only one set of parameters in this case, i.e., $w_{ij}(t-k) : \forall (i \neq j \in 1 \dots N) \wedge \forall k \in 1 \dots L$. The parametric model is required to be factorable and with this constraint we can then write:

$$w_{ij}^{n+1}(t-k) = w_{ij}^{n}(t-k) + \mu_n \frac{\partial}{\partial w_{ij}^{n}(t-k)} \log \left(\prod_{l=1}^{N} p_l(y_l^{n+1}; w_{ij}^{n}(t-k)) \right) \tag{III.1.23}$$

$$w_{ij}^{n+1}(t-k) = w_{ij}^{n}(t-k) + \mu_n \frac{p_i'(y_i^{n+1}; w_{ij}^{n}(t-k)) \prod_{l \neq i} p_l(y_l^{n+1}; w_{ij}^{n}(t-k))}{\prod_{l=1}^{N} p_l(y_l^{n+1}; w_{ij}^{n}(t-k))} \frac{\partial y_i}{\partial w_{ij}^{n}(t-k)}$$

$$w_{ij}^{n+1}(t-k) = w_{ij}^{n}(t-k) + \mu_n \frac{p_i'(y_i^{n+1}; w_{ij}^{n}(t-k))}{p_i(y_i^{n+1}; w_{ij}^{n}(t-k))} y_j(t-k) \tag{III.1.24}$$

The generic sequential parameter update algorithm of Equation III.1.24 has a number of important points. First, as the sample size tends to infinity, Equation III.1.24 will converge with probability one to an unbiased estimate of the true parameters of the underlying parametric model. This presupposes that the parametric model chosen is correct for the underlying latent data. We considered the previous work on blind separation of convolved sources and we now use the generic MLE (Equation III.1.24) to derive the second-order, fourth-order, and Infomax algorithms. Let us choose a parametric model for each marginal PDF based on the generalized Gaussian density. For this example we shall consider the generalized zero mean Gaussian form of PDF which, neglecting normalization constants, is given as:

$$p_i(y_i(t)) \propto \exp\left(-\frac{|y_i(t)|^s}{E\{|y_i(t)|^s\}}\right)$$
(III.1.25)

When the parameter value $s = 2$, then the generalized Gaussian density becomes the standard Gaussian density. We neglect the normalizing coefficients, as the derivative of the log PDF will be independent of the normalizing term. From Equation III.1.25 we have the general form of the derivative of the log probability density:

$$\frac{p'_i(y_i(t))}{p_i(y_i(t))} = \frac{-|y_i(t)|^{s-1} sign(y_i(t))}{E\{|y_i(t)|^s\}}$$
(III.1.26)

Using Equation III.1.26 in Equation III.1.24 we then have the generic form of:

$$w_{ij}^{n+1}(t - k) = w_{ij}^n(t - k) + \mu_n \frac{-|y_i(t)|^{s-1} sign(y_i(t))}{E\{|y_i(t)|^s\}} y_j(t - k)$$
(III.1.27)

when $s = 2$, Equation III.1.25 is the Gaussian distribution and we then have:

$$w_{ij}^{n+1}(t - k) = w_{ij}^n(t - k) - \mu_n \frac{y_i(t) y_i(t - k)}{E\{|y_i(t)|^2\}}.$$
(III.1.28)

This is precisely the temporal extension of second-order learning (Matsuoka, Masahiro, and Kawamoto,[147] Van Gerven[144]) for temporal anti-Hebbian learning. It is now clear that this weight update rule (Equation III.1.28) is seeking to fit the received data to a parametric model based on a product of marginal Gaussian distributions. Let us now reconsider the Hyperbolic-Cauchy distribution, which is defined as:

$$p_i(y_i(t)) = \frac{1}{2\sqrt{2}} \sec h^2\left(\frac{y_i(t)}{\sqrt{2}}\right).$$
(III.1.29)

The motivation in choosing this particular form of PDF is that the tails of the distribution are thicker than a standard Gaussian distribution. In cases where we are considering speech or noise, which have finite values of kurtosis, the tails of the distribution will be heavier than the normal distribution. Therefore, this may be a more suitable model than the standard Gaussian distribution for signals such as speech. Utilizing Equation III.1.29 in Equation III.1.24 and collecting all constant coefficients then gives:

$$w_{ij}^{n+1}(t - k) = w_{ij}^n(t - k) - \mu_n \tan h(y_i(t)) y_j(t - k).$$
(III.1.30)

It is interesting to note that Torkkola[148] arrives exactly at this update equation for a feedback structure similar to Figure III.1.12 by extending Bell and Sejnowski's[162] Infomax algorithm derivation, when the hyperbolic tangent activation function is considered for separation of convolved mixtures. The algorithm that Torkkola develops can now be considered from an MLE perspective where the parametric model for data fitting is chosen as a product of marginal Hyperbolic-Cauchy distributions. McDonald[163] was able to show that the PDF of stationary human speech could be approximated

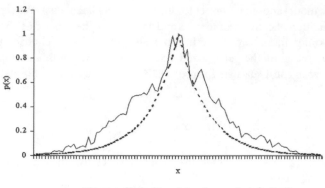

FIGURE III.1.14   Laplacian density and natural speech densities.

by the Laplacian density. Indeed, if we consider Figure III.1.14, it is seen that the Laplacian density fits the heavy-tailed distribution of natural speech more fully than either the Gaussian or the Hyperbolic-Cauchy distributions.

The Laplacian density is a special form of the generalized Gaussian where $s = 1$, and in this case it is a simple matter to see that the weight update equation will be given by:

$$w_{ij}^{n+1}(t - k) = w_{ij}^{n}(t - k)_i - \mu_n \frac{sign(y_i(t))y_j(t - k)}{E\{|y_i(t)|\}}. \tag{III.1.31}$$

It should be noted that Equation III.1.24 in its general form and the specific cases of Equations III.1.28, III.1.30, and III.1.31 are all forms of anti-Hebbian updates. The use of Equations III.1.28 and III.1.30 for separating artificially convolved mixtures of speech can now be considered in light of the maximum likelihood framework developed here. Since for speech, the Gaussian and Hyperbolic-Cauchy distributions capture similar amounts of the probability mass at the tails of the distribution, these distributions may then lead to almost similar sub-optimal parametric models. However, the form of the Laplacian PDF indicates that it may be an improved estimator when considering data generated by natural speech.

In this section we have developed update algorithms within a maximum likelihood context for separation of convolved mixtures of independent sources. The network architecture is a temporal extension of that proposed by Jutten and Herault[150] and Foldiak.[146] The temporal Infomax algorithm developed by Torkkola[148] has been shown to be a specific form of the more general update equation (Equation III.1.24).

1.7.5   Comparative Simulations Using Varying PDF Models

This section presents and discusses results from simulations carried out in separating speech from speech, and also speech from noise within a realistic acoustic environment. A major problem of recordings made in acoustic environments is the physical phenomenon of what is termed "multi-path" propagation of signals. For example, a microphone will receive a summation of the original transmitted source and delayed reflections. Recordings of natural speech or music in an anechoic chamber will not have the problem of multi-path propagation, and so only the delay between the microphones has to be considered. A more realistic acoustic environment is considered in these simulations. A standard living room with soft carpeting and furnishings was used (dimensions $8.5\,\mathrm{m} \times 6.0\,\mathrm{m} \times 2.5\,\mathrm{m}$). In this simulation, omni-directional microphones were placed 0.4 m apart and 0.5 m distant from a pair of loudspeakers in a square format. Loudspeakers were chosen for this simulation to ensure physical stationarity of the point sources.

If we consider Equations III.1.4 and III.1.5, clearly the aim of the learning is to identify the differential inverse filters, and any movement in the sources will cause a level of non-stationarity in the acoustic transfer functions to be estimated. The level of reverberation within the room is calculated using the standard T_{60} measure (Mackenzie[4]); an impulse is injected into the room and the time for the impulse response to decay by 60 dB is measured. This is measured to give a value of $T_{60} = 0.34$ sec, which is a typical value for standard domestic accommodation. It should be noted that the simple symmetric structure of sources and sensors reduced the complicating effect of reverberations within the room. To allow quantitative measurements of separation, each source was recorded separately on the direct and cross-coupled microphones. The simulations reported involve speech corrupted with competing speech and speech corrupted with wide-band masking noise. Each signal record was 8 sec long with a sampling rate of 11.025 kHz.

The MLE framework has the fundamental assumption that the sequences of observations are independent, i.e., no dependency between each sample of the data. However, as pointed out earlier (Figure III.1.4), natural speech can exhibit strong short-term temporal correlation, which has to be removed for the MLE assumptions to hold true. This can be achieved by a temporal pre-whitening operation where the received mixture of speech has temporal correlation removed. A simple moving average filter can be utilized in the whitening process. We shall consider the case where the signals are temporally pre-whitened initially. The effect that no pre-whitening of the mixtures has on the level and speed of separation will be considered. The signal-to-noise ratio (SNR) is used as a measure of the separation of the sources, which is:

$$SNR = -10 \log \left[\frac{E\{(y_i - s_i)^2\}}{E\{s_i^2\}} \right].$$

The original source power is given as $E\{s_i^2\}$ and the residual error power as $E\{(y_i - s_i)^2\}$.

It should be noted that, from a statistical perspective, temporal whitening is providing second-order temporal independence, and for each sample to be fully independent, information theoretic arguments are required. Comparing the PSD of whitened speech using a linear AR filter, one based on the hyperbolic tangent and the signum function show insignificant difference in form. The statistics of the driving process for speech are complex and nonstationary, and it appears that gross statistics are insufficient to capture the temporal subtleties of the signal. Nevertheless, second-order temporal whitening is sufficient for our purposes.

The simulation results of Figure III.1.15 show that as the number of epochs of self-adaptation increases, the separation performance for whitened speech is similar to that for non-whitened speech. Classical filter theory indicates that whitening will remove the eigenvalue disparity of the autocorrelation matrix and so give faster convergence of the LMS algorithm (Haykin[164]). The signals under consideration are natural speech propagated in a natural acoustic environment. The complex nature of the signals removes any potential convergence acceleration attributed to temporal whitening. Similar findings are reported in Deville and Charkani.[160]

Since multiple passes through the data are required, real-time on-line adaptation is still some way from realization. We also see that the Gaussian and Hyperbolic-Cauchy parametric models yield substantially similar performance in both cases. This confirms the findings of the previous section when comparing temporal linear anti-Hebbian learning and temporal Infomax learning. What is significant here is that the better-fitting Laplacian PDF model outperforms the other models considered with a 7-dB improvement for speech vs. speech. We should point out that the SNR values reported here are excellent, largely due to the simplicity of the recording geometry and the low levels of reverberation. As the levels of reverberation increase, the SNR performance drops significantly (Shields et al.[165]). Figure III.1.16 shows the separation performance when speech is contaminated with speech-shaped noise.

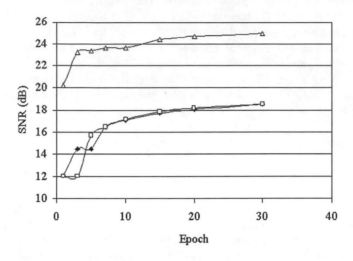

FIGURE III.1.15   Parametric model performance curves: speech vs. speech.

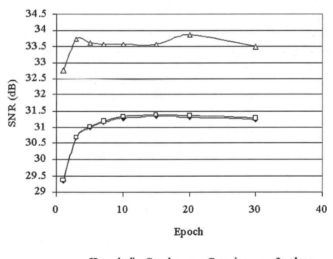

FIGURE III.1.16   Parametric model performance curves: speech vs. noise.

The other point to note regarding the SNR performance of the two simulations is that the performance for speech vs. competing speech is typically 10 dB less than for speech vs. noise. This may be largely due to the intermittent characteristics of natural speech; Nguyen Thi and Jutten,[151] and Van Gerven et al.[128] propose "nonpermanent" learning to help overcome this problem. The technique consists of tracking the output power of the signals and ceasing weight updates when the output power falls below a certain threshold.

The MLE perspective clarifies why simple linear decorrelation and Infomax give similar performance for separation of naturally occurring speech. Prior knowledge of the source signals can be used in choosing the parametric form of the network update algorithm as demonstrated by the use of the Laplace parametric model.

1.8   Conclusion

The current state of knowledge about human speech processing reveals some startling apparent conflicts that can have major implications for machine processing of speech, some of which are identified and discussed in this conclusion.

The development of present-day speech recognition technology is fundamentally based on HMM generative models, so evolutionary proposals (e.g., Smith, Gales, and Niranjan[166]) employing HMMs to create a natural distance measure for variable length sequences such as spoken words appear potentially beneficial. The application of SVMs to some important speech processing problems has also recently been explored, e.g., spoken digit recognition (Bazzi and Katebi[167]) and speaker identification (Fine, Navratil, and Gopinath[168]), with promising results being reported. However, there is strong justification for research into radically different approaches to ASR, speaker separation, and speech enhancement. These may require more of a top-down AI approach, concentrating on pattern extraction/processing or model-based schemes, but less dependent on phonemic information, in order to develop systems with performance as robust as that of humans.

A salutary reminder of the importance of the lower-level feature extraction to humans is that limited cochlear damage can seriously degrade the intelligibility of speech (assuming lip movements are hidden) in spite of:

- Knowledge of the location of the speaker
- The existence of considerable redundancy in speech
- The presence of many contextual cues
- In the case of an elderly person, a lifetime of auditory pattern recognition and language interpretation experience being available in the higher brain

So, in spite of the availability of the best pattern-processing engine we know (the brain), speech recognition is severely impeded by a moderate level of sensorineural hearing loss, which appears to cause damage to spectral resolution and magnitude sensitivity. Yet Lippman[169] showed that consonant intelligibility remains high even when speech energy in the mid-range (800–4000 Hz) is removed; and in tests with severely band-pass filtered speech (Warren et al.,[170] Stickney and Assmann[20]), humans with normal hearing were able to maintain high intelligibility scores (>95%). One would have assumed that the information available under these conditions would be even less than that available to many with a debilitating hearing deficit, and present-day ASR devices only achieve that sort of score with full bandwidth, low noise, low reverberation, speech signals.

Greenberg[171] emphasized the implications of spectral and temporal "incompleteness" for ASR. He also made the point that arguments emphasizing superior use by humans of redundancy in speech were often being used simply to bolster the prevalent bottom-up data-driven hierarchical framework of speech perception that has guided ASR developments. These redundancy arguments seem initially reasonable in that noise corruption may degrade some frequency bands more than others, and that by giving a higher weighting to information in the "cleaner" bands, a more robust estimate is obtained. Yet human hearing incorporates a simultaneous masking operation that effectively inserts notches into the wide-band spectrum of speech, on the basis of signal, *not speech*, amplitude, thus apparently reducing the breadth of information available. These considerations still bring one inevitably to the conclusion of Lipmann,[169] that the markedly superior speech recognition performance of humans under these circumstances suggests that they use a fundamentally different recognition process from modern ASR systems, including the new sub-band approaches.

Separation of speakers situated in realistic acoustic environments is a research topic with challenging unresolved problems. Anti-Hebbian learning with the addition of memory has been shown to be applicable to this "cocktail party" problem. The self-organization of a network with feedback, viewed from a MLE perspective, is equivalent to mutual information minimization when applied to the blind separation problem. This provides an overall generalization for many neural and adaptive filter techniques for source separation.

In the specific case of the "cocktail party" problem, it has been shown that the knowledge of the gross statistical characteristics of speech allows a more efficient algorithm than symmetric adaptive decorrelation or temporal Infomax. The MLE perspective also provides sound statistical justification for the ability of simple cross-correlation nulling to provide certain levels of separation of mixtures of speech. In the real-world case where the speakers are allowed to move, the transfer functions that comprise the "mixing" parameters become time-varying and have to be both estimated and tracked. The Kalman filter could be considered as a means of estimating the state densities where the mixing parameters define the state; however, the Kalman filter is only applicable to Gaussian tracking problems. One method recently proposed to deal with non-Gaussian source signals is based on a Monte Carlo numerical integration that employs particle filters (Everson and Roberts[172]). Samples drawn from the estimated posterior probability density of the mixing parameters, conditioned on the measured signals at each time point, are used in the numerical integration. Can such an approach throw any light on features of human hearing?

The processing system of the auditory brainstem involves many crossover paths between left and right channels at various positions in the hierarchy. It is known that some of these operations are associated with locating sound sources, but it is also likely that some are associated with signal extraction/enhancement processes that operate independently of location identification. Two channels are the minimum support for the system, but biological systems are strongly influenced by developmental and maintenance cost/benefit criteria, and two ears are all that many animals seem able to afford. Being outside Nature's evolutionary processes, engineering systems are in a position to improve on human/animal performance by employing more sensors and processors. The use of multiple microphones when employing correlation-based blind source separation of speech in realistic acoustic environments is being explored, and encouraging results have been reported (Parra and Spence[173]).

The literature implies the absence of a link between the ability to locate sound sources and the ability to reject undesired acoustic signals. That is, knowing where the sound sources are appears not to be used by humans in the speech separation/enhancement task, and in partial confirmation of this, many sufferers of hearing loss still have access to the low-frequency signals that support binaural lateralization. This could be viewed as undermining a principal objective of computational auditory scene analysis, and may place limits on its applicability.

The cross-fertilization of research ideas from signal processing and machine learning/AI is advancing the boundaries of what is possible in speech processing, but there are yet many fascinating unresolved problems, especially when considering realistic acoustic scenarios. Theoretically sound scientific research fueled by the results emerging from both these domains of investigation will yield improved methods of speech processing in the future. However, new members of the relevant signal processing and AI communities should recognize the importance of keeping themselves well informed on the current research results and discussions of the psycho-acousticians and neuro-physiologists investigating human hearing.

Acknowledgments

D.R. Campbell was assisted by grants from UK EPSRC (GR/L51652) and Defeating Deafness (246:PAI:DC) during the writing of this work. M. Girolami was assisted by grants from NCR and the British Library and Information Commission (RE/092) for work that is reported here.

References

1. Byrne, D. et al., An international comparison of long-term average speech spectra, *J. Acoust. Soc Am.,* 96(4), 2108–2120, 1994.
2. Houtgast, T. and Steeneken, H.J.M., The modulation transfer function in room acoustics as a predictor of speech intelligibility, *Acustica*, 28, 66–73, 1973.
3. Junqua, J.C., The influence of acoustics on speech production: a noise-induced stress phenomenon known as the Lombard reflex, *Speech Communication*, 20, 13–22, 1996.
4. Mackenzie, G.W., *Acoustics*, Focal Press, London, UK, 1964.
5. Allen, J.B., Cochlear modeling, *IEEE ASSP Magazine*, January, 3–29, 1985.
6. Lyon, R.F., A computational model of filtering, detection and compression in the cochlea, *ICASSP '82*, 1282–1285, 1982.
7. Lyon, R.F., A computational model of binaural localization and separation, *ICASSP '83*, 1148–1151, 1983.
8. Ghitza, O., Auditory nerve representation as a front-end for speech recognition in a noisy environment, *Computer Speech and Language*, 1, 109–130, 1986.
9. Ghitza, O., Auditory models and human performance in tasks related to speech coding and speech recognition, *IEEE Trans. Speech Audio. Proc.,* 2(1)II, 115–132, 1994.
10. Yang, X., Wang, K., and Shama, S.A., Auditory representations of acoustic signals, *IEEE Trans. Inf. Theory*, 38(2), 824–839, 1992.
11. Cooke, M.P., An explicit time-frequency characterization of synchrony in an auditory model, *Computer Speech and Language*, 6, 153–173, 1992.
12. Giguere, C. and Woodland, P.C., A computational model of the auditory periphery for speech and hearing research. I. Ascending path, *J. Acoust. Soc. Am.,* 95(1), 331–342, 1994.
13. Gagnon, L., A noise reduction approach for nonstationary additive interference, *ESCA ETRW on Speech Processing in Adverse Conditions*, Cannes-Mandelieu, France, 139–142, 1992.
14. Bourlard, H. and Dupont, S., Sub-band-based speech recognition, *Proc. ICASSP '97*, Munich, Germany, 1251–1254, 1997.
15. Hermansky, H., Should recognizers have ears? *Speech Comm.,* 25, 3–27, 1998.
16. Pickles, J.O., *An Introduction to the Physiology of Hearing*, Academic Press, San Diego, CA, 1992.
17. Moore, B.C.J. (Ed.), *Hearing*, Academic Press, San Diego, CA, 1995b.
18. Konishi, M., Listening with two ears, *Sci. American*, April, 34–41, 1993.
19. Moore, B.C.J., *Perceptual Consequences of Cochlear Damage*, Oxford University Press, Oxford, U.K., 1995a.
20. Stickney, G.S. and Assmann, P.F., Acoustic and linguistic factors in the perception of bandpass-filtered speech, *J. Acoust. Soc. Am.,* 109(3), 1157–1165, 2001.
21. Plomp, R., Binaural and monaural speech intelligibility of connected discourse in reverberation as a function of azimuth of a single competing sound source (speech or noise), *Acustica*, 34, 200–211, 1976.
22. Dermody, P., Human capabilities for speech processing in noise, *ESCA ETRW on Speech Processing in Adverse Conditions*, Cannes-Mandelieu, France, 11–19, 1992.

23. Gustafsson, H.A. and Arlinger, S.D., Masking of speech by amplitude-modulated noise, *J. Acoust. Soc. Am.*, 95(1), 518–529, 1994.

24. Ma, C. and O'Shaughnessy, D., The masking of narrowband noise by broadband harmonic complex sounds and implications for the processing of speech sounds, *Speech Communication*, 14(2), 103–118, 1994.

25. Van der Heijden, M. and Kohlrausch, A., The role of envelope fluctuations in spectral masking, *J. Acoust. Soc. Am.*, 97(3), 1800–1807, 1995.

26. Summerfield, A.Q. and Assmann, P., Auditory enhancement of speech perception, *The Psychophysics of Speech Perception*, Schouten, M.E.H. (Ed.), 140–150, 1987.

27. Watkins, A.J., Spectral transitions and a perceptual compensation for effects of transmission channels, *Proc. of Speech '88: 7th FASE Symp.*, Inst. of Acoustics, 711–718, 1988.

28. Darwin, C.J., McKeown, J.D., and Kirby, D., Compensation for transmission channel and speaker effects on vowel quality, *Speech Communication*, 8(3), 221–234, 1989.

29. Kollmeier, B. and Gilkey, R.H., Binaural forward and backward masking: evidence for sluggishness in binaural detection, *J. Acoust. Soc. Am.*, 87(4), 1709–1719, 1990.

30. Butler, R.A., On the relative usefulness of monaural and binaural cues in locating sound in space, *Psychon. Sci.*, 17(4), 245–246, 1969.

31. Koenig, A.H., Allen, J.B., Berkley, D.A., and Curtis, T.H., Determination of masking-level differences in a reverberant environment, *J. Acoust. Soc. Am.*, 61(5), 1374–1376, 1977.

32. Butler, R.A. and Flannery, R., The spatial attributes of stimulus frequency and their role in monaural localization of sound in the horizontal plane, *Perception and Psychophysics*, 28(5), 449–457, 1980.

33. Oldfield, SR. and Parker, S.P.A., Acuity of sound localisation: a topography of auditory space. III. Monaural hearing conditions, *Perception*, 15, 67–81, 1986.

34. Wickesberg, R.E. and Oertel, D., Delayed, frequency-specific inhibition in the cochlear nuclei of mice: a mechanism for monaural echo suppression, *J. Neuroscience*, 10(6), 1762–1768, 1990.

35. Brown, G.J. and Cooke, M., Computational auditory scene analysis, *Comput. Speech Lang.*, 8(4), 297–336, 1994.

36. Durlach, N.I., Binaural signal detection: equalization and cancellation theory, in *Foundations of Modern Auditory Theory*, Tobias, J.V. (Ed.), Vol. II, Academic Press, London, 1972.

37. Carhart, R., Tillman, T.W., and Johnson, K.R., Release of masking for speech through interaural time delay, *J. Acoust. Soc. Am.*, 42(1), 124–138, 1967.

38. Carhart, R., Tillman, T.W., and Johnson, K.R., Effects of interaural time delays on masking by two competing signals, *J. Acoust. Soc. Am.*, 43(6), 1223–1230, 1968.

39. Hine, J.E., Martin, R.L., and Moore, D.R., Free-field binaural unmasking in ferrets, *Behavioural Neuroscience*, 108(1), 196–205, 1994.

40. Feng, A.S., Information processing in the auditory brainstem, *Current Opinion in Neurobiology*, 2, 511–515, 1992.

41. Jeffres, L.A., Binaural signal detection: Vector theory, in *Foundations of Modern Auditory Theory*, Tobias, J.V. (Ed.), Vol. II, Academic Press, London, 1972.

42. Bernstein, L.R. and Trahiotis, C., Discrimination of interaural envelope correlation and its relation to binaural unmasking at high frequencies, *J. Acoust. Soc. Am.*, 91(1), 306–316, 1992.

43. Allen, J.B., How do humans process and recognize speech? *IEEE Trans. Speech and Audio Process.*, 2(4), 567–577, 1994.

44. Fletcher, H., *Speech and Hearing*, Van Nostrand, New York, 1929.

45. Schneider, B.A., Dale, B., and Trehub, S.E., Binaural unmasking in infants, *J. Acoust. Soc. Am.*, 83(3), 1124–1132, 1988.

46. Moncur, J.P. and Dirks, D., Binaural and monaural speech intelligibility in reverberation, *J. Speech and Hearing Res.*, 10, 186–195, 1967.

47. Durlach, N.I., Gabriel, K.J., Colburn, H.S., and Trahiotis, C., Interaural correlation discrimination. II. Relation to binaural unmasking, *J. Acoust. Soc. Am.*, 79(5), 1548–1557, 1986.

48. Culling, J.F. and Summerfield, Q., Perceptual separation of concurrent speech sounds: absence of across-frequency grouping by common interaural delay, *J. Acoust. Soc. Am.*, 98, 837–846, 1995.
49. Bronkhurst, A.W. and Plomp, R., The effect of head-induced interaural time and level differences on speech intelligibility in noise, *J. Acoust. Soc. Am.*, 83(4), 1508–1516, 1988.
50. Wightman, F.L. and Kistler, D.J., The dominant role of low-frequency interaural time differences in sound localization, *J. Acoust. Soc. Am.*, 91(3), 1648–1661, 1992.
51. Schneider, B., Moraglia, G., and Jepson, A., Binocular unmasking: an analogue to binaural unmasking? *Science*, 243, 1479–1481, 1989.
52. Widrow, B. and Stearns, S.D., *Adaptive Signal Processing*, Prentice-Hall, Englewood Cliffs, NJ, 1985.
53. Bodden, M. and Blauert, J., Separation of concurrent speech signals: a cocktail party processor for speech signals, *ESCA ETRW on Speech Processing in Adverse Conditions*, Cannes-Mandelieu, France, 147–150, 1992.
54. Banks, D., Localisation and separation of simultaneous voices with two microphones, *IEE Proc.-I*, 140(4), 229–234, 1993.
55. Hoffman, M.W., Trine, T.D., Buckley, K.M., and Van Tasell, D.J., Robust adaptive microphone array processing for hearing aids: realistic speech enhancement, *J. Acoust. Soc. Am.*, 96(2), 1, 759–770, 1994.
56. Deller, J.R., Proakis, J.G., and Hansen, J.H.L., *Discrete-Time Processing of Speech Signals*, Macmillan, New York, 1993.
57. Gold, B. and Morgan, N., *Speech and Audio Signal Processing*, Wiley, New York, 2000.
58. Engebretson, A.M., Benefits of digital hearing aids, *IEEE Engineering in Medicine and Biology*, April/May, 238–248, 1994.
59. Gong, Y., Speech recognition in noisy environments: a survey, *Speech Communication*, 16, 261–291, 1995.
60. Cole, R. et al., The challenge of spoken language systems: research directions for the nineties, *IEEE Trans. Speech and Audio Process.*, 3(1), 1–21, 1995.
61. Kates, J.M., Toward a theory of optimal hearing aid processing, *J. Rehab. Res. Dev.*, 30(1), 39–48, 1993.
62. Lim, J.S. and Oppenheim, A.V., Enhancement and bandwidth compression of noisy speech, *Proc. IEEE*, 67(12), 1586–1604, 1979.
63. Moir, T.J., Campbell, D.R., and Dabis, H.S., A polynomial approach to optimal and adaptive filtering with application to speech enhancement, *IEEE Trans. Signal Process.*, 39(5), 1221–1224, 1991.
64. Van Compernolle, D., Ma, W., and Van Diest, M.M., Speech recognition in noisy environments with the aid of microphone arrays, *EUROSPEECH '89*, Paris, France, 2, 657–660, 1989.
65. Mokbel, C., Barbier, L., and Chollet, G., Adapting a HMM speech recogniser to noisy environments, *ESCA ETRW on Speech Processing in Adverse Conditions*, Cannes-Mandelieu, France, 211–214, 1992.
66. Hirsch, H.G., Intelligibility improvement of noisy speech for people with cochlear implants, *ESCA ETRW on Speech Processing in Adverse Conditions*, Cannes-Mandelieu, France, 69–72, 1992.
67. Wang, K. and Shamma, S., Self-normalization and noise-robustness in early auditory representations, *IEEE Trans. Speech Processing*, 2(3), 421–435, 1994.
68. Allen, J.B., Berkley, D.A., and Blauert, J., Multimicrophone signal processing technique to remove room reverberation from speech signals, *J. Acoust. Soc. Am.*, 62(4), 912–915, 1977.
69. Gilloire, A., Experiments with sub-band acoustic echo cancellers for teleconferencing, *ICASSP '87*, 2141–2144, 1987.
70. Goulding, M.M. and Bird, J.S., Speech enhancement for mobile telephony, *IEEE Trans. Vehic. Technol.*, 39(4), 316–326, 1990.

71. Walker, G., Byrne, D., and Dillon, H., The effects of multichannel compression/expansion amplification on the intelligibility of nonsense syllables in noise, *J. Acoust. Soc. Am.*, 76(3), 746–757, 1984.

72. White, M.W., Compression systems for hearing aids and cochlear prostheses, *J. Rehab. Res.*, 23(1), 25–39, 1986.

73. Plomp, R., The negative effect of amplitude compression in multichannel hearing aids in the light of the modulation-transfer function, *J. Acoust. Soc. Am.*, 83(6), 2322–2327, 1988.

74. Kollmeier, B., Speech enhancement by filtering in the loudness domain, *Acta Otolaryngol (Stockh), Suppl.*, 469, 207–214, 1990.

75. Stone, M.A. and Moore, B.C.J., Spectral feature enhancement for people with sensorineural hearing impairment: effects on speech intelligibility and quality, *J. Rehab. Res. Dev.*, 29(2), 39–56, 1992.

76. Baer, T., Moore, B.C.J., and Gatehouse, S., Spectral contrast enhancement of speech in noise for listeners with sensorineural hearing impairment: effects on intelligibility, quality and response times, *J. Rehab. Res. Dev.*, 30(1), 49–72, 1993.

77. Yund, E.W. and Buckles, K.M., Multichannel compression hearing aids: effect of number of channels on speech discrimination in noise, *J. Acoust. Soc. Am.*, 97(2), 1206–1222, 1995a.

78. Ming, J. and Jack Smith, F., Union: a new approach for combining sub-band observations for noisy speech recognition, *Speech Communication*, 34, 41–55, 2001.

79. Morris, A., Hagen, A., Glotin, H., and Bourlard, H., Multi-stream adaptive evidence combination for noise robust ASR, *Speech Communication*, 34, 25–40, 2001.

80. Yund, E.W. and Buckles, K.M., Enhanced speech perception at low signal-to-noise ratios with multichannel compression hearing aids, *J. Acoust. Soc. Am.*, 97(2), 1224–1240, 1995b.

81. Toner, E. and Campbell, D.R., Speech-enhancement-based conceptually on auditory processing, *ESCA ETRW on Speech Processing in Adverse Conditions*, Cannes-Mandelieu, France, 151–154, 1992.

82. Weinstein, E., Feder, M., and Oppenheim, A.V., Multi-channel signal separation by decorrelation, *IEEE Tran. Speech and Audio Process.*, 4(1), 405–413, 1993.

83. Silverman, H.F., Some analysis of microphone arrays for speech data acquisition, *IEEE Trans. Acoust. Speech and Signal Process.*, 35(12), 1699–1712, 1987.

84. Silverman, H.F. and Kirtman, S.E., A two-stage algorithm for determining talker location from microphone array data, *Computer Speech and Language*, 6, 129–152, 1992.

85. Grenier, Y., A microphone array for car environments, *ICASSP '92*, I, 305–308, 1992.

86. Peterson, P.M., Durlach, N.I., Rabinowitz, W.M., and Zurek, P.M., Multimicrophone adaptive beamforming for interference reduction in hearing aids, *J. Rehab. Res. Dev.*, 24(4), 103–110, 1987.

87. Zurek, P.M., Greenberg, J.E., and Peterson, P.M., Sensitivity to design parameters in an adaptive-beamforming hearing aid, *Proc. ICASSP '90*, 2, 1129–1132, 1990.

88. Greenberg, J.E. and Zurek, P.M., Evaluation of an adaptive beamforming method for hearing aids, *J. Acoust. Soc. Am.*, 91(3), 1662–1676, 1992.

89. Soede, W., Bilsen, F.A., and Berkhout, A.J., Assessment of a directional microphone array for hearing-impaired listeners, *J. Acoust. Soc. Am.*, 94(2), pt. 1, 799–808, 1993.

90. Soede, W., Berkhout, A.J., and Bilsen, F.A., Development of a directional hearing instrument based on array technology, *J. Acoust. Soc. Am.*, 94(2), pt. 1, 785–798, 1993.

91. Stadler, R.W. and Rabinowitz, W.M., On the potential of fixed arrays for hearing aids, *J. Acoust. Soc. Am.*, 94(3), pt. 1, 1332–1342, 1993.

92. Van Hoesel, R.J.M. and Clark, G.M., Evaluation of a portable two-microphone adaptive beam-forming speech processor with cochlear implant patients, *J. Acoust. Soc. Am.*, 97(4), 2498–2503, 1995.

93. Bloom, P.J. and Cain, G.D., Evaluation of two-input speech dereverberation techniques, *ICASSP '82*, 164–167, 1982.

94. Yamada, H., Wang, H., and Itakura, F., Recovering of broad-band reverberant speech signal by sub-band MINT method, *ICASSP '91*, Toronto, Canada, 969–972, 1991.

95. Ferrara, E.R. and Widrow, B., Multichannel adaptive filtering for signal enhancement, *IEEE Trans. Acoust. Speech and Sig. Process.*, 29(3), 766–770, 1981.

96. Griffith, L.J. and Jim, C.W., An alternative approach to linearly constrained adaptive beamforming, *IEEE Trans. Antennas and Propag.*, AP-30, 27–34, 1982.

97. Farrel, K., Mammone, R.J., and Flanagan, J.L., Beamforming microphone arrays for speech enhancement, *ICASSP '92*, San Francisco, CA, I, 285–288, 1992.

98. Oh, S., Viswanathan, V., and Papamichalis, P., Hands-free voice communication in an automobile with a microphone array, *ICASSP '92*, San Francisco, CA, I, 281–284, 1992.

99. Chazan, D., Medan, Y., and Shvadron, U., Noise cancellation for hearing aids, *IEEE Trans. Acoust. Speech and Sig. Proc.*, 36(11), 1697–1705, 1988.

100. Van Compernolle, D., Hearing aids using binaural processing principles, *Acta Otolaryngol (Stockh), Suppl.*, 469, 76–84, 1990.

101. Peterson, P.M., Wei, S., Rabinowitz, W.M., and Zurek, P.M., Robustness of an adaptive beamforming method for hearing aids, *Acta Otolaryngol (Stockh), Suppl.*, 469, 85–90, 1990.

102. Gudvangen, S. and Flockton, S.J., Modelling of acoustic transfer functions for echo cancellers, *IEE Proc. Vis. Image Signal Process.*, 142(1), 47–51, 1995.

103. Rabiner, L.R., Crochiere, R.E., and Allen, J.B., FIR system modeling and identification in the presence of noise and with band-limited inputs, *IEEE Trans. Acoust. Speech Sig. Process.*, ASSP-26(4), 319–333, 1978.

104. Cowan, C.F.N., Performance comparisons of finite linear adaptive filters, *IEE Proc. F*, 134(3), 211–215, 1987.

105. Hassibi, B., Sayed, A.H., and Kailath, T., Optimality of the LMS algorithm, *IEEE Trans. Signal Proc.*, 44(2), 267–280, 1996.

106. Eweda, E., Comparison of RLS, LMS, and Sign algorithms for tracking randomly time-varying channels, *IEEE Trans. Sig. Process.*, 42(11), 2937–2944, 1994.

107. Strube, H.W., Separation of several speakers recorded by two microphones (cocktail-party processing), *Signal Processing*, 3, 355–364, 1981.

108. Dabis, H.S., Moir, T.J., and Campbell, D.R., Speech enhancement by recursive estimation of differential path transfer functions, *ICSP '90*, Beijing, PRC, 345–348, 1990.

109. Brey, R.H., Robinette, M.S., Chabries, D.M., and Christiansen, R.W., Improvement in speech intelligibility in noise employing an adaptive filter with normal and hearing-impaired subjects, *J. Rehab. Res. Dev.*, 24(4), 75–86, 1987.

110. Weiss, M., Use of an adaptive noise canceller as an input preprocessor for a hearing aid, *J. Rehab. Res. Dev.*, 24(4), 93–102, 1987.

111. Lu, M. and Clarkson, P.M., The performance of adaptive noise cancellation systems in reverberant rooms, *J. Acoust. Soc. Am.*, 93(2), 1122–1135, 1993.

112. Wallace, R.B. and Goubran, R.A., Improved tracking adaptive noise canceller for nonstationary environments, *IEEE Trans. Sig. Process.*, 40(3), 700–703, 1992.

113. Le Bouquin-Jeannes, R., Faucon, G., Azirani, A.A., and Ehrmann, F., Speech enhancement using sub-band decomposition and comparison with full-band techniques, *Signal Processing VII: Theories and Applications*, M. Holt, C. Cowan, P. Grant, and W. Sandham (Eds.), 1206–1209, 1994.

114. Guy, R.W. and Abdou, A., A measurement system and method to investigate the directional characteristics of sound fields in enclosures, *Noise Control J.*, 42(1), 8–18, 1994.

115. Kollmeier, B., Peissig, J., and Hohmann, V., Binaural noise-reduction hearing aid scheme with real-time processing in the frequency domain, *Scand. Audiol., Suppl.*, 38, 28–38, 1993.

116. Gagnon, L. and McGee, W.F., Speech enhancement using resonator filter banks, *ICASSP '91*, Toronto, Canada, 981–984, 1991.

117. Hansen, J.H.L. and Clements, M.A., Constrained iterative speech enhancement with application to speech recognition, *IEEE Trans. Signal Process.*, 39(4), 795–805, 1991.

118. Bateman, D.C., Bye, D.K., and Hunt, M.J., Spectral contrast normalisation and other techniques for speech recognition in noise, *ICASSP '92*, San Francisco, CA, I, 241–244, 1992.

119. Lippman, R., Speech recognition by machines and humans, *Speech Communication*, 22, 1–15, 1997.

120. Campbell, D.R., Moir, T.J., and Dabis, H.S., Multivariable polynomial matrix formulation of adaptive noise cancelling, *Signal Processing*, 26, 177–183, 1992.

121. Shields, P. and Campbell, D.R., A multi-microphone sub-band adaptive speech enhancement system employing diverse sub-band processing, *Speech Communication*, 25, 165–175, 1998.

122. Le Bouquin-Jeannes, R. and Faucon, G., Study of a voice activity detector and its influence on a noise reduction system, *Speech Communication*, 16, 245–254, 1995.

123. Agaiby, H. and Moir, T.J., Knowing the wheat from the weeds in noisy speech, *Eurospeech 97*, Rhodes, Greece, 1119–1122, 1997.

124. Toner, E. and Campbell, D.R., Speech enhancement using sub-band intermittent adaption, *Speech Communication*, 12, 253–259, 1993.

125. Mahalanobis, A., Song, S., Mitra, S.K., and Petraglia, M.R., Adaptive FIR filters based on structural sub-band decomposition for system identification problems, *IEEE Trans. Circs. Syst.—II: Analog and Digital Signal Processing*, 40(6), 375–381, 1993.

126. Gilloire, A. and Vetterli, M., Adaptive filtering in sub-bands with critical sampling: analysis, experiments, and application to acoustic echo cancellation, *IEEE Trans. Sig. Process.*, 40(8), 1862–1875, 1992.

127. Knecht, W.G., Schenkel, M.E., and Moschytz, G.S., Neural network filters for speech enhancement, *IEEE Trans. Speech and Audio Proc.*, 3(6), 433–438, 1995.

128. Van Compernolle, D. and Van Gerven, S., Signal separation in a symmetric adaptive noise canceler by output decorrelation, *ICASSP '92*, IV, 221–224, 1992.

129. Van Gerven, S., Van Compernolle, D., Nguyen-Thi, H.L., and Jutten, C., Blind separation of sources: a comparative study of a second- and fourth-order solution, *Signal Processing VII: Theories and Applications*, M. Holt, C. Cowan, P. Grant, and W. Sandham (Eds.), 1153–1156, 1994.

130. Haykin, S., *Neural Networks, A Comprehensive Foundation*, Prentice-Hall, Englewood Cliffs, NJ, 1999.

131. Vapnik, V., *The Nature of Statistical Learning Theory*, Springer-Verlag, Berlin, 1995.

132. Muller, K.R., Mika, S., Ratsch, G., Tsuda, K., and Scholkopf, B., An introduction to kernel-based learning algorithms, *IEEE Trans. Neural Networks*, 12(2), 181–201, 2001.

133. Scholkopf, B., Mika, S., Burges, C.J.C., Knirock, P., Muller, K.R., Ratsch, G., and Smola, A., Input space vs. feature space in kernel-based methods, *IEEE Trans. Neural Networks*, 10(5), 1000–1017, 1999.

134. Jaakkola, T. and Haussler, D., Exploiting generative models in discriminative classifiers, in *Adv. Neural Information Processing Systems* 11, M.S. Kearns, S.S. Solla, and D.A. Cohn (Eds.), MIT Press, 1999.

135. Hebb, D.O., *The Organization of Behavior: A Neuropsychological Theory*, Wiley, 1949.

136. Comon, P., Independent component analysis, a new concept? *Signal Processing*, 36, 287–314, 1994.

137. Hertz, J., Krogh, A., and Palmer, R.G., *Introduction to the Theory of Neural Computation*, Addison-Wesley, 1991.

138. Kohonen, T., *Self-Organising Maps*, Springer Verlag, 1995.

139. Carpenter, G. and Grossberg, S., *Pattern Recognition by Self-Organising Neural Networks*, MIT Press, 1991.

140. Bourland, H. and Morgan, N., *Connectionist Speech Recognition — A Hybrid Approach*, Kluwer Academic Press, 1994.

141. Wan, E. and Nelson, A., Networks for speech enhancement, in *Handbook of Neural Networks for Speech Processing*, S. Katagiri (Ed.), Artech House, Boston, 1998.

142. Oppenheim, A.V. and Schafer, R.W., *Discrete-Time Signal Processing*, Prentice-Hall, ISBN: 0-13-216771-9, 1989.

143. Barnett, S., *Matrices in Control Theory*, Van Nostrand, ISBN: 0-442-00581-4, 1971.

144. Chan, D.C.B., Godshill, S.J., and Rayner, P.J.W., Multi-channel Multi-tap Signal Separation By Output Decorrelation, Cambridge University, CUED/F-INFENG/TR 250, ISSN 0951–9211, 1996.

145. Van Gerven, S., Adaptive Noise Cancellation and Signal Separation with Applications to Speech Enhancement., Ph.D. thesis, Katholieke Universiteit Leuven, ISBN 90-5682-025-7, 1996.

146. Foldiak, P., Adaptive network for optimal linear feature extraction. *IEEE/INNS Int. J. Conf. Neural Networks*, 1, pp. 401–405. Washington, D.C.: Institute of Electrical and Electronics Engineering, San Diego, 1989.

147. Matsuoka, K., Masahiro, O., and Kawamoto, M., A neural net for blind separation of nonstationary signals, *Neural Networks*, 8, 411–419, 1995.

148. Torkkola, K., Blind separation of convolved sources based on information maximisation, *IEEE Workshop on Neural Networks for Signal Processing, NNSP '96*, Kyoto, Japan, 1996.

149. Platt, J.C. and Faggin, F., Networks for the separation of sources that are superimposed and delayed, *Neural Information Processing Systems*, 4, 730–737, 1992.

150. Jutten, C. and Herault, J., Blind separation of sources. Part 1. An adaptive algorithm based on neuromimetic architecture, *Signal Processing*, 24, 1–10, 1991.

151. Nguyen Thi, H.L. and Jutten, C., Blind source separation for convolutive mixtures, *Signal Processing*, 45(2), 209–229, 1995.

152. Yellin, D. and Weinstein, E., Multichannel signal separation: methods and analysis, *IEEE Trans. Signal Processing*, 44(1), 106–118, 1996.

153. Principe, J.C., Wang, C., and Wu, H.S., Temporal decorrelation using teacher forcing anti-Hebbian learning and its application in adaptive blind source separation, *NIPS '96, Blind Signal Processing Workshop*, http://www.bip.riken.go.jp/absl/back/nips96ws/nips96ws.html, 1996.

154. Gorokhov, A. and Loubaton, P., Second-order blind identification of convolutive mixtures with temporally correlated sources: a subspace method, *Signal Processing VII, Theories and Applications*, Trieste, Italy, Elsevier, 1996.

155. Lindgren, U., Sahlin, H., and Broman, H., Source separation using second-order statistics, *Signal Processing VII, Theories and Applications*, Trieste, Italy, Elsevier, 1996.

156. Lambert, R., Multichannel Blind Deconvolution: FIR Matrix Algebra and Separation of Multipath Mixtures. Ph.D. thesis, University of Southern California, 1996. Available at http://home.socal.rr.com/russdsp/Mydis.zip.

157. Cichocki, A., Amari, S.I., and Cao, J., Blind separation of delayed and convolved signals with self-adaptive learning rate, *Int. Symp. Nonlinear Theory and Applications*, 229–232, 1996.

158. Lee, T.W., Bell, A.J., and Orgmeister, R., Blind source separation of real world signals, in *Proc. IEEE/ICNN, Int. Conf. Neural Networks*, 4, 2129–2134, 1997.

159. Charkani, N. and Deville, Y., Optimisation of the asymptotic performance of time-domain convolutive source separation algorithms, *Proc. European Symp. Artificial Neural Networks*, pp. 273–278, ISBN 2-9600049-7-3, 1997.

160. Deville, Y. and Charkani, N., Analysis of the stability of time domain source separation algorithms for convolutively mixed signals, *Proc. Int. Conf. Acoustics, Speech and Signal Processing*, 3, 1835–1839, 1997.

161. Cox, D.R. and Hinkley, D.V. *Theoretical Statistics*, Chapman & Hall, 1974.
162. Bell, A. and Sejnowski, T., An information maximization approach to blind separation and blind deconvolution, *Neural Computation*, 7, 1129–1159, 1995.
163. McDonald, R., Signal-to-noise and idle channel performance of differential pulse code modulation systems—Particular applications to voice signals, *BSTJ*, 45, 1123–1151, 1966.
164. Haykin, S., *Adaptive Filter Theory*, 3rd ed., Prentice-Hall, Englewood Cliffs, NJ, 1996.
165. Shields, P.W., Girolami, M., Campbell, D.R., and Fyfe, C., Adaptive processing schemes inspired by binaural unmasking for enhancement of speech corrupted with noise and reverberation, *Proc.1st European Workshop on Neuromorphic Systems*, World Scientific, 1997.
166. Smith, N., Gales, M., and Niranjan, M., Data-Dependent Kernels in SVM Classification of Speech Patterns. CUED/F-INFENG/TR.387, Cambridge University Engineering Dept. 2001.
167. Bazzi, I. and Katebi, D., Using support vector machines for spoken digit recognition, *Proc. 6th Int. Conf. Spoken Language Processing*, Beijing, China, 2000.
168. Fine, S., Navratil, J., and Gopinath, R., A hybrid GMM/SVM approach to speaker identification, *Proc. Int. Conf. Acoustics, Speech and Signal Processing*, vol. 1, 2001.
169. Lippman, R., Accurate consonant perception without mid-frequency speech energy, *IEEE Trans. Speech and Audio Processing*, 4(1), 66–69, 1996.
170. Warren, R.M., Riener, K.R., Bashford, J.A., and Brubaker, B.S., Spectral redundancy: intelligibility of sentences heard through narrow spectral slits, *Perception and Psychophysics*, 57(2), 175–182, 1995.
171. Greenberg, S., Understanding speech understanding: towards a unified theory of speech perception, *Proc. ESCA Workshop on Auditory Basis of Speech Perception*, Keele, U.K., 1–8, 1996.
172. Everson, R.M. and Roberts, S.J., Particle filters for nonstationary ICA, *Adv. Independent Components Analysis*, M. Girolami (Ed.), Springer, 23–41, 2000.
173. Parra, L. and Spence, C., Separation of non-stationary natural signals, in *Independent Components Analysis: Principles and Practice*, R. Everson and S. Roberts (Eds.), Cambridge University Press, 2001.
174. Martin, M. (Ed.) *Speech Audiometry*, 2nd ed., Whurr Publ. Ltd., London, U.K., 1997.

2

Fuzzy Theory, Methods, and Applications in Nonlinear Signal Processing

Kenneth E. Barner
University of Delaware

2.1   Introduction

Signal processing theory and practice have been dominated by linear methods that satisfy the principle of superposition. Linear signal processing enjoys the rich theory of linear systems and, in many applications, linear signal processing algorithms prove to be optimal. Most importantly, linear filters are inherently simple to implement, which is perhaps the dominant reason for their widespread use in practice. Although linear filters will continue to play an important role in signal processing, nonlinear filters are emerging as viable alternative solutions. The major force behind this paradigm shift is the growth in challenging applications that require the use of increasingly sophisticated signal processing algorithms. At the same time, the ongoing advances of computers and digital signal processors, in terms of speed, size, and cost, make the implementation of sophisticated algorithms practical and cost-effective.

Unlike linear signal processing, nonlinear signal processing lacks a unified and universal set of tools for analysis and design. Hundreds of nonlinear signal processing algorithms have been proposed. Most of the proposed methods, although well tailored for a given application, are not generally applicable. While nonlinear signal processing is a dynamic, rapidly growing field, a large class of nonlinear signal processing algorithms can be studied with fundamentals that are well formulated. In this chapter, we view the filtering problem from a Maximum Likelihood (ML) approach. It is shown that the ML optimization leads directly to the class of linear filters for signals with the Gaussian statistics and to the class of nonlinear weighted median filters for signals with double exponential, or Laplacian, distributions.

The ML development that leads to the class of weighted median filters shows that this class of filters operates jointly on spatial* and rank (SR) order information. The joint utilization of SR information has proved advantageous for two primary reasons: (1) spatial ordering can be used to exploit correlations between neighboring samples and (2) rank order can be used to isolate outliers and ensure robust behavior. Although the exploitation of SR ordering information in nonlinear filtering algorithms has yielded good results, traditional ordering information is based on a crisp relationship. Such crisp relations yield no information on important quantities such as sample spread or diversity.

The simple relaxation of the ordering relation from a crisp (binary) operator to a more general affinity (real valued) operator leads to the concept of fuzzy SR orderings. Thus, fuzzy SR orderings not only relate spatial and rank orderings, but also contain information on sample spread (affinity). Powerful fuzzy nonlinear filtering algorithms can be realized by embedding fuzzy SR ordering information into the filter structure. Such filters can be simply realized as (1) generalizations of existing nonlinear filters that employ fuzzy, rather than crisp, ordering relations, (2) generalizations of linear filters that embed fuzzy SR ordering information into the traditional weighted sum filter structure, or (3) new filter structures specifically designed to exploit fuzzy SR information.

To illustrate the advantages of fuzzy SR ordering information, consider the application of filtering in the time–frequency (TF) plane.[35] A common problem with TF representations is the presence of undesirable cross-terms that must be filtered out, while preserving the desired auto-components. Figure III.2.1 depicts a comparison of the TF distribution of a bat echo-location chirp and several methods designed to minimize the cross-components. As an inspection of the resulting TF plots shows, the method incorporating fuzzy SR information (center affine filter, discussed in Section 2.4, Figure III.2.1(f)) clearly produces the best results compared to more traditional methods.

The remainder of this chapter theoretically motivates the use of fuzzy SR ordering information, details the theory of fuzzy ordering and fuzzy order statistics, and develops several classes of fuzzy nonlinear filters based on these concepts. Specifically, the classes of affine filters and fuzzy weighted median filters are developed and applied to a wide range of signal processing and communications problems. These topics are covered in this chapter as follows. Section 2.2 begins with a theoretical discussion of ML estimation and formally develops the concept of SR ordering. The crisp ordering relation is relaxed in Section 2.3, which develops the concepts of sample affinity and the resulting fuzzy SR ordering. Fuzzy filter generalizations are developed in Section 2.4, where we focus on the fuzzy weighted median and affine filter classes. In the case of affine filters, the discussion is limited to the two important median affine and center affine filter subclasses. Section 2.5 presents the results of applying the developed filters to several important signal processing and communications problems, including robust frequency selective filtering, Synthetic Aperture Radar image filtering, TF domain filtering, multiresolution signal representations, image smoothing, image zooming, and multiuser detection. Finally, conclusions are drawn in Section 2.6.

---

*The ordering is spatial in two-dimensional signals cases (i.e., images) and temporal for one-dimensional time sequences.

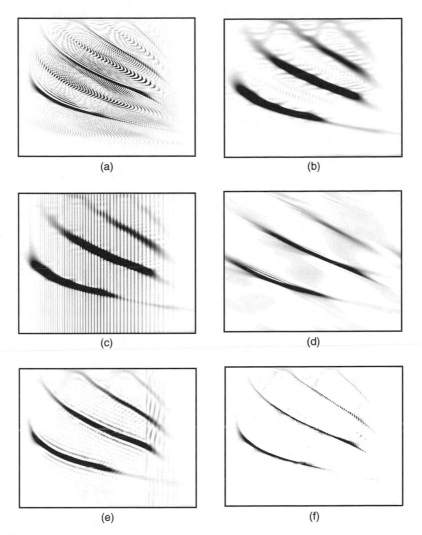

FIGURE III.2.1   Time-frequency representations of the bat echo-location chirp using (a) Wigner distribution, (b) smoothed pseudo Wigner distribution, (c) Choi-Williams distribution, (d) Baraniuk–Jones method 1, (e) Baraniuk–Jones method 2, and (f) fuzzy SR information-based center affine filtered WD. (From Arce, G.R. and Hasan, S.R., Elimination of interference terms of the discrete Wigner distribution using nonlinear filtering, *IEEE Trans. Signal Proc.*, 48, 2321–2331, Aug. 2000. With permission.)

2.2   Maximum Likelihood Estimation and Spatial–Rank Ordering

2.2.1   ML Estimation

To motivate the development of theoretically sound signal processing methods, consider first the modeling of observation samples. In all but trival cases, nondeterministic methods must be used. As most signals have random components, probability-based models from a powerful set of modeling methods. Accordingly, signal processing methods have deep roots in satistical estimation theory.

Consider a set of N observation samples. In most applications, the observation samples are captured by a moving window centered at some position **n**, where we consider the general case of a vector index to account for multidimensional signals. Such samples will be denoted as

$\mathbf{x}[\mathbf{n}] = [x_1][\mathbf{n}], x_2[\mathbf{n}], \ldots, x_N[\mathbf{n}]]^T$. For notational convenience, we will drop the index $\mathbf{n}$, unless necessary for clarity.

Assume now that we model samples as independent and identically distributed (i.i.d.). Each observation sample is then characterized by the common probability density function (pdf) $f_\beta(x)$, where β is the mean, or location, of the distribution. Often β is information carrying and unknown, and thus must be estimated. The ML estimate of the location is achieved by maximizing, with respect to β, the probability of observing $x_1, x_2, \ldots, x_N$. For i.i.d. samples, this results in:

$$\hat{\beta} = \arg \max_{\beta} \prod_{i=1}^{N} f_\beta(x_i). \qquad (\text{III.2.1})$$

Thus, the value of β that maximizes the product of the pdfs consitutes the ML estimate.

The degree to which the ML estimate accurately represents the location is, to a large extent, dependent on how accurately the model distribution represents the true distribution of the observation process. To allow for a wide range of sample distributions, the commonly assumed Gaussian distribution can be generalized by allowing the exponential rate of tail decay to be a free parameter. This results in the *generalized Gaussian* density function:

$$f_\beta(x) = c e^{-(|x-\beta|/\sigma)^p}, \qquad (\text{III.2.2})$$

where p governs the rate of tail decay, $c = p/(2\sigma\Gamma(1/p))$, and $\Gamma(\cdot)$ is the Gamma function. This includes the standard Gaussian distribution as a special case ($p = 2$). For $p < 2$, the tails decay slower than in the Gaussian case, resulting in a heavier tailed distribution. Of particular interest is the case $p = 1$, which yields the double exponential, or Laplacian, distribution:

$$f_\beta(x) = \frac{1}{2\sigma} e^{-|x-\beta|/\sigma}. \qquad (\text{III.2.3})$$

The ML criteria can be applied to optimatlly estimate the location of a set of N samples distributed according to the generalized Gaussian distribution, yielding:

$$\hat{\beta} = \arg \max_{\beta} \prod_{i=1}^{N} c e^{-(|x-\beta|/\sigma)^p} = \arg \min_{\beta} \sum_{i=1}^{N} |x_i - \beta|^p. \qquad (\text{III.2.4})$$

Determining the ML estimate is thus equivalent to minimizing:

$$G_p(\beta) = \sum_{i=1}^{N} |x_i - \beta|^p, \qquad (\text{III.2.5})$$

with respect to β. For the Gaussian case ($p = 2$), this reduces to the sample mean, or average:

$$\hat{\beta} = \arg \min_{\beta} G_2(\beta) = \frac{1}{N} \sum_{i=1}^{N} x_i. \qquad (\text{III.2.6})$$

A much more robust estimator is realized if the underlying sample distribution is taken to be the heavy-tailed Laplacian distribution ($p = 1$). In this case, the ML estimator of location is given by the value β that minimizes the sum of least absolute deviations:

$$G_1(\beta) = \sum_{i=1}^{N} |x_i - \beta|, \qquad (\text{III.2.7})$$

which can easily be shown to be the sample median:

$$\hat{\beta} = \arg \max_{\beta} G_1(\beta) = \text{MED}[x_1, x_2, \ldots, x_N]. \qquad (\text{III.2.8})$$

The sample mean and median thus play analogous roles in location estimation: while the mean is associated with the Gausssian distribution, the median is related to the Laplacian distribution, which has heavier tails and provides a better model for many signals (i.e., images), as well as those contaminated with impulsive outliers.[1-4]

Although the median is a robust estimator that possesses many optimality properties, the performance of the median filter is limited by the fact that it is spatially blind. That is, all observation samples are treated equally regardless of their location within the observation window. This limitation is a direct result of the i.i.d. assumption made in the filter development. A much richer class of filters is realized if this assumption is relaxed to the case of independent, but not identically distributed, samples.

Consider the generalized Gaussian distribution case where the observation samples have a common location parameter β, but where each x_i has a (possibly) unique scale parameter σ_i. Incorporating the unique scale parameters into the ML criteria yields a location estimate given by value of β minimizing:

$$G_p(\beta) = \sum_{i=1}^{N} \frac{1}{\sigma_i^p} |x_i - \beta|^p. \qquad (\text{III.2.9})$$

In the special case of the standard Gaussian distribution ($p = 2$), the ML estimate reduces to the normalized weighted average:

$$\hat{\beta} = \arg\min_{\beta} \sum_{i=1}^{N} \frac{1}{\sigma_i^2}(x_i - \beta)^2 = \frac{\sum_{i=1}^{N} w_i \cdot x_i}{\sum_{i=1}^{N} w_i}, \qquad (\text{III.2.10})$$

where $w_i = 1/\sigma_i^2 > 0$. In the heavier-tailed Laplacian distribution special case ($p = 1$), the ML estimate reduces to the weighted median (WM), originally introduced over a hundred years ago by Edgemore,[5] and defined as:

$$\hat{\beta} = \arg\min_{\beta} \sum_{i=1}^{N} \frac{1}{\sigma_i} |x_i - \beta| = \text{MED}[w_1 \diamond x_1, w_2 \diamond x_2, \ldots, w_N \diamond x_N], \qquad (\text{III.2.11})$$

where $w_i = 1/\sigma_i > 0$ and $\diamond$ is the replication operator defined as $w_i \diamond x_i = \overbrace{x_i, x_i, \ldots, x_i}^{w_i \text{ times}}$.

A trival, but yet very important, special case of the weighted median filter (as well as the weighted sum filter) is the identity operator. Assuming the samples constitute an ordered (temporal or spatial) set from an observed process, let $\delta_c = \frac{N+1}{2}$ be the index of the center observation sample. Then it is easy to see that:

$$x_{\delta_c} = \text{MED}[w_1 \diamond x_1, w_2 \diamond x_2, \ldots, w_N \diamond x_N] \qquad (\text{III.2.12})$$

for $w_{\delta_c} = 1$ and $w_i = 0$ for $i = \neq \delta_c$. Thus, the weighted median has two important special cases: (1) the standard median filter, which operators strictly on rank order information, and (2) the identity filter, which operators strictly on spatial order.

To illustrate the importance of these cases, and the corresponding orderings upon which they are based, consider the filtering of a moving average (MA) process corrupted by Laplacian noise. Figure III.2.2 shows the correlation between the desired MA process and the filter outputs, for the identity and median cases, as a function of SNR in the corrupted observation. The figure shows that for high SNRs, the identity filter output (central observation sample) has the highest correlation with the desired output, while for low SNRs the median has the highest correlation. Thus, this simple example illustrates the importance of spatial order in high SNR cases and rank order in low SNR cases.

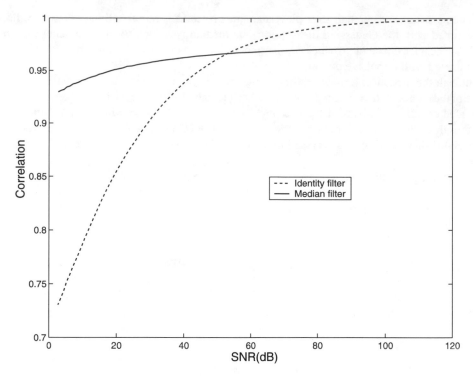

FIGURE III.2.2   Correlation coefficient between a MA process and the identity and (window size 5) median filter outputs as a function of SNR in the Laplacian noise corrupted observation.

The general formulation of the weighted median filter attempts to exploit both spatial ordering, through repetition of samples, and rank ordering, through median selection. The filter is thus able to exploit spatial correlations among neighboring samples and limit the influence of outliers. A more formal consideration of spatial and rank ordering can be obtained by considering the full ordering relations betwccn obscrvcd samples.

2.2.2   Spatial–Rank Ordering

To formally relate the spatial ordering and rank ordering of samples in a signal processing application, consider again the typical case in which an observation window passes over an observation sequence in a predefined scanning pattern. At each location $\mathbf{n}$ the observation window covers N samples, which can be indexed according to their spatial location and written in vector form:

$$\mathbf{x}_\ell[\mathbf{n}] = [x_1[\mathbf{n}], x_2[\mathbf{n}], \ldots, x_N[\mathbf{n}]]^T. \tag{III.2.13}$$

The subscript ℓ has now been added to explicitly indicate that the samples are indexed according to their natural spatial order within the observation image. A second natural ordering of the observed samples is rank order, which yields the order statistics of the observation samples:

$$x_{(1)}[\mathbf{n}] \le x_{(2)}[\mathbf{n}] \le \cdots \le x_{(N)}[\mathbf{n}]. \tag{III.2.14}$$

Writing the order statistics in vector form yields the rank order observation vector:

$$\mathbf{x}_L[\mathbf{n}] = [x_{(1)}[\mathbf{n}], x_{(2)}[\mathbf{n}], \ldots, x_{(N)}[\mathbf{n}]]^T. \tag{III.2.15}$$

Again, the spatial location of the observation window is only to be shown explicitly when needed for clarity. Thus, we write the spatial order and rank order observation vectors as simply $\mathbf{x}_\ell$ and $\mathbf{x}_L$.

A crisp, or binary, relation between the samples of two sets A and B can be denoted by a crisp membership function $\mu_C(a, b) : A \times B \mapsto \{0, 1\}, a \in A, b \in B$. Note that both the spatial and rank ordered samples consititute the same set, $X = \{x_1, \ldots, x_N\} = \{x_{(1)}, \ldots, x_{(N)}\}$. Thus, to relate the spatial and rank orderings of the samples in X, we can define the SR matrix:

$$\mathbf{R} = \begin{bmatrix} R_{1,(1)} & \cdots & R_{1,(N)} \\ \vdots & \ddots & \vdots \\ R_{N,(1)} & \cdots & R_{N,(N)} \end{bmatrix} \tag{III.2.16}$$

where $R_{i,(j)} = \mu_C(x_i, x_{(j)})$ and:

$$\mu_C(x_i, x_{(j)}) = \begin{cases} 1 & \text{if } x_i \text{ has rank } j \, (x_i \leftrightarrow x_{(j)}) \\ 0 & \text{otherwise} \end{cases} \tag{III.2.17}$$

This crisp relation produces a binary relation matrix, i.e., $R_{i,(j)} \in \{0, 1\}$.

The matrix $\mathbf{R}$ contains the full joint SR information of the observation set X. Thus, $\mathbf{R}$ can be used as a transformation between the two orderings and to extract the marginal vectors $\mathbf{x}_\ell$ and $\mathbf{x}_L$. The transformations yielding the spatial and rank order indices are given by:

$$\mathbf{s} = [1 : N] \, \mathbf{R} \quad \text{and} \quad \mathbf{r}^T = \mathbf{R} \, [1 : N]^T, \tag{III.2.18}$$

where $[1 : N] = [1, 2, \ldots, N]$, and $\mathbf{s} = [s_1, s_2, \ldots, s_N]$ and $\mathbf{r} = [r_1, r_2, \ldots, r_N]$ are the spatial and rank order index vectors, respectively, i.e., $x_{s_j} \leftrightarrow x_{(j)}$ and $x_i \leftrightarrow x_{(r_i)}$ for $i, j - 1, 2, \ldots, N$. Similarly, the spatial and rank ordered samples are related by:

$$\mathbf{x}_\ell = \mathbf{R}\mathbf{x}_L \text{ and } \mathbf{x}_L^T = \mathbf{x}_\ell \mathbf{R}^T. \tag{III.2.19}$$

As an illustrative example, suppose a three-sample observation window is used and a particular spatial order observation vector is given by $\mathbf{x}_\ell = [10, 1, 2]$. This results in the SR matrix:

$$\mathbf{R} = \begin{bmatrix} 0 & 0 & 1 \\ 1 & 0 & 0 \\ 0 & 1 & 0 \end{bmatrix}, \tag{III.2.20}$$

from which we can obtain the spatial and rank orders indexes $\mathbf{s} = [1, 2, 3]\mathbf{R} = [2, 3, 1]$, $\mathbf{r}^T = \mathbf{R}[1, 2, 3]^T = [3, 1, 2]^T$, and the spatial and rank order samples $\mathbf{x}_\ell = \mathbf{R}[1, 2, 10]^T = [10, 1, 2]^T$, $\mathbf{x}_L^T = [10, 1, 2]\mathbf{R} = [1, 2, 10]$.

The SR matrix fully relates the spatial and rank orderings of the observed samples. Thus, the structure of the SR matrix captures spatial correlations of the data (spatial order information) and indicates which samples are likely to be outliers and which samples are likely to be reliable (rank order information). To illustrate the structure of the SR matrix for typical signals, and to show how this structure changes with the underlying signal characteristics, the statistics of the SR matrix are examined for two types of signals.

Consider first a one-dimensional statistical sequence consisting of an MA process. This process is generated by simple FIR filtering of white Gausssian noise. In the case of white noise samples

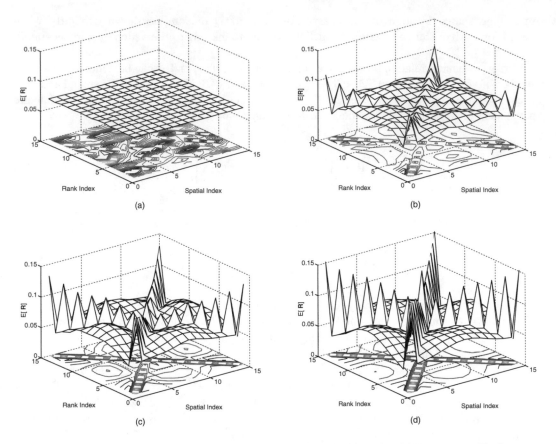

FIGURE III.2.3 Expected SR matrices for low-pass FIR filtered white noise (MA) processes. The low-pass filter cutoff frequencies in the examples are (a) $\omega_c = 1.0$ (no filtering), (b) $\omega_c = 0.33$, (c) $\omega_c = 0.25$, and (d) $\omega_c = 0.20$. (From Barner K.E. and Hardie, R.C., Spatial-rank order selection filters, in *Nonlinear Image Processing.* (Mitra S.K. and Sicuranza, G., Eds.), San Diego, CA: Academic Press, 2001. With permission.)

(no filtering), all spatial–rank order combinations are equally likely and the expected SR matrix is uniform:

$$
E[\mathbf{R}] = \begin{bmatrix} \frac{1}{N} & \cdots & \frac{1}{N} \\ \vdots & \ddots & \vdots \\ \frac{1}{N} & \cdots & \frac{1}{N} \end{bmatrix}, \tag{III.2.21}
$$

looseness-1 where $E[\cdot]$ denotes the expectation operator. Figure III.2.3(a) shows $E[\mathbf{R}]$ for the white noise case when the window size is $N = 15$. As expected, all spatial-order pairs are equally likely. Applying a low-pass FIR filter with a cutoff frequency of $\omega_c = 0.33$ to the noise in order to generate an MA process, yields a time sequence with a greater concentration of low-frequency power. This greater concentration of low-frequency power gives the time series increasingly sinusoidal structure, which is reflected in the resulting SR matrix, as shown in Figure III.2.3(b). As the figure shows, extreme samples are most likely to be located in the first or last observation window location and monotonic observations are more likely than other onservations. Decreasing the cutoff of the FIR filter to $\omega_c = 0.25$ and $\omega_c = 0.20$ increases the sinusoidal nature of the time domain sequence and increases the structure of the corresponding SR matrices, as illustrated in Figures III.2.3(c) and (d).

Similar results hold for images. Consider the original and a contaminated Gaussian noise-corrupted version of the well-known image Lenna, which is utilized in Section 2.5 and shown in Figure III.2.28.

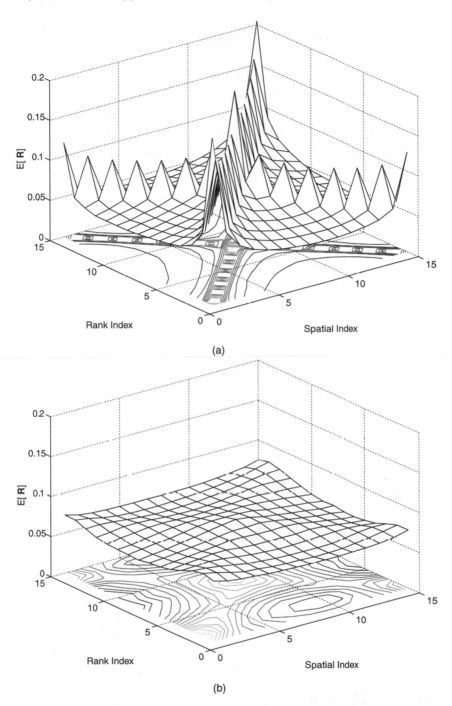

FIGURE III.2.4   Expected SR matrix for the image (a) Lenna and (b) Lenna corrupted with contaminated Gaussian noise. (From Barner K.E. and Hardie, R.C., Spatial-rank order selection filters, in *Nonlinear Image Processing*. (Mitra S.K. and Sicuranza, G., Eds.), San Diego, CA: Academic Press, 2001. With permission.)

The expected SR matrix for an $N = 15$ one-dimensional observation window passing over these images are shown in Figure III.2.4. As expected, the SR matrix for the original image has a structure that corresponds to the underlying image. The heavily corrupted noisy image, however, has lost much of the underlying structure as all spatial–rank order pairs are nearly equally likely. Thus, just as additive noise tends to decorrelate samples and flatten the power spectral density (PSD) of a signal, it also tends to flatten the expected SR matrix. Appropriate filtering can help restore the SR structure to that of the underlying signal.

2.3 Sample Affinity and Fuzzy Spatial–Rank Ordering

2.3.1 Sample Affiinity

The spatial and rank orderings of samples provides valuable information charactering the observed samples. Strict use of traditional ordering, however, can often lead to suboptimal performance. For instance, strict use of rank ordering can be misleading, as it is often assumed that extreme rank samples should be discarded. Maximum and minimum samples, however, are not necessarily outliers and may, in fact, be information-bearing samples that should not merely be discarded. Thus, traditional ordering methods disregard such important measures as sample value, spread, and diversity.

More general relations between samples can be realized by adopting a fuzzy set theory approach. Consider a real valued fuzzy membership function:

$$\mu_{\tilde{R}}(a, b) : A \times B \mapsto [0, 1] | a \in A, b \in B \tag{III.2.22}$$

that describes the relations between the samples in the sets A and B. The relation function $\mu_{\tilde{R}}(a, b)$ can be any shape that reflects the most relevant information between samples for the problem at hand. Thus, a wide range of fuzzy membership functions can be defined. In fact, the crisp relation function $\mu_c(a, b)$ can be viewed as a special case of the fuzzy membership function. For the ordering and filter generalizations considered here, we base the membership on the concept of affinity, and accordingly impose the following conditions on the membership function:

1. $\lim_{|a-b| \to 0} \mu_{\tilde{R}}(a, b) = 1$
2. $\lim_{|a-b| \to \infty} \mu_{\tilde{R}}(a, b) = 0$
3. $\mu_{\tilde{R}}(a_1, b_1) \geq \mu_{\tilde{R}}(a_2, b_2)$, for $|a_1 - b_1| \leq |a_2 - b_2|$

The intuitive justification for these conditions is that two identical samples should have high affinity, or relation 1, while infinitely distant samples should have low affinity, or relation 0. Additionally, the affinity between samples should increase as the distance between them decreases. Many membership functions, such as rectangular and triangular membership functions, satisfy the constraints. Here we utilized the commonly used Gaussian membership function:

$$\mu_G(a, b) = e^{-(a-b)^2/2\sigma^2}, \tag{III.2.23}$$

where $\sigma > 0$ controls the spread of the membership functions.

To illustrate the effect of the Gaussian membership function, consider again the example $\mathbf{x}_\ell = [10, 1, 2]^T$. The relation between $x_3 = x_{(2)}$ and each of the observation samples is illustrated in Figure III.2.5 for memberships functions with spread $\sigma = 2$ and $\sigma = 5$. The figure shows that the similarly valued samples x_2 and x_3 have a strong relation in both cases, but that the relation between the more distant x_1 and x_3 samples is significantly reduced as σ is decreased.

2.3.2 Fuzzy Spatial–Rank Ordering

The affinity-based relations between all observation samples can be represented in a fuzzy SR matirx as a straightforward generalization of the crisp SR matrix.[4,6–8] Simply replacing the crisp relation with the more general fuzzy relation yields:

$$\tilde{\mathbf{R}} = \begin{bmatrix} \tilde{R}_{1,(1)} & \cdots & \tilde{R}_{1,(N)} \\ \vdots & \ddots & \vdots \\ \tilde{R}_{N,(1)} & \cdots & \tilde{R}_{N,(N)} \end{bmatrix}, \tag{III.2.24}$$

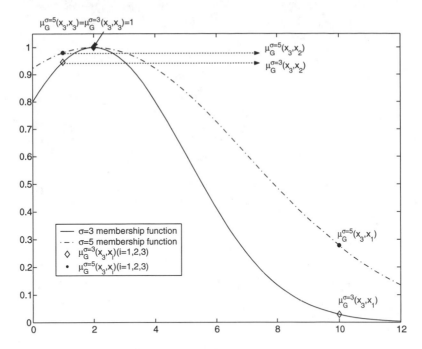

FIGURE III.2.5 The affinity-based relation between $x_3 = 2$ and the elements of $\mathbf{x}_\ell = [10, 1, 2]$ for membership function spread parameters $\sigma = 3$ and $\sigma = 5$.

where $\tilde{R}_{i,(j)} = \mu_{\tilde{R}}(x_i, x_{(j)})$. This fuzzy relation produces a real-valued relation matrix, i.e., $\tilde{R}_{i,(j)} \in [0, 1]$. In this context, $\mu_{\tilde{R}}(x_i, x_{(j)})$, and equivalently $\tilde{R}_{i,(j)}$, denotes the degree to which x_i and $x_{(j)}$ are related.

The real-domain fuzzy rank and spatial vectors $\tilde{\mathbf{r}}$ and $\tilde{\mathbf{s}}$ are now defined in an analogous manner to their crisp counterparts:

$$\tilde{\mathbf{s}} = [1 : N]\tilde{\mathbf{R}} \quad \text{and} \quad \tilde{\mathbf{r}}^T = \tilde{\mathbf{R}}[1 : N]^T. \tag{III.2.25}$$

Similarly, the crisp and fuzzy spatial and rank order vectors are given by $\tilde{\mathbf{x}}_\ell = \tilde{\mathbf{R}}\mathbf{x}_L$ and $\tilde{\mathbf{x}}_L^T = \mathbf{x}_\ell^T \tilde{\mathbf{R}}$.

A careful inspection of the fuzzy terms indicates that the ranges of values have been increased beyond that of their crisp counterpart. Specifically, $\tilde{x}_i \tilde{x}_{(j)} \in [x_{(1)}, \sum_{l=1}^{N} x_l]$, and $\tilde{r}_i, \tilde{s}_i \in [1, \sum_{l=1}^{N} l]$. To yield more intuitive values, these terms can be restricted to the same range as their crisp counterpart by normalizing the rows or columns of the fuzzy SR matrix. Since the rows and columns correspond to space and rank respectively, we designate $\tilde{\mathbf{R}}^\ell$ and $\tilde{\mathbf{R}}^L$ to be the row (spatial) and column (rank) normalized fuzzy SR matrices, respectively.

The normalized fuzzy rank and spatial index vectors are now given by $\tilde{\mathbf{r}}^T = \tilde{\mathbf{R}}^\ell[1 : N]^T$ and $\tilde{\mathbf{s}} = [1 : N]\tilde{\mathbf{R}}^L$. Similarly, $\tilde{\mathbf{x}}^\ell = \tilde{\mathbf{R}}^\ell \mathbf{x}_L$ and $\tilde{\mathbf{x}}_L^T = \mathbf{x}_\ell^T \tilde{\mathbf{R}}^L$. Given this normalization, the appropriate bounds hold $\tilde{r}_i, \tilde{s}_{(j)} \in [1, N]$ and $\tilde{x}_i, \tilde{x}_{(j)} \in [\tilde{x}_{(1)}, x_{(N)}]$. Carrying out the matrix expressions for a single term yields the following expressions for $\tilde{r}_i$ and $\tilde{x}_{(j)}$:

$$\tilde{r}_i = \frac{\sum_{j=1}^{N} j \tilde{R}_{i,(j)}}{\sum_{j=1}^{N} \tilde{R}_{i,(j)}}, \tag{III.2.26}$$

and

$$\tilde{x}_{(j)} = \frac{\sum\limits_{i=1}^{N} x_i \tilde{R}_{i,(j)}}{\sum\limits_{i=1}^{N} \tilde{R}_{i,(j)}}. \tag{III.2.27}$$

Thus, $\tilde{r}_i$ is a normalized weighted sum of the integers $1, 2, \ldots, N$ and $\tilde{x}_{(j)}$ is a normalized weighted sum of the samples $x_1, x_2, \ldots, x_N$. The weights in each case are the affinity relations between samples.

To illustrate the value in utilizing fuzzy relations, consider again the example $\mathbf{x}_\ell^T = [10, 1, 2]$ and Gaussian membership function ($\sigma = 3$). The fuzzy SR matrix in this case is:

$$\tilde{\mathbf{R}} = \begin{bmatrix} 0.0111 & 0.0286 & 1.0000 \\ 1.0000 & 0.9460 & 0.0111 \\ 0.9460 & 1.0000 & 0.0286 \end{bmatrix}, \tag{III.2.28}$$

and the normalized fuzzy SR matrices are:

$$\tilde{\mathbf{R}}^\ell = \begin{bmatrix} 0.0107 & 0.0275 & 0.9618 \\ 0.5110 & 0.4834 & 0.0057 \\ 0.4791 & 0.5065 & 0.0145 \end{bmatrix} \tag{III.2.29}$$

and

$$\tilde{\mathbf{R}}^L = \begin{bmatrix} 0.0057 & 0.0145 & 0.9618 \\ 0.5110 & 0.4791 & 0.0107 \\ 0.4834 & 0.5065 & 0.0275 \end{bmatrix}. \tag{III.2.30}$$

The resulting fuzzy rank vector is $\tilde{\mathbf{r}} = [2.9512, 1.4947, 1.5354]$. The resulting fuzzy space and rank ordered sample vectors are $\tilde{\mathbf{x}}_\ell = [9.6840, 1.5344, 1.6367]^T$ and $\tilde{\mathbf{x}}_L = [1.5344, 1.6367, 9.6840]^T$, respectively. The comparison of $\tilde{\mathbf{r}}$ and $\tilde{\mathbf{x}}_\ell$ with their crisp counterparts is illustrated in Figure III.2.6. As the figure shows, $\tilde{x}_1 \approx x_1$ and $\tilde{r}_1 \approx r_1$. This is a result of the fact that $x_1 \gg x_2, x_3$ in relation to the spread of the membership function. Conversely, since x_2 and x_3 are similarly valued, and thus highly related, $\tilde{r}_2 \approx \tilde{r}_3 \approx \frac{r_2+r_3}{2}$ and $\tilde{x}_2 \approx \tilde{x}_3 \approx \frac{x_2+x_3}{2}$. Thus, fuzzy ranking and order-statistic reflect not only the ordering of samples but also their spread. These concepts and properties are considered further in the following section.

2.3.3 Properties of Fuzzy Spatial–Rank Ordering

The fuzzy SR matrix, indices, and samples possess powerful properties that are useful in relating sample orderings and values. Several of the most important properties are highlighted below. A fuller discussion of properties is given in References 4 through 8.

Super-ordination property — *The fuzzy SR matrix is a superset of the crisp SR matrix, containing information on the spread of the samples in addition to the crisp SR information.*

This property arises from the fact that a fuzzy relation can be mapped to its crisp counterpart through thresholding: $R_{i(j)} = T_1(\tilde{R}_{i,(j)})$, where $T_\delta(a)$ is the thresholding operation that yields 1 if $a \geq \delta$ and 0 otherwise. Extending this element-wise to the fuzzy SR matrix yields $\mathbf{R} = T_1(\tilde{\mathbf{R}})$,

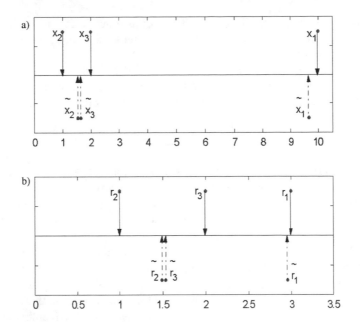

FIGURE III.2.6 Comparison of crisp and fuzzy (a) samples and (b) ranks for the observation $\mathbf{x}_\ell = [10, 1, 2]^T$.

showing that the fuzzy SR matrix can be reduced to it crisp counterpart by thresholding.* In case of a row (column) normalized fuzzy SR matrix, elements are thresholded with δ set to the maximum row (column) element.

Reduction property — *As the membership function spread decreases, fuzzy relations, ranks, and order statistics reduce to their crisp counterparts. Conversely, as the membership function spread increases to infinity, all fuzzy relations become equal, fuzzy ranks converge to the median, and fuzzy order statistics converge to the sample mean.*

This property describes the limiting behavior, with respect to membership function spread of the fuzzy SR matrix. It is easy to see $\lim_{\sigma \to 0} \tilde{\mathbf{R}} = \mathbf{R}$. Conversely, $\sigma \to \infty$ indicates infinite spread of the membership function and the realtion between all samples is equal, i.e., $\tilde{R}_{i,(j)} = 1$ for $i, j = 1, 2, \ldots, N$. For a row (column) normalized fuzzy matrix, $\lim_{\sigma \to \infty} \tilde{R}_{i,(j)} = \frac{1}{N}$. From this it is easy to see that $\lim_{\sigma \to 0} \tilde{\mathbf{r}} = \mathbf{r}$ and $\lim_{\sigma \to \infty} r_i = \frac{N+1}{2}$. Also, $\lim_{\sigma \to 0} \tilde{x}_{(j)} = x_{(j)}$ and $\lim_{\sigma \to \infty} \tilde{x}_{(j)} = \tilde{x} = \frac{1}{N} \sum_l^N x_l$.

Spread-sensitivity property — *The inter-sample spacing is captured in the fuzzy SR matrix.*

If the membership function covers the range $[0, x_{(N)} - x_{(1)}]$, and is continuous and continuously decreasing (as a function of $|x_i - x_j|$), then the sample spread can be exactly determined from $\tilde{\mathbf{R}}$. That is, $|x_i - x_j| = \mu_R^{-1}(\tilde{R}_{i,(r_j)})$. Thus, the observation vector can be recovered from $\tilde{\mathbf{R}}$ to within a constant. It is easy to see that the observation vector can also be recovered from the normalized SR matrix.

This property explains why the fuzzy SR matrix captures sample spread information. To illustrate, consider two strictly increasing sequences: one linear, the other exponential. While their corresponding crisp SR matrices are diagonal identiy matrices, concealing the sample spacing, the

---

*This assumes all observation values are unique. For the case of equally valued samples, stable sorting is utilized to assign each sample a unique crisp rank index, and this rank can be extracted from the fuzzy SR matrix by simply preserving the original time order of equally valued samples.

fuzzy SR matrices reflect the sample spread. This is illustrated in Figure III.2.7 for a sequence of length $N = 20$. From the figure, it is clear that the greater separation between the later exponential samples is captured in the fuzzy SR matrix.

Distribution of fuzzy order statistics — As Equation III.2.27 shows, the fuzzy order statistics are weighted sums of the observation samples (or the observation order statistics). Thus, the distribution of $\tilde{x}_{(j)}$ can be formed as an appropriately weighted sum of the order statistic distributions:

$$f_{\tilde{x}_{(i)}}(x) = \sum_{j=1}^{N} \omega_j f_{x_{(j)}}(x) \tag{III.2.31}$$

where the ω_j terms are the appropriate membership function and sample spread based weights, and $f_{x_{(i)}}(\cdot)$ and $f_{x_{(i)}}(\cdot)$ are the distribution of the ith fuzzy and crisp order statistics, respectively. Clearly, the limiting case $\lim_{\sigma \to 0} f_{\tilde{x}_{(j)}}(\cdot) = f_{\tilde{x}_{(j)}}(\cdot)$ and $\lim_{\sigma \to \infty} f_{\tilde{x}_{(j)}}(\cdot) = f_{\bar{x}}(\cdot)$ hold. For other cases, the extent of the membership function spread determines how much weighted averaging of the crisp order statistic distribution is realized in the determination of $f_{\tilde{x}_{(j)}}(\cdot)$.

As an example of the role played by the membership function spread, consider the case i.i.d. zero mean, unit variance Gaussian observation samples with $N = 9$. The distribution of $\tilde{x}_{(j)}$ is shown in Figure III.2.8 for several membership function spread values. The figure illustrates that as σ grows, and increased averaging is introduced, the variance of the fuzzy order statistics is reduced and their means are shifted to the center of the original distributions. Thus, the membership function spread controls the migration of $\tilde{x}_{(j)}$ from the crisp order statistic $x_{(j)}$ to the sample mean $\bar{x}$.

Clustering property — *Samples from a finite number of populations yield fuzzy samples (fuzzy ranks) that are averages of the observation samples (crisp ranks) within the individual population.*

Let the samples be from L populations with $\mu(x_i, x_k) = 1$ for $i, k \in I_l, \mu(x_i, x_k) = 0$ for $i \in I_l, k \in I_m, l \neq m$ and $\cup_{l=1}^{L} I_l = \{1, 2, \ldots, N\}$. Also, let p_i be the population index of x_i, i.e., $x_i \in I_{p_i}$. In this case, the fuzzy relation between samples in different populations is 0 and the relation between samples in the same population I_l is, in the normalized case, $\frac{1}{\|I_l\|}$ where $\|I_l\|$ denotes the number of samples in I_ℓ. This results in the fuzzy time and rank samples:

$$\tilde{x}_i = \tilde{x}_{r_i} = \frac{1}{\|I_{p_i}\|} \sum_{k \in I_{p_i}} x_k \tag{III.2.32}$$

and

$$\tilde{r}_i = \frac{1}{\|I_{p_i}\|} \sum_{k \in I_{p_i}} r_k \tag{III.2.33}$$

Note that the fuzzy rank indices are obtained by averaging crisp ranks which, for the case of equally valued samples, form an integer subsequence, $r_{min}, r_{min} + 1, \ldots, r_{max}$, where r_{min} and r_{max} are the minimum and maximum rank of samples in a given population. Averaging such a subsequence results in a value that is always a multiple of $\frac{1}{2}$. Thus, fuzzy ranks are typically such that $\tilde{r}_i \in \{1, 1\frac{1}{2}, 2, 2\frac{1}{2}, \ldots, N\}$.

Relaxing the membership relations between samples in the above property, we see that the fuzzy relations introduce averaging among similarly valued samples, where similarly is determined by the membership function shape and spread.

2.4 Fuzzy Filter Definitions

Having established the concepts of affinity and fuzzy SR orderings, these concepts can now be adopted into filtering structures. One method for accomplishing the inclusion is through the simple

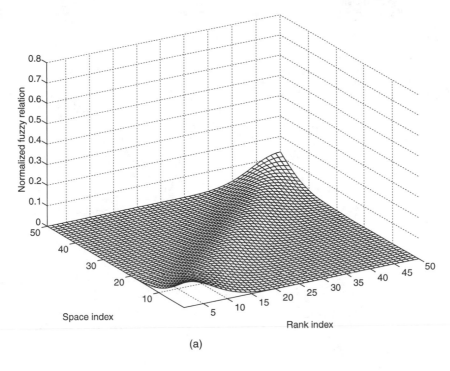

(a)

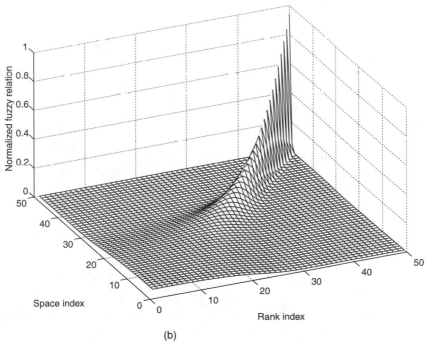

(b)

FIGURE III.2.7   The row normalized fuzzy SR matrices for (a) linearly increasing and (b) exponentially increasing sequences.

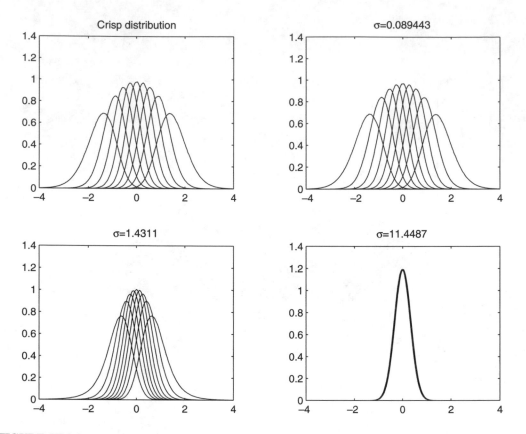

FIGURE III.2.8   The distribution of the crisp and fuzzy order statistics of i.i.d. zero mean Gaussian samples with variance 4 for the $N = 9$ case.

modification of established filtering algorithms. Thus, existing algorithms that have proved useful can be updated to include affinity or fuzzy ordering information. The additional degrees of freedom introduced by fuzzy methods, and thus the consideration of sample spread by the filtering algorithm, lead to improved performance.

The fuzzy generalization of two broad classes is considered here. First, the general class of affine filters is established.[9] Affine filters are realized as a simple extension of the broad class of weighted sum filters in which affinity, or sample spread, weighting is introduced in the weighted sum output. The second class of filters established is the fuzzy generalization of weighted median filters.[7,8] Fuzzy weighted median filters are realized by simply inserting fuzzy samples into the standard weighted median filter formulation. As is demonstrated in Section 2.5, the affinity and fuzzy generalizations lead to improved performance.

2.4.1   Affine Filters

Affine filters are realized by including an affinity weighting, to a specified reference point, in a standard weighted sum filter. Two important subclasses of affine filters are the median affine and center affine filter classes. The median affine filter class is established by utilizing the median sample as the reference point, and introducing the affinity weighting into the standard weighted sum filter. The center affine filter class is realized by setting the central observation sample as the reference point, and introducing the affinity weighing into the weighted sum of order statistics filter.

2.4.1.1 Median Affine Filters

The standard linear FIR filter has been successfully applied to many problems. Indeed, by formulating the filter output as a weighted sum of spatially, or temporally, ordered samples, important filter characteristics are realized, such as frequency selectivity. Linear FIR filters, however, performed poorly in the presence of outliers. This performance can be improved if a simple valid reference point can be established and the validity of each sample measured with respect to the reference point. This is the motivation behind the median affine filter.

The median operator is robust, and can thus serve to establish a valid reference point for many signal statistic cases. Thus, the standard weighted sum FIR filter given in Equation III.2.9 can be made more robust by weighting each sample according to its affinity to the median reference point. Utilizing again $\delta = \frac{N+1}{2}$ as the central index, the median affine filter is defined as:

$$\text{MAFF}[\mathbf{x}] = \frac{\sum\limits_{i=1}^{N} w_i \tilde{R}_{i,(\delta)} x_i}{\sum\limits_{i=1}^{N} |w_i| \tilde{R}_{i,(\delta)}}, \tag{III.2.34}$$

where the w_i's are the filter weights and $\tilde{R}_{i,(\delta)}$ is the affinity of the ith observation with respect to the median reference point, $\text{MED}[\mathbf{x}] = x_{(\delta)}$.

The filter structure in Equation III.2.34 weights each observation twice: first, according to its reliability, and second, according to its natural order (Figure. III.2.9). Median affine estimates are therefore based on observations that are both reliable and favorable due to their natural order. Observations which fail to meet either, or both, criteria have only a limited influence on the estimate.

By varying the dispersion of the affinity (or membership) function, certain properties of the median affine filter can be stressed: while a large value of σ emphasizes the linear properties of the filter, a small value puts more weight on its median properties. Of special interest are the limiting cases. For $\sigma \to \infty$, the affinity function is constant on its entire domain. The estimator, therefore, weights all observations merely according to their natural order, i.e.:

$$\lim_{\sigma \to \infty} \text{MAFF}[\mathbf{x}] = \frac{\sum\limits_{i=1}^{N} w_i x_i}{\sum\limits_{i=1}^{N} w_i}, \tag{III.2.35}$$

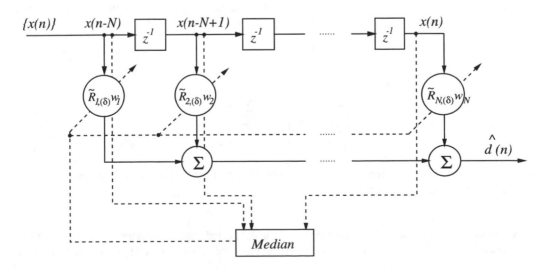

FIGURE III.2.9 Structure of the (unnormalized) median affine filter.

and the median affine estimator reduces to a normalized linear filter. In contrast, for $\sigma \to 0$, the affinity function shrinks to an impulse at MED($\mathbf{x}$). Thus, the constant weights w_i are disregarded and the estimate is equal to the median, i.e.:

$$\lim_{\sigma \to 0} \text{MAFF}[\mathbf{x}] = \text{MED}[\mathbf{x}]. \tag{III.2.36}$$

In addition to the limiting cases, the median affine filter includes important filters, such as the MTM filter,[10] as subclasses.

The median affine filter also possesses several desirable properties,[9] including: (1) data-adaptiveness, (2) translation invariance, (3) the ability to suppress impulses, and (4) the ability to preserve signal trends and discontinuities. These properties lend understanding to the filtering process and help explain the improved performance achieved through the incorporation of sample affinity into the filter structure.

2.4.1.2 Center Affine Filters

A second important subclass of affine filters is the class of center affine filters. While the median serves as an acceptable reference point for certain applications, the observation sample located (spatially) in the center of the observation window is particularly important in other applications. Thus, the class of center affine filters is based on the concept of affinity to the central observation sample.

Center affine filters are related to median affine filters through a simple change in ordering. That is, rather than a generalization of the linear filter based on affinity to the median, the center affine filter is a generalization of the order-statistic weighted sum (L) filter based on affinity to the central spatial sample. Accordingly, the output of the center affine filter is given by:

$$\text{CAFF}[\mathbf{x}] = \frac{\displaystyle\sum_{i=1}^{N} w_{(i)} \tilde{R}_{\delta,(i)} x_{(i)}}{\displaystyle\sum_{i=1}^{N} |w_{(i)}| \tilde{R}_{\delta,(i)}}, \tag{III.2.37}$$

where the $w_{(i)}$'s are the filter coefficients and $\tilde{R}_{i,(\delta)}$ is the affinity of the ith order statistic ($x_{(i)}$) with respect to the central (spatial) sample reference point, x_δ.

The center affine filter weights each order-statistic according to its affinity to the central observation sample, and according to its rank. The filter output, therefore, is based mainly on those order statistics that are simultaneously close to the central observation sample, and preferable due to their rank order. Note that, as opposed to the median affine filters, here the temporal weights $\tilde{R}_{\delta,(i)}$ are time-varying and the rank-order weights $w_{(i)}$ are constant. Like the median affine filter, the center affine filter with nonnegative weights reduces to its basic structures at the limits of the dispersion parameter σ:

$$\lim_{\sigma \to 0} \text{CAFF}[\mathbf{x}] = x_\delta$$

and

$$\lim_{\sigma \to \infty} \text{CAFF}[\mathbf{x}] = \frac{\displaystyle\sum_{i=1}^{N} w_{(i)} x_{(i)}}{\displaystyle\sum_{i=1}^{N} w_{(i)}}. \tag{III.2.38}$$

Thus, the center affine filter reduces to the identity filter and the L-filter, with coefficients $w_{(i)}$, for $\sigma \to 0$ and $\sigma \to \infty$, respectively.

2.4.1.3 Optimization

To appropriately tune the performance of an affine filter, the filter coefficients can be optimized. Two approaches to optimizing the parameters are presented. First, a simple suboptimal procedure that addresses only the affinity spread function is presented. A more comprehensive optimization procedure that addresses both the filter weights and affinity spread function is then presented. This optimization is based on a stochastic adaptive procedure. In both cases, the presented methods address the optimization of the median affine filter. The methods can be applied to the design of center affine filters by simply interchanging corresponding quantities.

A simple and intuitive design procedure can be derived from the fact that the median affine filter behaves like a linear filter for $\sigma \to \infty$. Setting σ to a large initial value allows the use of the multitude of linear filter design methods to find the w_i coefficients of the median affine filter. Holding the w_i's constant, the filter performance can, in general, be improved by gradually reducing the value of σ until a desired level of robustness is achieved. During the actual filtering process, σ is fixed. Since this process strengthens the median-like properties, while weakening the influence of the FIR filter weights, this procedure is referred to as the *medianization* of a linear FIR filter.

The median affine filter can be adaptively optimized under the mean square error (MSE) criteria in an approach that has been applied to related filter structures, such as radial basis functions.[11,12] Consider first the optimization of σ for a fixed set of filter coefficients. To simplify the notation, let $\gamma = \sigma^2$. Then, under the MSE criteria, the cost function to be minimized is:

$$J(\gamma) \triangleq E[e^2] = E[(d - \hat{d})^2], \tag{III.2.39}$$

where $e = d - \hat{d}$ is the filtering error, and $E[\cdot]$ stands for the statistical expectation operator. The optimization problem can be stated as the minimization of $J(\gamma)$, where γ is restricted to nonnegative, real-valued numbers. Due to the nonlinearity of the median affine estimate, this is a nonlinear optimization problem.

Although finding a closed form solution is intractable, an iterative LMS type approach can be adopted.[9,13] Under this approach, γ is indexed and updated according to:

$$\gamma(n+1) = \gamma(n) - \mu_\gamma \frac{\partial J}{\partial \gamma}(n), \tag{III.2.40}$$

where μ_γ is the appropriately chosen step size. For the case of positive valued weights, differentiating $J(\gamma)$ with respect to γ, substituting in the above, and performing some simplification, yields the update:

$$\gamma(n+1) = \gamma(n) + \mu_\gamma \frac{(d(n) - \hat{d}(n))}{\gamma^2(n)} \left(\sum_{i=1}^{N} w_i \tilde{R}_{i,(\delta)}(x_i(n) - \hat{d}(n))(x_i(n) - x_{(\delta)}(n))^2 \right). \tag{III.2.41}$$

At each iteration the positively constrained can be enforced by a simple maximum operator, $\gamma(n+1) = \max\{\gamma(n+1), 0\}$.

Adopting a similar approach for the optimization of the filter weights, in which $J(\gamma)$ is differentiated with respect to the filter weights, yields in the update:

$$w_i(n+1) = w_i(n) - \mu_w \frac{\partial J}{\partial w_i}(n) \tag{III.2.42}$$

for $i = 1, 2, \ldots, N$, where μ_w is the step size and:

$$\frac{\partial J}{\partial w_i}(n) = -(d(n) - \hat{d}(n)) \left[\tilde{R}_{i,(\delta)}(n) \sum_{k=1}^{N} w_k \tilde{R}_{k,(\delta)}(n)(x_i(n) - x_k(n)) \right]. \tag{III.2.43}$$

Both iteration updates are computationally simple and can be modified to address the case of negative filter weights. Additionally, they can be applied in an alternating procedure to yield a globally optimal set of affine-filter parameters.

2.4.2 Fuzzy Weighted Median Filters

2.4.2.1 Fuzzy Median Filters

The median filter is widely used due to its effectiveness in preserving signal structures, i.e., edges and monotonic regions, while smoothing noise. In many practical applications, however, it is advantageous to introduce some weighted averaging of samples, particularly among similarly valued samples. In such cases the fuzzy median can be utilized. The fuzzy median filter is realized by simply replacing the crisp order statistics with their fuzzy counterparts,[6-18] and is thus defined as:

$$\text{FMED}[\mathbf{x}] = \tilde{x}_{(\delta)} = \frac{\sum_{i=1}^{N} x_i \tilde{R}_{i,(\delta)}}{\sum_{i=1}^{N} \tilde{R}_{i,(\delta)}}, \tag{III.2.44}$$

where $\tilde{R}_{i,(\delta)}$ is the fuzzy relation between x_i and $x_{(\delta)}$. Note that this is simply a special case of the median affine filter in which all spatial order samples are giving equal weight, $w_i = 1$ for $i = 1, 2, \ldots, N$.

To illustrate the advantages of using fuzzy techniques in a median structure, consider the filtering of an ideal step edge corrupted by additive Gaussian noise. Figure III.2.10 shows the ensemble average of the outputs of crisp and fuzzy median filters with window size of $N = 9$ operating on the Gaussian noise-corrupted step signal. It can be seen that while the crisp median filter smoothes the additive noise, it also introduces significant edge smoothing. The fuzzy median, in contrast, preserves the edge transition while performing more effective noise smoothing.

This improved performance can be attributed to the fact that the fuzzy median is an unbiased estimator at all window locations spanning the edge transition (as well as in strictly uniform regions).[7] The crisp median, however, is biased at all window locations that span the edge transition and only unbiased in strictly uniform regions. It can also be shown that variance of the fuzzy median estimator is less than its crisp counterpart. In nonideal cases, such that as depicted in Figure III.2.10, some smoothing does occur due to imperfect relation between samples on a common side of the edge transition and some nonzero relation among samples on the opposite side of the transitions.

As a second example, consider the filtering of a chirp image. Figure III.2.11 shows the original, Gaussian noise-corrupted, median filtered, and fuzzy median filtered chirp images. The results show that, like in the one-dimensional case, the fuzzy extension to the median filter improves the edge preservation and noise smoothing capabilities.

2.4.2.2 Fuzzy Weighted Median Filters

Recall that the median filter can be generalized by weighting each of the spatially ordered samples prior to sorting and median selection. This results in the expression given in Equation III.2.11, and repeated here for convenience:

$$\text{WMED}[\mathbf{x}] = \text{MED}[w_1 \diamond x_1, w_2 \diamond x_2, \ldots, w_N \diamond x_N], \tag{III.2.45}$$

where $\diamond$ is the replication operator. Note that although the discussion presented in Section 2.2 restricted the replication weights to positive integer values, the filter can easily incorporate real valued (positive and negative) weights utilizing the procedures in Reference 14.

The weighted median filter can be generalized to the fuzzy weighted median filter by simply replacing the crisp observation samples with their fuzzy counterparts:

$$\text{FWMED}[\mathbf{x}] = \text{MED}[w_1 \diamond \tilde{x}_1, w_2 \diamond \tilde{x}_2, \ldots, w_N \diamond \tilde{x}_N]. \tag{III.2.46}$$

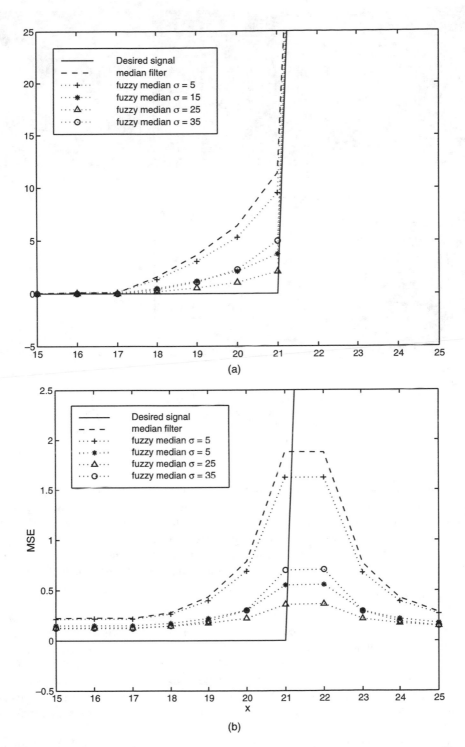

FIGURE III.2.10 (a) Step signal of height 100, corrupted by Gaussian noise of variance 10, filtered with median filter and fuzzy median filter and (b) the resulting mean square filtering error.

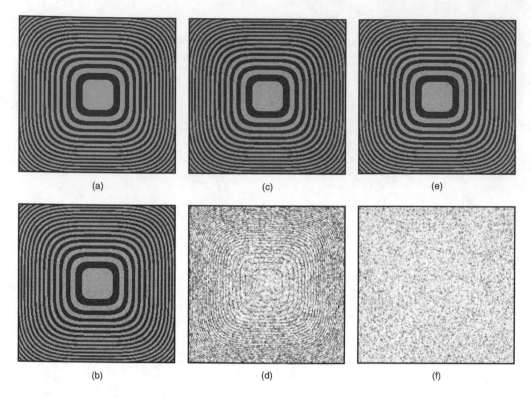

FIGURE III.2.11   Filtered chirp images and the resulting scaled error images: (a)–(b) Original image and the image corrupted by additive Gaussian noise with variance 100, (c)–(d) median (RMSE = 7.0, MAE = 5.3), and (e)–(f) fuzzy median (RMSE = 4.1, MAE = 3.1) filter. Each filter utilized a 3 × 3 observation window.

As it can be shown that the fuzzy order statistics and the fuzzy spatial order samples constitute the same set, this definition of the fuzzy weighted median is equivalent to first weighting the (crisp) spatial order samples, and then selecting the fuzzy median from this expanded set. Under both equivalent definitions, the fuzzy weighted median is able to exploit spatial correlations (through spatial weighting), limit the influence of outliers (through ranking and median selection), and introduce selected weighted averaging of similarly valued samples (through the use of fuzzy samples).

2.4.2.3   Optimization

The optimization of the fuzzy weighted median filter follows a similar approach to that presented for the class of affine filters. In the fuzzy weighted median filter case, however, the error criteria chosen is the mean absolute error (MAE) criteria. This is due to the fact that median and weighted median filters are typically optimized under this criteria since it arises naturally out of the ML development under the Laplacian assumption. Additionally, median type filters are typically applied to signals with heavy-tailed distributions. Utilizing criteria such as the MSE for such signals tends to overemphasize the influence of outliers in the optimization procedure, and thus lower power-error criteria are typically employed.

Consider first the optimization of the fuzzy order statistics, from which the optimization of the fuzzy median follows directly. Under the MAE criteria, the cost to be minimized is $J(\gamma) = E(|d - \hat{d}|)$, where the optimization of the jth order statistic is achieved by setting $\hat{d} = \tilde{x}_{(j)}$ as the estimate of a desired signal d. Differentiating this cost criteria with respect to γ, substituting into a stochastic gradient-based algorithm, and replacing the expectation operator with instantaneous estimates yields:

$$\gamma(n+1) = \gamma(n) - \mu_\gamma \frac{\partial J(\gamma)}{\partial \gamma}(n) \tag{III.2.47}$$

$$= \gamma(n) + \mu_\gamma \frac{sgn(d(n) - \tilde{x}_j(n))}{\gamma^2(n)} \frac{\sum_{i=1}^{N} \tilde{R}_{i,(j)}(x_i(n) - \tilde{x}_j(n))(x_i(n) - x_{(j)}(n))^2}{\sum_{i=1}^{N} \tilde{R}_{i,(j)}} \tag{III.2.48}$$

where μ_γ is the step size and, as before, a positivity constraint is placed on γ.

The optimization of the fuzzy median follows directly from the above by simply setting $j = \delta = \frac{N+1}{2}$. Thus, in the fuzzy median case there is only a single parameter to optimize. In the case of a FWM filter, the spatial weights and membership function spread parameter must be optimized. The optimization of the WM filter is often carried out under the MAE criteria utilizing a gradient-based algorithm.[15,16] Such an optimization yields the following weight update expression:

$$w_j(n+1) = w_j(n) + \mu_\omega e(n) sgn(w_j(n)) sgn[sgn(w_j(n))x_j(n) - \hat{d}(n)]. \tag{III.2.49}$$

As in the previous case, this optimization can be combined with the iterative membership function optimization expression and applied in an alternating fashion. While this does not guarantee globally optimal results, the procedure is simple, computationally efficient, and has yielded good results.

2.5 Applications

The introduced concepts and filters can be applied to a wide range of applications. Indeed, since the affinity and fuzzy concepts are easily embedded into existing methods, yielding additional degrees of freedom and improved results, nearly all problems can be successfully addressed through affinity or fuzzy based methods. To give an appreciation of the performance gains that can be achieved through more general methods, results are presented for affine filters, fuzzy ordering, and weighted median filtering applied to a broad range of applications. First, affine filters are applied to robust frequency selective filtering, multiresolution signal decomposition, inverse synthetic aperture radar image filtering, and time-frequency filtering for cross-term elimination. Fuzzy weighted median filters are then applied to noise smoothing of images and image zooming. Finally, fuzzy ranking is applied to signal detection in a communications problem.

2.5.1 Affine Filters

Affine order-statistic filters have a broad spectrum of potential applications due to their wide range of filter characteristics and their flexibility. The different structures that the filter can take on lend themselves to different problems. The first problem considered is robust frequency selective filtering, in which temporal (spatial) order of samples must be exploited to gain frequency selectivity, while some rank ordering should be considered to ensure robust behavior. This problem is effectively addressed through median affine filtering. These same characteristics allow median affine filters to be effectively employed in multiresolution signal decomposition problems, which are considered next. The following three problems focus on cases where the central observation sample plays a crucial role, and are thus problems effectively addressed through center affine filtering.

2.5.1.1 Robust Frequency-Selective Filtering

Filters that are jointly robust and frequency selective are of great interest. While several robust low-pass filters exist, the design of robust band-pass and high-pass filters remains a challenging task. The robust median affine filters can be easily designed to yield robust, frequency-selective

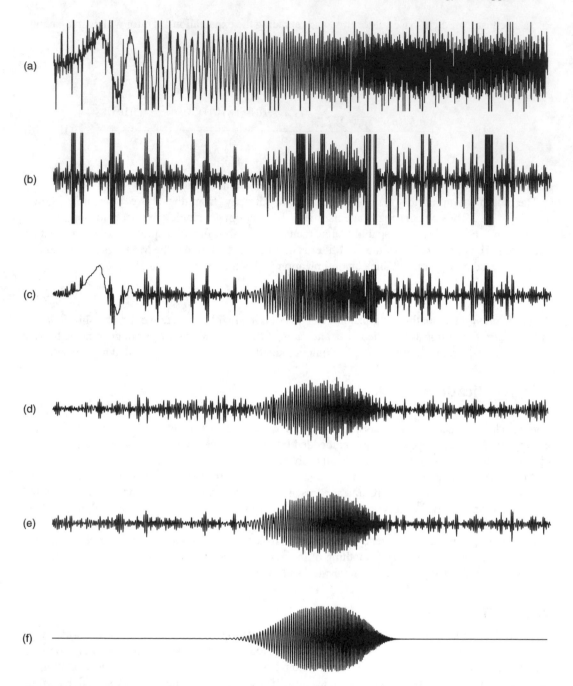

FIGURE III.2.12   (a) The noise-corrupted chirp and the result of the (b) linear FIR bandpass, (c) FIR-WOS, (d) $L\ell$, and (e) median affine filters. The desired output is (f) and MSE and MAE results are reported in Table III.2.1. (From Flaig, A., Arce, G.R., and Barner, K.E., Affine order statistic filters: a data-adaptive filtering framework for nonstationary signals, *IEEE Trans. Signal Proc.*, 46, 2101–2112, 1998. With permission.)

behavior. Comparisons to the frequency-selective FIR-WOS[17] and $L\ell$ filters[18–20] show that median affine bandpass filters can achieve a better performance with significantly fewer coefficients.

The performance of the compared filters is illustrated using a 2048-sample quad-chirp (sinusoidal waveform with quadratically increasing frequency) additively contaminated by α-stable noise ($\alpha = 1.2$) to simulate an impulsive environment[21] (Figure III.2.12). The corresponding mean square

TABLE III.2.1   Comparison of Filter Performance on
Noise-Corrupted Chirp Shown in Figure III.2.12

Filter	MSE	MAE
Identity	1.07×10^6	23.72
FIR	1.72×10^5	33.05
FIR-WOS	0.0989	0.2028
$L\ell$	0.0205	0.1085
Median affine	0.0128	0.0862

and absolute errors are given in Table III.2.1. As expected, the impulsive noise is detrimental to the linear filter. The FIR-WOS filter is able to trim the impulses, at the cost of varying attenuation in the passband. Due to the employed WOS structure, however, this filter is not able to stop very low frequencies, requiring additional measures to complete the desired bandpass operation. The $L\ell$ filter is clearly more robust, but suffers from the blind rejection of the extreme order statistics, which results in artifacts over most of the frequency spectrum. This effect is even stronger for higher signal-to-noise ratios. The median affine filter preserves the desired frequency band well, while strongly attenuating the impulsive noise, which is reflected in a MSE that is roughly half of that achieved by the $L\ell$ filter. By comparing the linear estimate to that of the median affine filter, it can be seen that the latter behaves exactly like the linear filter whenever no impulses are present. Finally, note that the FIR-WOS and $L\ell$ filters utilize $2N + 1$ and N^2 coefficients, respectively. The median affine filter, in contrast, requires only $N + 1$ coefficients.

2.5.1.2   Multiresolution Representations

Multiresolution representations have been widely accepted as important analysis and processing tools that have been applied to a wide array of problems. The most widely used multiresolution procedures are based on wavelets, which enable the decomposition of a signal into the sum of a lower resolution (or coarse) signal plus a detail signal. Each coarse approximation in turn can be decomposed further, yielding a coarser signal and a detail signal at that resolution.[22,23]

It has been shown[24] that the calculations of the wavelet decomposition can be accomplished using quadrature mirror filters (QMF) (Figure III.2.13). Take $\text{Wav}_{low}[\cdot]$ and $\text{Wav}_{high}[\cdot]$ to be the 1D low-pass and high-pass filtering operations applied on each of the rows of a 2D data set. Consider **X** to be a 2D data matrix. Then **X** can be decomposed into its approximation and detail coefficients as follows:

$$\mathbf{X}_a^1 = ((\downarrow 2)\text{Wav}_{low}[((\downarrow 2)\text{Wav}_{low}[\mathbf{X}])^T])^T \tag{III.2.50}$$

$$\mathbf{X}_h^1 = ((\downarrow 2)\text{Wav}_{low}[((\downarrow 2)\text{Wav}_{high}[\mathbf{X}])^T])^T \tag{III.2.51}$$

$$\mathbf{X}_v^1 = ((\downarrow 2)\text{Wav}_{high}[((\downarrow 2)\text{Wav}_{low}[\mathbf{X}])^T])^T \tag{III.2.52}$$

$$\mathbf{X}_d^1 = ((\downarrow 2)\text{Wav}_{high}[((\downarrow 2)\text{Wav}_{high}[\mathbf{X}])^T])^T \tag{III.2.53}$$

In these expressions, $\mathbf{X}_a^1$ is the first level approximation of **X**, and $\mathbf{X}_h^1$, $\mathbf{X}_v^1$, and $\mathbf{X}_d^1$ correspond to horizontal, vertical, and diagonal details, respectively. It is also possible to perfectly reconstruct the original signal from the approximation and detail coefficients.

The multiresolution structure can be relaxed to allow decompositions based on nonlinear filters. As a simple modification, a one-dimensional median filter can be used in the multiresolution structure

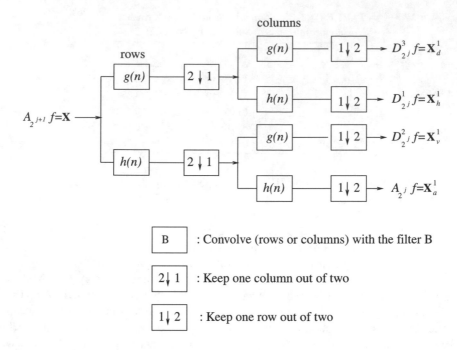

$$B \quad : \text{Convolve (rows or columns) with the filter B}$$

$$2{\downarrow}1 \quad : \text{Keep one column out of two}$$

$$1{\downarrow}2 \quad : \text{Keep one row out of two}$$

FIGURE III.2.13   The quadrature mirror filter multiresolution wavelet decomposition structure. (From Asghar, M. and Barner, K.E., Nonlinear multiresolution techniques with applications to scientific visualization in a haptic environment, *IEEE Trans. Visualization and Computer Graphics*, 7, 76–93, Mar. 2001. With permission.)

shown in Figure III.2.13. In this case, the low-pass filter $h(n)$ is replaced by the median filter. Since there is no high-pass median filter, some modification of the multiresolution structure is required. Instead of high-pass filtering the data, the detail signal can be obtained by subtracting the median filtered data from the original data before decimation (Figure III.2.14). In this case, the level one decomposed signal and detail signals are given by:

$$\mathbf{X}_a^1 = ((\downarrow 2)\, \text{MED}[((\downarrow 2)\, \text{MED}[\mathbf{X}])^{\mathrm{T}}])^{\mathrm{T}} \tag{III.2.54}$$

$$\mathbf{X}_v^1 = ((\downarrow 2)((\downarrow 2)\, \text{MED}[\mathbf{X}]^{\mathrm{T}} - \text{MED}[(\downarrow 2)\, \text{MED}[\mathbf{X}]^{\mathrm{T}}]))^{\mathrm{T}} \tag{III.2.55}$$

$$\mathbf{X}_h^1 = ((\downarrow 2)\, \text{MED}[(\downarrow 2)(\mathbf{X} - \text{MED}[\mathbf{X}])^{\mathrm{T}}])^{\mathrm{T}} \tag{III.2.56}$$

$$\mathbf{X}_d^1 = ((\downarrow 2)((\downarrow 2)(\mathbf{X} - \text{MED}[\mathbf{X}])^{\mathrm{T}} - \text{MED}[(\downarrow 2)(\mathbf{X} - \text{MED}[\mathbf{X}])^{\mathrm{T}}]))^{\mathrm{T}} \tag{III.2.57}$$

Perfect reconstruction is possible from the approximation and detail components only if the data is not decimated after the filtering operation. Decimation, however, is advantageous in many applications, as it reduces the size of the dataset.

Linear multiresolution techniques tend to smooth the edges in signals, while nonlinear techniques, such as median filters, are known to preserve edges. A median filter, however, completely removes details with spatial span smaller than half of the window size. Conversely, linear filters do not take into account the spread of samples, and thus spread the effect of a pulse to its neighboring samples. The affine median filter provides a bridge between these two extremes and provides additional flexibility to the user in that a single parameter is required to specify the degree of filter nonlinearity. This flexibility allows the filter to behave as a linear filter, as a median filter, or as a hybrid filter

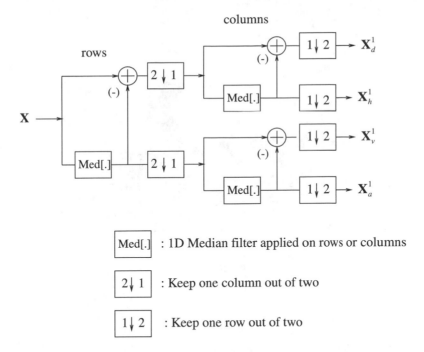

Med[.]   : 1D Median filter applied on rows or columns

2↓ 1   : Keep one column out of two

1↓ 2   : Keep one row out of two

FIGURE III.2.14   The nonlinear filter multiresolution structure used to generate the nonlinear decomposition. (From Asghar, M. and Barner, K.E., Nonlinear multiresolution techniques with applications to scientific visualization in a haptic environment, *IEEE Trans. Visualization and Computer Graphics*, 7, 76–93, Mar. 2001. With permission.)

combining the properties of the linear and nonlinear filters. Employing the median affine filter in the same structure yields:

$$\mathbf{X}_a^1 = ((\downarrow 2)\, \mathrm{MAFF}[((\downarrow 2)\, \mathrm{MAFF}[\mathbf{X}])^T])^T \tag{III.2.58}$$

$$\mathbf{X}_v^1 = ((\downarrow 2)((\downarrow 2)\, \mathrm{MAFF}[\mathbf{X}]^T - \mathrm{MAFF}[(\downarrow 2)\, \mathrm{MAFF}[\mathbf{X}]^T]))^T \tag{III.2.59}$$

$$\mathbf{X}_h^1 = ((\downarrow 2)\, \mathrm{MAFF}[(\downarrow 2)(\mathbf{X} - \mathrm{MAFF}[\mathbf{X}])^T])^T \tag{III.2.60}$$

$$\mathbf{X}_d^1 = ((\downarrow 2)((\downarrow 2)(\mathbf{X} - \mathrm{MAFF}[\mathbf{X}])^T - \mathrm{MAFF}[(\downarrow 2)(\mathbf{X} - \mathrm{MAFF}[\mathbf{X}])^T]))^T \tag{III.2.61}$$

The DEM data used to illustrate the performance of the multiresolution techniques is the Lake Charles dataset. The actual size of this dataset is 1200×1200. However, for illustration purposes, only certain sections of the dataset are rendered at one time. A size 150×150 portion of the dataset is shown in Figure III.2.15. The approximations of the Lake Charles dataset produced by the level one biorthogonal wavelet, median filter, affine median filter, and scale invariant affine median filter[25] decompositions are shown in Figure III.2.16. As the figure shows, the original dataset has a number of sharp edges that are smeared in the wavelet decomposed surface. In comparison, the median-based decomposition contains edges that preserve sharp transitions. The hybrid behavior of the affine median filter produces a decomposition that smoothes fine details while preserving edge integrity.

The difference in the behavior of the multiresolution techniques is more marked in the level two decomposed approximations, which are shown in Figure III.2.17. The figure shows that the median-based multiresolution decomposition perfectly preserves large-scale edge features while completely removing small details. At the other extreme, the wavelet decomposition significantly smoothes fine details as well as large-scale edge features. The affine median decomposition can be varied between

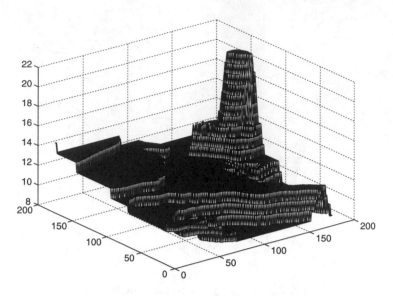

FIGURE III.2.15   Original Lake Charles (150 × 150) DEM dataset.

these limiting cases through control over the spread parameter γ. This hybrid response allows edges to be well preserved in the decomposition while simultaneously smoothing fine details, as the figures show.

In determining the behavior of a decomposition method, it is instructive to examine the detail signals $\mathbf{X}_h^i$, $\mathbf{X}_v^i$, and $\mathbf{X}_d^i$. Utilizing the multiresolution structure in Figure III.2.14, the level one detail signals of the Lake Charles decomposition are shown in Figure III.2.18. The figure shows that the detail signals in the linear decomposition contain significant power due to the edge smoothing of the wavelet transform. In contrast, the edge-preserving properties of the median filter are evident in the identically zero detail coefficients along most edges in the median decomposition. The detail signals show that the affine median decomposition has better edge preservation character than the linear decomposition, while also smoothing details in a more desirable fashion than the median.

As a final example, consider the case in which the observed dataset is corrupted by independent, additive noise. Specifically, let 10% of the samples in the Lake Charles dataset be corrupted by zero mean, unit variance Gaussian noise. The additive noise can either be attributed to the sensing, transmission, and storage of the data, or can be interpreted as small-scale detail. In either case, the noise terms should be significantly attenuated in the decomposition approximations. Figure III.2.19 shows a realization of the corrupted Lake Charles dataset, and the resulting level one approximations for the wavelet, median, and scale invariant affine median decompositions. As the figure shows, the wavelet decomposition spreads the noise terms across neighboring samples while only modestly reducing the magnitude. The median decomposition, in contrast, completely eliminates the noise terms. The noise is nearly eliminated in the affine median decomposition, resulting in an approximation with a slight texture indicating the presence of the noise/details in the observed dataset. Moreover, the large-scale edges are well preserved in the affine median decomposition, yielding an approximation with the most desirable characteristics.

These examples illustrate that the QMF decomposition structure can be easily extended to incorporate nonlinear filters. Furthermore, the affine median-based decomposition has advantages over linear and median filter-based decompositions of clean and noise corrupted datasets.[25,26] In particular, the affine median-based decomposition is controlled by a single parameter that governs the level of detail smoothing, resulting in a flexible multiresolution decomposition that is effective in smoothing fine detail, while simultaneously preserving large-scale features and sharp edge transitions.

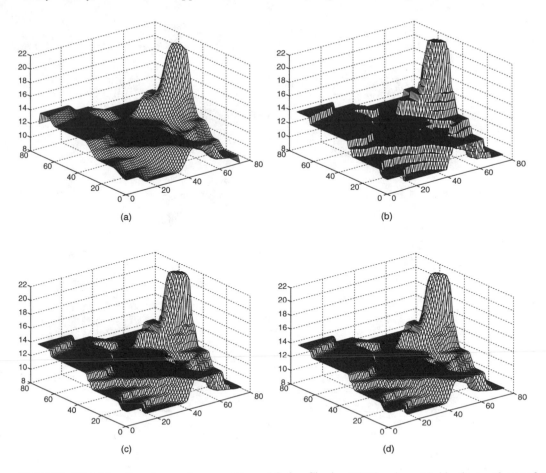

(a) (b)

(c) (d)

FIGURE III.2.16   Level one decomposition of Lake Charles DEM using: (a) biorthogonal wavelet, (b) median, (c) affine median, and (d) scale invariant affine median methods. (From Asghar, M. and Barner, K.E., Nonlinear multiresolution techniques with applications to scientific visualization in a haptic environment, *IEEE Trans. Visualization and Computer Graphics*, 7, 76–93, Mar. 2001. With permission.)

2.5.1.3 Inverse Synthetic Aperture Radar Image Filtering

Inverse synthetic aperture radar (ISAR) has attracted increasing interest in the context of target classification due to its high resolution.[27] ISAR images emerge from the mapping of the reflectivity density function of the target onto the range-Doppler plane. Difficulties in target identification arise from the fact that radar backscatters from the target are typically embedded in heavy clutter noise. For a proper target classification, it is therefore necessary to remove the noise without altering the target backscatters. From a signal processing point of view, ISAR images constitute nonstationary two-dimensional signals. The target backscatters are represented as pulse-like features in non-Gaussian noise.

We compare the performance of a center affine filter to that of a WOS filter and an $L\ell$ filter on the ISAR image* depicted in Figure III.2.20(a). This image is a 128×128, 8 bits/pixel intensity image of a B-727. A pair of synthetic images, Figure III.2.20(b), is used to optimize each of the filters. Figure III.2.21 shows the $L\ell$, WOS, and center affine filtering outputs and errors. An examination of the figure shows that the WOS filter eliminates the noise well, but blurs details of the plane. The

---

*Data provided by Victor C. Chen, Airborne Radar, Radar Division Naval Research Laboratory, Washington, D.C. 20375.

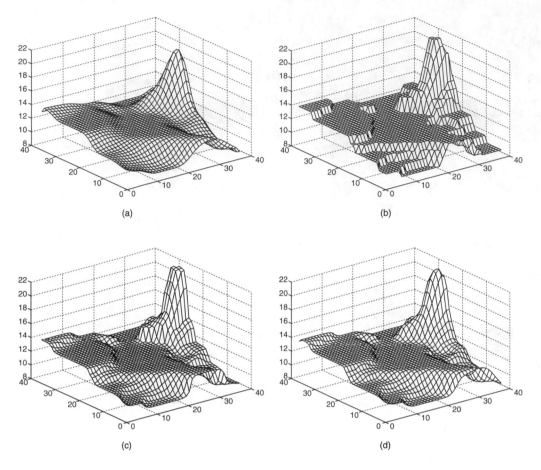

FIGURE III.2.17   Level two decomposition of Lake Charles DEM using: (a) biorthogonal wavelet, (b) median, (c) affine median, and (d) scale invariant affine median methods. (From Asghar, M. and Barner, K.E., Nonlinear multiresolution techniques with applications to scientific visualization in a haptic environment, *IEEE Trans. Visualization and Computer Graphics*, 7, 76–93, Mar. 2001. With permission.)

$L\ell$ filter preserves the plane much better, but is not very effective in removing the clutter noise. The center affine filter removes the background noise to a large extent, while preserving the plane in all its details. The superior performance of the center affine filter can be explained by its affinity-based sample preference. While operating in the background noise, all observations are close to the center sample. Thus, the center affine filter behaves like an L filter smoothing the clutter noise. When encountering backscatter from the plane, the center affine filter considers only those pixels with intensity similar to the center sample, thus preserving the plane details.

2.5.1.4   Time-Frequency Cross-Term Filtering

Quadratic time-frequency representations (TFRs) are powerful tools for the analysis of signals.[28] Among quadratic TFRs, the Wigner distribution (WD), which for continuous time is given by:

$$WD_x(t, f) = \int_\tau x\left(t + \frac{\tau}{2}\right) x^*\left(t - \frac{\tau}{2}\right) e^{-j2\pi f \tau} d\tau, \qquad (III.2.62)$$

satisfies a number of desirable mathematical properties and features optimal time-frequency concentration.[29–33] By using the WD, more subtle signal features may be detected, especially for signals having short length and high time-frequency variation. However, despite the desirable

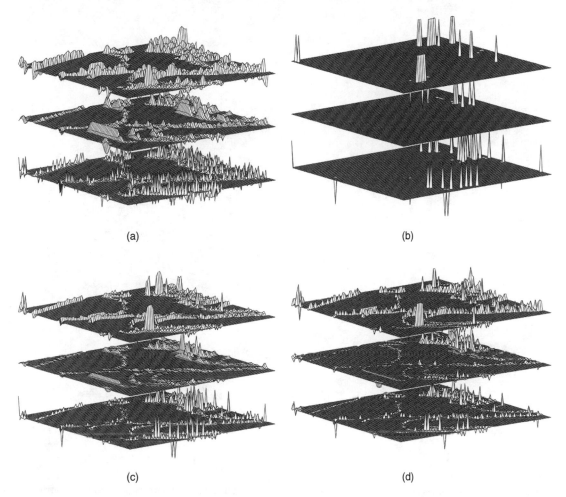

(a)

(b)

(c)

(d)

FIGURE III.2.18 Level one horizontal (top), vertical (middle), and diagonal (bottom) detail signals for the Lake Charles dataset. Decomposition method: (a) biorthogonal wavelet, (b) median, (c) affine median, and (d) scale invariant affine median. (From Asghar, M. and Barner, K.E., Nonlinear multiresolution techniques with applications to scientific visualization in a haptic environment, *IEEE Trans. Visualization and Computer Graphics*, 7, 76–93, Mar. 2001. With permission.)

properties of the Wigner distribution, its use in practical applications has often been limited by the presence of cross-terms. For instance, the Wigner distribution of the sum of two signals $x(t) + y(t)$:

$$WD_{x+y}(t, f) = WD_x(t, f) + 2\Re e(WD_{x,y}(t, f)) + WD_y(t, f) \qquad (\text{III.2.63})$$

has a cross-term $2\Re e(WD_{x,y}(t, f))$ in addition to the two auto-components, where the cross-Wigner distribution is defined as:

$$WD_{x,y}(t, f) = \int_{-\infty}^{+\infty} x\left(t + \frac{\tau}{2}\right) y^*\left(t - \frac{\tau}{2}\right) e^{-j2\pi f\tau} d\tau. \qquad (\text{III.2.64})$$

The cross or interference terms are often a serious problem in practical applications, especially if a WD outcome is to be visually analyzed by a human analyst. Cross-terms lie between two auto-components and are oscillatory, with their frequencies increasing with increasing distance in time-frequency between the two auto-components.[32,34,35] For real-valued band-pass signals, the

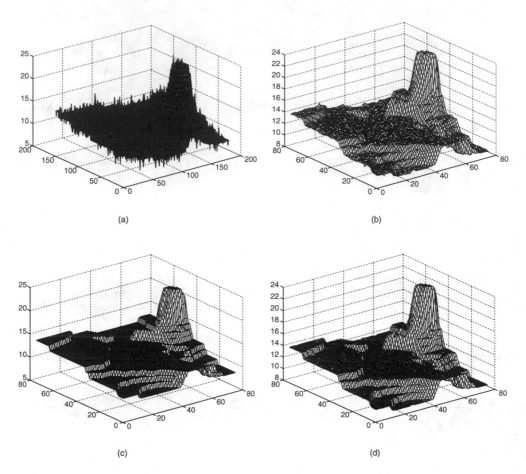

FIGURE III.2.19   Level one approximations of the (a) Lake Charles dataset with 10% of samples corrupted by Gaussian noise. Decomposition methods: (b) biorthogonal wavelet, (b) median, and (c) scale invariant affine median. (From Asghar, M. and Barner, K.E., Nonlinear multiresolution techniques with applications to scientific visualization in a haptic environment, *IEEE Trans. Visualization and Computer Graphics*, 7, 76–93, Mar. 2001. With permission.)

Wigner distribution of the analytic signal is generally used because removal of the negative frequency components also eliminates cross-terms between positive and negative frequency components. Although helpful in eliminating cross-terms, cross-terms between multiple components in the analytic signal still make interpretation difficult. Since cross-terms have oscillations of relatively high frequency, they can be attenuated by means of a smoothing operation which corresponds to the convolution of the WD with a 2D smoothing kernel. For most methods, the smoothing tends to produce the following effects: (1) a (desired) partial attenuation of the interference terms; (2) an (undesired) broadening of signal terms, i.e., a loss of time-frequency concentration; and (3) a (sometimes undesired) loss of some of the mathematical properties of the WD (e.g., Wigner distribution preserves the time and frequency marginals of a signal[28]).

The interference terms of the WD are a consequence of the WD's bilinear (or quadratic) structure; they occur in the case of multicomponent signals and can be identified mathematically with quadratic cross-terms.[28,29,32] According to the quadratic superposition law, the WD of an N-component signal consists of N signal terms and $\frac{N(N-1)}{2}$ ITs. Each signal component generates a signal term, and each pair of signal components generates an IT. While the number of signal terms grows linearly with the

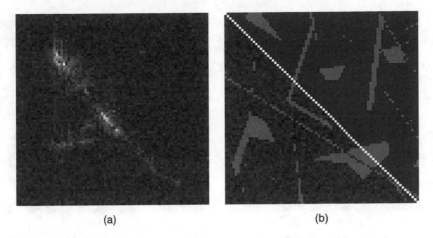

(a)   (b)

FIGURE III.2.20 (a) ISAR image of B-727 and (b) portion of synthetic original and noisy training images used for optimization.

number N of signal components, the number of ITs grows quadratically with N.[32] This is shown in Figure III.2.22.

The goal of filtering in the WD plane is thus to reduce, or eliminate, the cross-terms while preserving the auto terms. To achieve this result, a center affine filter can be applied to the WD. To take advantage of the structure of the filtering problem, the following modifications can be made to the standard center affine filtering operation:[35]

1. The absolute values of the samples, $|W_{m_1,m_2}|$, rather than their actual values, W_{m_1,m_2}, are utilized to calculating the respective affinities. Accordingly, the reference point of the affinity function is set as the absolute value of the center sample, $|W_c|$.
2. The affinity spread parameter γ is made proportional to the local variance at the observation window location. Thus, a higher value of γ is obtained, in general, for an observation window centered in a cross-term region than for a window centered in an auto-term region.

While (1) may not affect the affinities of samples in the auto-term region (as most of the samples are already positive), it will drastically change the affinities of samples in the cross-term region. Both positive and negative samples will get higher affinities; more samples will enter the estimate and thus, the filter behavior will change toward a linear low-pass filter. On the other hand, (2) will further ensure that more samples (including both positive and negative samples) get higher affinities in the cross-term region while assigning very low affinities to most of the samples in the auto-term region. This scenario is depicted in Figure III.2.23.

To evaluate the cross-term reduction, we consider the special case of a signal composed of the sum of time-frequency shifted signals $x(t) = c_1 x_1(t) + c_2 x_2(t)$ where $x_i(t) = x_0(t - t_i)e^{j2\pi f_i t}$, for $i = 1, 2$. Consider the case where $x_0(t)$ is a Gaussian atom. Our particular interest in a Gaussian shaped atom derives from the fact that it has good locality (energy concentration) in the joint $t - f$ domain, and its WD is non-negative.[36] Figure III.2.24 demonstrates the center affine filtered WD of the signal composed of two Gaussian atoms.

In order to demonstrate the relative performance of the center affine filtered WD, a comparison of the experimental results provided by several time-frequency representations is shown. Consider first a 128-point test signal comprised of one parabolic chirp, one sinusoidal pulse, and two parallel Gaussian pulses:

$$
\begin{aligned}
x(n) \quad = \quad & r_{1,64}(n)(e^{-\frac{j\pi n}{8}} + e^{-\frac{j3\pi}{8}(n+0.5n^2+0.33n^3)}) \\
& + r_{75,124}(n)e^{-\frac{(n-99)^2}{100}}(e^{\frac{j\pi n}{4}} + e^{\frac{j3\pi}{4}})
\end{aligned}
\tag{III.2.65}
$$

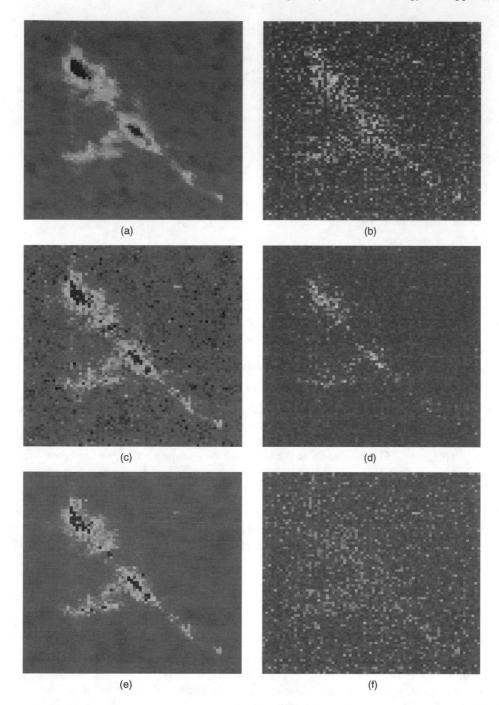

FIGURE III.2.21   Filter output and difference image for filters operating on the ISAR image in Figure III.2.20(a). WOS filter (a) output and (b) difference image, $L\ell$ filter (c) output and (d) difference image, and center affine filter (e) output and (f) difference image. (From Flaig, A., Arce, G.R., and Barner, K.E., Affine order statistic filters: a data-adaptive filtering framework for nonstationary signals, *IEEE Trans. Signal Proc.*, 46, 2101–2112, 1998. With permission.)

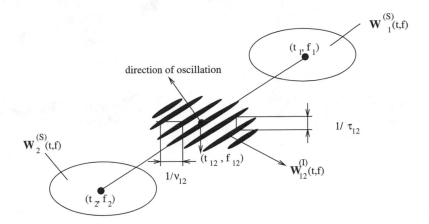

FIGURE III.2.22 Interference geometry of Wigner distribution. (From Arce, G.R. and Hasan, S.R., Elimination of interference terms of the discrete Wigner distribution using nonlinear filtering, *IEEE Trans. Signal Proc.*, 48, 2321–2331, Aug. 2000. With permission.)

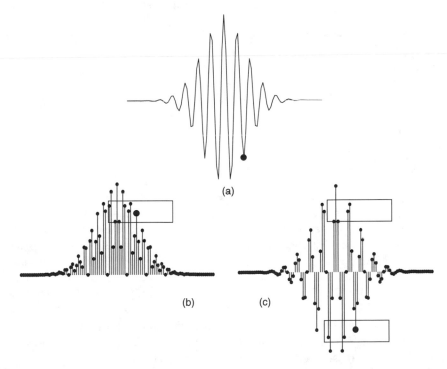

FIGURE III.2.23 Nonlinear filtering mechanism of a sample in a cross-term component: (a) center sample point (dot) to be filtered in a cross-term, (b) samples within square window represent samples affine to the center sample (absolute values), and (c) filtering of original samples according to their affinities (samples within the two square windows enter in the filtering operation). (From Arce, G.R. and Hasan, S.R., Elimination of interference terms of the discrete Wigner distribution using nonlinear filtering, *IEEE Trans. Signal Proc.*, 48, 2321–2331, Aug. 2000. With permission.)

with $r_{a,b}(n)$ the gating function:

$$r_{a,b}(n) = \begin{cases} 1, & a \le n \le b \\ 0, & \text{otherwise} \end{cases} \qquad (III.2.66)$$

(a)   (b)

FIGURE III.2.24   (a) Discrete Wigner distribution of a signal composed of two Gaussian atoms and (b) center affine filtered distribution.

Figure III.2.25(a) shows the Wigner distribution of the signal having auto components well localized but numerous high-amplitude oscillating cross-terms. Figure III.2.25(b) shows the pseudo smoothed Wigner distribution (using a 13-point Gaussian time smoothing window and 31-point Gaussian frequency smoothing window) of the test signal, which reduces the cross-terms by both frequency and time direction smoothing. The interpretation of this signal is much easier, but the component localization becomes coarser. Figure III.2.25(c) shows the Choi-Williams distribution of the given signal with kernel width $\sigma = 1$. Again, most of the cross-terms are minimized at the cost of reduced localization. Obvious problems are also visible, especially when the signal components overlap in time and frequency. Figure III.2.25(d) shows the results of the Baraniuk–Jones method 1.[37] This method fails to track the smoothly varying parabolic chirp and, as a result, that component appears as two connected linear chirps. Figure III.2.25(e) shows the representation given in Reference 38. In this scheme a time-adaptive radially Gaussian kernel is used. The results are adequate, although some loss in auto-component localization occurs. In addition, the computation cost is high as the kernel is computed in a local window sliding over the signal. Figure III.2.25(f) shows the center affine filtered TFR. An almost complete reduction in cross terms is attained without losing the resolution and localization provided by the Wigner distribution.

As a second example, consider a signal consisting of three Gaussian atoms: $B(t - T) + B(t) + B(t + T)$. The two outer atoms generate cross-terms overlapping with the center atom. As illustrated in Figure III.2.26(a), the local variance of the WV distribution around the center Gaussian atom is approximately equal to the local variance of the cross-term-only regions. To illustrate the interference term canceling ability of the center affine filter with respect to the standard median filter, we apply each filter to the original WVD. The results for the 7×7 median, 9×9 median, and center affine filters are shown in Figure III.2.26(b–d). As the figure shows, the result of applying a median filter to the WV distribution is not adequate, since although the cross-terms are somewhat reduced, significant distortion is introduced into the auto term components. The center affine filter, in contrast, significantly reduces the cross-terms while preserving the auto terms.

As a final practical example, consider the processing of protein sequence data. The amino acid sequence of the protein may be considered as a 20-symbol alphabet sequence, or it may be considered as a sequence of numerical values reflecting various physiochemical aspects of the amino acids, i.e., hydrophobicity, bulkiness, or electron-ion interaction potential (EIIP).[39] Analysis of the numerical representation of sequences often identifies characteristic patterns that may be too weak to be detected as patterns of symbols.[40] It has been shown[40,41] that proteins of a given family have a common characteristic frequency component related to their function. However, since frequency analysis alone contains no spatial information, there is no indication as to which residues contribute to the frequency components. Through the use of the Wigner-Ville time-frequency method, information

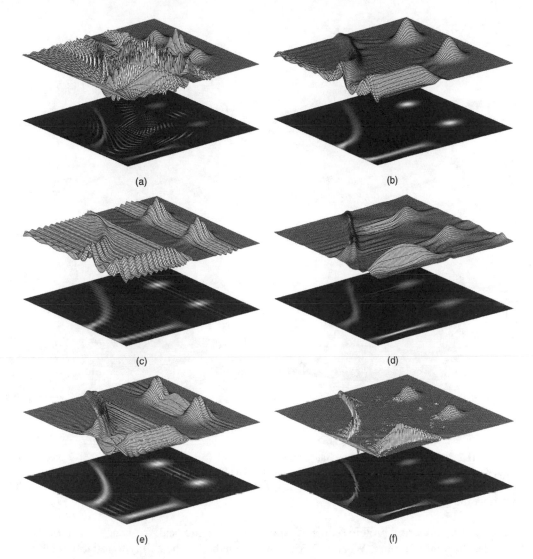

FIGURE III.2.25   Time-frequency representation of the signal given by Equation III.2.65 using (a) Wigner distribution, (b) pseudo smoothed Wigner distribution, (c) Choi–Williams distribution with spread factor $\sigma = 1$, (d) Baraniuk–Jones distribution (method 1), (e) Baraniuk–Jones (method 2), and (f) center affine filtering. (From Arce, G.R. and Hasan, S.R., Elimination of interference terms of the discrete Wigner distribution using nonlinear filtering, *IEEE Trans. Signal Proc.*, 48, 2321–2331, Aug. 2000. With permission.)

involving secondary structure and biologically active sites can be retained. This result is illustrated using the fibroblast growth factors class of proteins.[42]

Fibroblast growth factors constitute a family of proteins that affect the growth, migration, differentiation, and survival of certain cells. The amino acid sequence of a basic fibroblast growth factor (FGF) for human is 155 amino acids (or residues) long. The amino acid representation of the sequence can be converted from an alphabetic string to a numerical signal by using the EIIP for each amino acid in the sequence.

The Resonant Recognition Model (RRM)[40] is a physicomathematical model that analyzes the interaction of a protein and its target using digital signal processing methods. One application of this model involves prediction of a protein's biologically active sites. In this technique, a Fourier transform

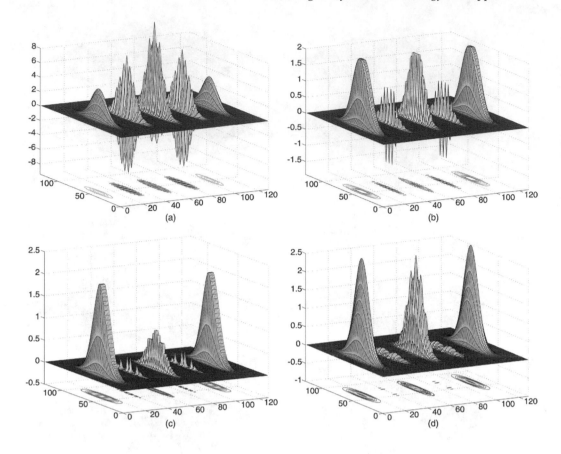

FIGURE III.2.26 (a) WVD of three Gaussian atoms and the (b) 7×7 median, (c) 9×9 median, and (d) center affine filter outputs. (From Arce, G.R. and Hasan, S.R., Elimination of interference terms of the discrete Wigner distribution using nonlinear filtering, *IEEE Trans. Signal Proc.*, 48, 2321–2331, Aug. 2000. With permission.)

is applied to the numerical protein sequence, and a characteristic peak frequency is determined for a particular protein's function. What is lacking in this method is the ability to reliably identify the individual amino acids that contribute to that peak frequency. The TFR of the amino acid sequence, however, preserves both the frequency information along with the spatial relationships. The discrete WV TFR of the human basic FGF is shown in Figure III.2.27. As illustrated in the previous examples, cross-terms make interpretation of this representation difficult. After application of the center affine filter, Figure III.2.27(b), cross-terms are minimized and primary terms are retained. The retained terms correspond to the activation sites and characteristic frequency component. Indeed, experiments have shown that the potential cell attachment sites of FGFs are between residues 46–48 and residues 88–90, and the characteristic frequency has been shown in the literature as 0.4512,[43] which is in agreement with the presented results.[42]

2.5.2 Fuzzy Ordering and Fuzzy Median Filters

Fuzzy ordering and fuzzy order statistic-based methods can also be applied to a wide range of problems. Here we consider three representative applications in which the fuzzy weighted median and fuzzy ordering show advantages over their traditional crisp counterparts. Specifically, the fuzzy weighted median filter is applied to image smoothing and image zooming. In the image smoothing

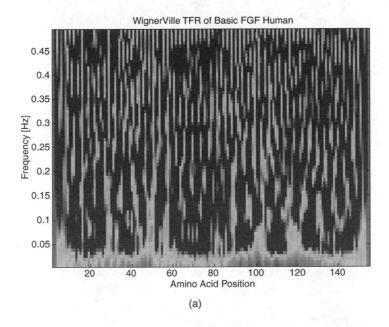

(a)

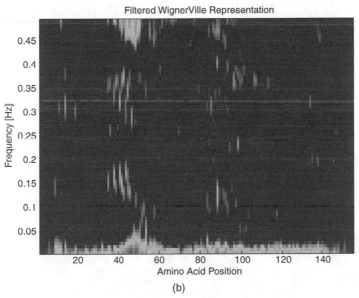

(b)

FIGURE III.2.27 (a) Wigner-Ville TFR of Basic FGF Human and (b) the center affine filtered representation indicating the activation sites.

case, the use of fuzzy samples in the weighted median structure leads to improved performance, especially when background noise with a Gaussian distribution is considered. In such cases, the fuzzy weighted median preserves image details, (i.e., edges), while removing outlying samples and appropriately smoothing background noise. The fuzzy weighted median also shows improved results over traditional linear and crisp weighted median filters in image zooming, or interpolation. Interpolations based on fuzzy methods show improved edge representations, while avoiding blocking artifacts in uniform regions. Lastly, fuzzy ranking is utilized in a multiuser detection scheme. The detection results show that the fuzzy ranking concept yields improved results compared to crisp ranking schemes, which do not consider the spread of samples.

2.5.2.1 Image Smoothing

Noise smoothing is an important component in many applications. This is particularly true in image processing, where sensor irregularities, atmosphere interference, or transmission/storage errors often introduce noise into the image capture, transmission, or display processes. Additionally, images are nonstationary signals within which important visual cues are represented by sharp transitions, i.e., edges. Consequently, linear filters do not yield satisfactory results in image noise filtering applications. The development of nonlinear images smoothing filters has thus become an active area of research.

The median filter and its generalizations have, perhaps, gathered the most attention in addressing noise smoothing in images.[17,45–51] The weighted median filter has been widely researched as an effective filter for removing noise, especially noise with heavy-tailed distributions. The median and weighted median filters have not, however, shown good performance in processing Gaussian noise contaminated images. The median and weighted median filter structures can be effectively utilized to process a wide range of corrupting noise processes if the more general fuzzy realizations are utilized.

The performance gain realized through the fuzzy generalizations is evaluated here for the fuzzy median and fuzzy weighted median filter cases.[4] The filters are compared to their crisp counterparts in a noise smoothing application where we consider two cases: an image corrupted by varying levels of (1) impulsive noise and (2) contaminated Gaussian noise. The presented results utilize the images Lenna and Albert, Figure III.2.28. Corrupted versions of the Lenna image are used as the input to each of the filters considered. Optimization is performed utilizing Lenna and noise with the same statistics as the observation image. The robustness of the algorithms, with respect to changing image statistics, is demonstrated by applying the filters optimized on Albert to the image Lenna. In all cases, a 5×5 observation window is utilized.

Consider first the case of salt and pepper impulsive noise, where the probability of noise contamination is p and varies from 0.05 to 0.45. Figure III.2.29(a) shows the MAE for each of the filters operating on the impulse corrupted versions of Lenna as a function of the noise contamination probability. As the figure shows, the fuzzy and crisp weighted median filters have nearly identical performance. This similarity in performance is a result of the optimization that yields a membership spread parameter of $\lambda = 0.01$, which is very small compared to the range of signal values, [0, 255]. Thus, in the impulsive noise case, the optimal weighted median filter utilizes crisp order statistics as

(a)   (b)

FIGURE III.2.28 512×512 8-bit/pixel grey-scale images: (a) Lenna and (b) Albert.

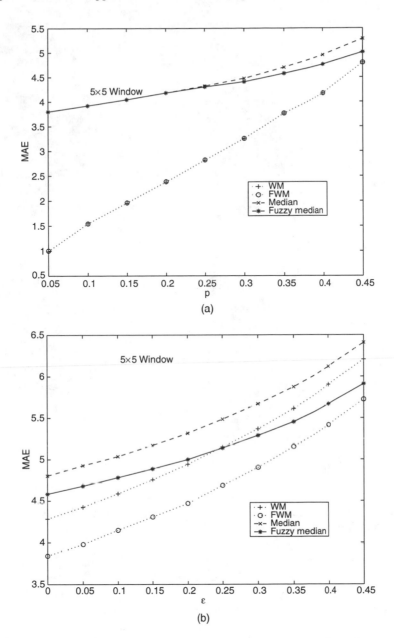

FIGURE III.2.29   MAE for the median, fuzzy median, weighted median, and fuzzy weighted median filters operating on the image Lenna corrupted by (a) impulsive noise with impulse probability p and (b) $\Phi(10, 100, \epsilon)$ contaminated Gaussian noise.

the output. This is an intuitive result since in the impulsive case, samples are either unaltered or bear no information. Thus, if the filter is centered on a sample that is not corrupted, it is best to simply perform the identity operation and weighted averaging has no advantage. If the filter is centered on a corrupted sample, then this sample contains no information about the original signal, and the best output selection is a crisp order statistics that can preserve local structure. Thus, there is no advantage to averaging samples and the optimization procedure correctly yields a crisp filter. Similar results are obtained in the median case.

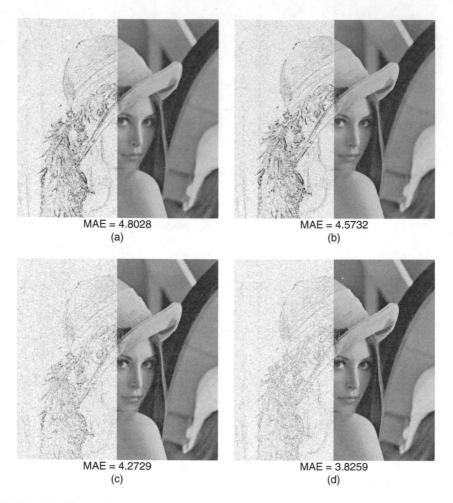

MAE = 4.8028
(a)

MAE = 4.5732
(b)

MAE = 4.2729
(c)

MAE = 3.8259
(d)

FIGURE III.2.30   Filtered image and the scaled difference for Lenna processed by the (a) median, (b) fuzzy median, (c) WM, and (d) FWM filters.

Consider next the performance of the filters operating in a contaminated Gaussian noise environment. In this case, the observation image is corrupted by $\Phi(10, 100, \epsilon)$ contaminated Gaussian noise, where ϵ indicates that mixing proportioned of two Gaussian processes with variance 10 and 100. The MAE performance of each of the filters operating in the contaminated Gaussian environment is shown in Figure III.2.29(b) as a function of the contamination parameter for ϵ in the range 0 to 0.45. In this case, there is a clear separation in the performance of the crisp and fuzzy filters. Selected weighted averaging among similarly valued samples is advantageous in this case, since there is always background Gaussian noise, regardless of the mixture parameter value. The figure shows a fairly consistent performance advantage for the fuzzy filters over the entire range of contamination. The performance gain is primarily due to two factors: the fuzzy versions of the filters (1) smooth the background noise more effectively and (2) preserve edges better in the presence of additive noise than their crisp counterparts.

A subjective evaluation of the filter outputs also illustrates the superior performance of the fuzzy filter generalizations. Figure III.2.30 shows the output images for each of the filters in the $\Phi(10, 100, 0)$ case. The right half of each image in the figure shows the output of a particular filter, while the left half is an error image. As the images show, the fuzzy filter generalizations yields

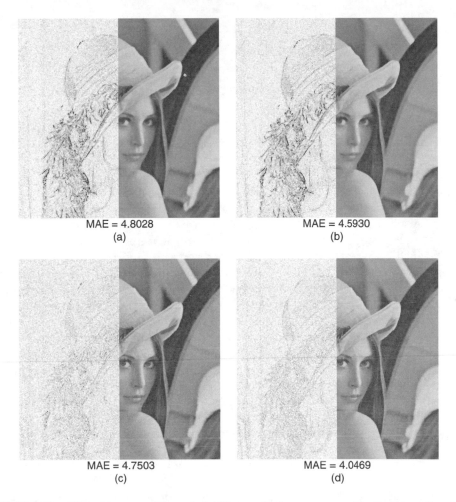

MAE = 4.8028
(a)

MAE = 4.5930
(b)

MAE = 4.7503
(c)

MAE = 4.0469
(d)

FIGURE III.2.31   Filtered image and the scaled difference for Lenna processed by the (a) median, (b) fuzzy median, (c) WM, and (d) FWM filters. Fuzzy median, WM and FWM filters are optimized using Albert.

superior noise smoothing performance while simultaneously preserving more filter structure than the crisp filter structures.

To evaluate the robustness of the filters to changing image statistics, we consider the filtering of the image Lenna by filters optimized on Albert. Figure III.2.31 shows the results of this operation for the same noise case as considered in Figure III.2.30. A comparison of the two figures shows that little loss in performance is realized by optimizing on a different image. In fact, although the resulting MAE is slightly higher in this case, the results for the Albert optimized filters appear visually superior. This is most likely due to the fact that Albert contains a significant number of edges, and therefore the optimization yields filters that are designed to place a high priority on edge preservation. When applied to the Lenna image, this results in outputs with superior edge preservation compared to the Lenna optimized filters, which place a somewhat reduced priority to edge preservation, due to the fairly uniform nature of the Lenna image.

2.5.2.2   Image Zooming

The resizing of signals is an important problem in many applications. This is particularly important for image data, as many forms of information are represented in images. Moreover, the rapid growth

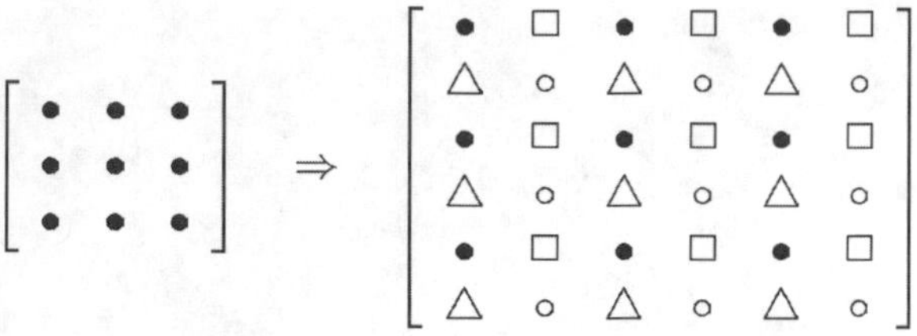

$$
\begin{bmatrix} \bullet & \bullet & \bullet \\ \bullet & \bullet & \bullet \\ \bullet & \bullet & \bullet \end{bmatrix}
\Rightarrow
\begin{bmatrix}
\bullet & \square & \bullet & \square & \bullet & \square \\
\triangle & \circ & \triangle & \circ & \triangle & \circ \\
\bullet & \square & \bullet & \square & \bullet & \square \\
\triangle & \circ & \triangle & \circ & \triangle & \circ \\
\bullet & \square & \bullet & \square & \bullet & \square \\
\triangle & \circ & \triangle & \circ & \triangle & \circ
\end{bmatrix}
$$

FIGURE III.2.32   Polyphase interpolation. Left: Original image; original pixels denoted by "•". Right: Zoomed image; interpolated pixels denoted by "△", "□", and "○".

of communications systems, such as the Internet, has made the exchange and display of image-based information widely available. Within such systems, images are often stored or transmitted at one size and displayed at another. As an example, images are often transmitted at a reduced pixel resolution, to save storage or bandwidth, and then zoomed, or interpolated, to a higher pixel density for display. Such operations have traditionally been performed utilizing linear operators. Linear operations, however, do not yield desirable results, especially for nonstationary signals such as images. Improved zooming results can be realized through the utilization of nonlinear interpolators. In particular, we show that the fuzzy weighted median filter produces superior zooming compared to linear and traditional weighted median filters.

Numerous structures exist for zooming, but most are based on first inserting zero-valued pixels between existing pixels and then interpolating the appropriate values of the inserted pixels from the surrounding samples. As an example, consider an algorithm that zooms an image by a factor of two. In this case, an empty array is constructed with twice the number of rows and columns as the original image. The original pixels are inserted into the array in alternating rows and columns, as indicated in Figure III.2.32. In the figure, pixel values from the original image are designated by a "•". The "new" pixels to be interpolated are indicated by the symbols "△", "□", and "○". The unique symbol given to each pixel results from the differing set of neighbors for each case. Since there are four cases, including the original image pixels, this approach is referred to as polyphase interpolation.

In the bilinear approach, the average of the two original image pixels neighboring a "△" sample is used to set the pixel value of these samples. A similar two-neighbor average is used to set the "□" pixel values. Lastly, the four original samples neighboring each "○" pixels are averaged to set the values of these remaining samples.

A similar polyphase approach utilizing weighted median, or fuzzy weighted median, filters can be developed as a straightforward extension to the linear approach.[4,15] In this case, the algorithm first interpolates "○" pixels as the weighted median of the original pixels at its four corners. As each of these samples are equally distant to the sample to be estimated, they are considered equally reliable and each assigned a weight of 1. The remaining pixels are determined by taking the weighted median of the four neighboring samples for each case. More specifically, for each "□" pixel, the two original pixels to its left and right are assigned weight 1, while the two interpolated "○" pixels above and below are assigned weight 0.5. This lower weighting reflects the fact that the interpolated samples are less reliable than the original pixels. Similarly, the original pixels above and below each "△" pixel are assigned weight 1, and the interpolated pixels to the right and left are assigned weight 0.5. The choice of 0.5 for the weights is chosen to reflect the reliability of the estimated samples. It should be noted that due to the structure of the weighted median, and the number of samples utilized, any assignment of weight values between zero and one for the estimated samples yields identical results.

Fuzzy weighted median filter can be implemented in the above-mentioned algorithm by simply replacing the pixels used for interpolation by their fuzzy counterpart. Thus, the interpolation process can be illustrated as follows. Consider an image represented by an array of pixel values denoted as $\{a_{i,j}\}$. Then the fuzzy weighted median polyphase interpolation performs the following mapping:

$$
\begin{bmatrix} a_{1,1} & a_{1,2} & a_{1,3} \\ a_{2,1} & a_{2,2} & a_{2,3} \\ a_{3,1} & a_{3,2} & a_{3,3} \end{bmatrix} \Rightarrow
\begin{bmatrix}
x_{1,1}^{\bullet} & x_{1,1}^{\square} & x_{1,2}^{\bullet} & x_{1,2}^{\square} & x_{1,3}^{\bullet} & x_{1,3}^{\square} \\
x_{1,1}^{\triangle} & x_{1,1}^{\circ} & x_{1,2}^{\triangle} & x_{1,2}^{\circ} & x_{1,3}^{\triangle} & x_{1,3}^{\circ} \\
x_{2,1}^{\bullet} & x_{2,1}^{\square} & x_{2,2}^{\bullet} & x_{2,2}^{\square} & x_{1,3}^{\bullet} & x_{2,3}^{\square} \\
x_{2,1}^{\triangle} & x_{2,1}^{\circ} & x_{2,2}^{\triangle} & x_{2,2}^{\circ} & x_{1,3}^{\triangle} & x_{2,3}^{\circ} \\
x_{3,1}^{\bullet} & x_{3,1}^{\square} & x_{3,2}^{\bullet} & x_{3,2}^{\square} & x_{3,3}^{\bullet} & x_{3,3}^{\square} \\
x_{3,1}^{\triangle} & x_{3,1}^{\circ} & x_{3,2}^{\triangle} & x_{3,2}^{\circ} & x_{3,3}^{\triangle} & x_{3,3}^{\circ}
\end{bmatrix}, \qquad (\text{III.2.67})
$$

where $a_{i,j}$ is the value of the pixel in the ith row and jth column of the original image. The pixels in the interpolated image are determined as:

$$
x_{i,j}^{\bullet} = a_{i,j}, \qquad (\text{III.2.68})
$$

$$
x_{i,j}^{\circ} = \text{MED}\,[\tilde{a}_{i,j}^{\circ}, \tilde{a}_{i+1,j}^{\circ}, \tilde{a}_{i,j+1}^{\circ}, \tilde{a}_{i+1,j+1}^{\circ}], \qquad (\text{III.2.69})
$$

$$
x_{i,j}^{\square} = \text{MED}\,[\tilde{a}_{i,j}^{\square}, \tilde{a}_{i,j+1}^{\square}, 0.5 \diamond \tilde{x}_{i-1,j}^{\square}, 0.5 \diamond \tilde{x}_{i+1,j}^{\square}], \qquad (\text{III.2.70})
$$

$$
x_{i,j}^{\triangle} = \text{MED}\,[\tilde{a}_{i,j}^{\triangle}, \tilde{a}_{i+1,j}^{\triangle}, 0.5 \diamond \tilde{x}_{i,j-1}^{\triangle}, 0.5 \diamond \tilde{x}_{i,j+1}^{\triangle}], \qquad (\text{III.2.71})
$$

In the above samples, $\tilde{a}_{i,j}^{\circ}$, $\tilde{a}_{i,j}^{\square}$, and $\tilde{a}_{i,j}\triangle$, are the fuzzy samples based on observation windows centered at locations $\circ$, $\square$, and $\triangle$, respectively. Note that for the same pixel location, these samples can take on different values, i.e., $\tilde{a}_{i,j}^{\circ}$, $\tilde{a}_{i,j}^{\square}$, and $\tilde{a}_{i,j}^{\triangle}$ may not be equal to each other. This is due to the fact that the observation window is centered at a different location in each case, and each fuzzy sample is set according to the relationship among the samples in the given window.

To illustrate the zooming performance of the bilinear, weighted median, and fuzzy weighted median methods, consider the zooming of the image area outlined in Figure III.2.33(a). The results of zooming this area by a factor of two, utilizing each method, is shown in Figure III.2.33(a–d). As the images show, the linear method introduces high frequency distortions along the edges. This is clearly seen by examining the white leg of the chair. The weighted median reduces this distortion somewhat, but introduces blocking artifacts, which can be seen in smooth areas such as facial regions. The fuzzy weighted median, in contrast, produces sharp edges with minimal artifacts. See again the leg of the chair. Also, no blocking artifacts are seen in the uniform regions.

Fuzzy weighted median filters thus offer improved performance in the interpolation of non-stationary signals (i.e., images). Visually important cues, such as edges, are well represented in the interpolated signal. Additionally, the blocking artifacts that are present in the output of weighted median filters are noticeably absent in the fuzzy weighted median output.

2.5.2.3　Multiuser Detection

Rank-ordering plays an important role not only in filtering, but also in nonparametric detection. In this application, the significance of rank-ordering arises from the fact that the sample ranks are uniformly distributed independent of the distribution of the pertaining samples, implying that detection algorithms based on ranks can be designed that are robust to varying sample statistics. This is very important if little or nothing is known about the noise (or signal) statistics, or if the noise environment is changing rapidly, as is often the case in wireless communications. The majority of the detection algorithms developed to date have been based on crisp ranks that contain only very coarse information about the observation samples, and thus lead to detection schemes that exhibit low efficiency when compared to optimal techniques. The use of fuzzy ranks can help to overcome

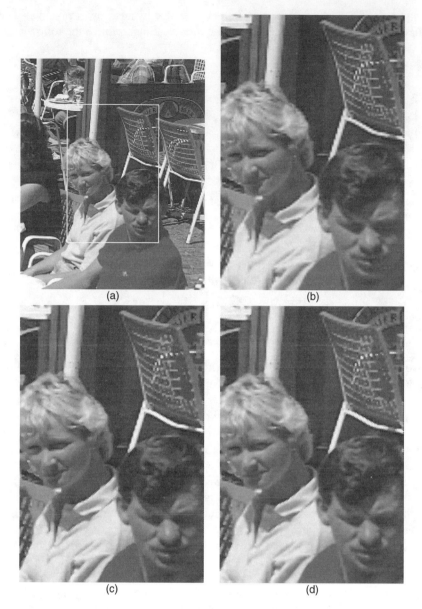

FIGURE III.2.33 (a) Original image with area of interest outlined and zoomed areas of interest utilizing the (b) bilinear, (c) weighted median, and (d) fuzzy weighted median methods.

this drawback. We illustrate this through the use of Fuzzy Rank Order (FRO) detectors[8,52–54] for fast frequency hopping multiple access networks.

In a frequency-hopping spread spectrum network, K users transmit sequences of L tones called frames. Each tone is chosen from a set of Q possible tones. At the receiver, the signals of all users are superimposed. The frames of the received signal are non-coherently detected at Q possible frequencies for each of the L time slots. The results are arranged in a $Q \times L$ received matrix, where the rows represent the frequencies and the columns represent the time slots. To detect user k, the rows are de-spread with the kth user's spreading sequence, yielding a decoded matrix with entries $x_{i,j}$. The decoded matrix will contain one row at the level of the desired user's symbol. This row is referred

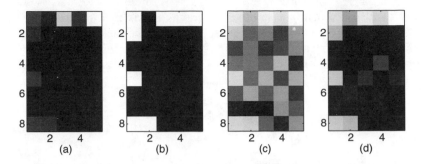

FIGURE III.2.34 (a) Decoded matrix **x**, (b) thresholded matrix, (c) crisp rank matrix, and (d) fuzzy rank matrix, for $Q = 8, L = 5$, and SNR $= 25$ dB. The first row corresponds to the correct decision. (From Flaig, A., Barner, K.E., and Arce, G.R., Fuzzy ranking: theory and applications, *Signal Proc. Special Issue on Fuzzy Processing*, 80, 1017–1036, June 2000. With permission.)

to as the correct row. The samples in the correct row correspond to the desired signal plus noise, and the entries in the remaining rows are due to noise only, where the noise is due to a combination of background noise and contributions from interfering users (multiple access interference). Denoting the cumulative distribution of the samples in the correct row and the spurious rows by F_{S+N} and F_N, and assuming that all elements in the received matrix are mutually independent,[55] the decision as to which one of the Q rows is the correct row can be put in the framework of a Q-ary hypothesis test:

$$H_k : x_{k,\ell} \sim F_{S+N}, \ell = 1, 2, \dots, L \text{ and}$$
$$x_{i,j} \sim F_N, \quad j = 1, 2, \dots, L$$
$$i = 1, \dots, k - 1, k + 1, \dots, Q \tag{III.2.72}$$

Thus, the samples in the row under test have to be compared to the remaining pooled data in some fashion.

The conventional detector forms a binary detection matrix by detecting any energy (above the noise floor) in a given time-frequency slot in **x** as a hit, regardless of the actual amount of energy present. To obtain a decision as to which symbol was transmitted, the number of hits is determined, and the row with the maximum number of hits is chosen to be the correct row. Figure III.2.34(a) depicts the energy samples seen in a typical received matrix, for $Q = 8, L = 5$, and SNR $= 25$ dB, by gray tones, where white is assigned to the maximum energy sample and black is assigned to the element with the smallest value. In this example, the first row is the correct row. This is reflected in the received matrix by two clear hits on the third and fifth hop. Aside from those correct hits, there are also somewhat weaker hits scored by interfering users in row two, five, and eight. Figure III.2.34(b) shows the corresponding threshold matrix, where black (0) and white (1) represent nonhits and hits, respectively. Due to the hard decision, the second hit in the first row is irrevocably lost.

The hard-decision majority vote (HDMV) method described above works well in high SNR environments,[56,57] where samples that originated from signal plus noise and samples that stem from noise only are easily distinguished. A notable loss in efficiency, however, is incurred in low SNR environments.

Decision rules based on rank ordering are motivated by the observation[55] that the cumulative distributions satisfy:

$$F_{S+N}(x) \geq F_N(x) \tag{III.2.73}$$

for all x. That is, the samples in the correct row are *stochastically larger* than the samples in the remaining rows. The maximum rank sum receiver (MRSR)[58] exploits this property by rank-ordering the samples in the decoded matrix $\mathbf{x}$ and forming a rank matrix $\mathbf{r} = \{r_{k,\ell}\}$, whose elements are the crisp ranks of the corresponding entries in $\mathbf{x}$, i.e., $r_{k,\ell}$ is the rank of $x_{k,\ell}$. Thus, according to Equation III.2.73 hits are typically assigned high ranks, whereas noise samples are assigned low ranks. The decision is made by summing the ranks across each row, and choosing the row with the largest sum as the correct row:

$$H_k : \max_k \sum_{\ell=1}^{L} r_{k,\ell} \qquad\qquad \text{(III.2.74)}$$

The main shortcoming of this approach is that the real-valued energy samples in $\mathbf{x}$ are replaced by integer ranks that reflect only the relative values of the received energies but not their spread. As an example, consider the integer rank matrix pertaining to $\mathbf{x}$, Figure III.2.34(c), where bright tones correspond to high ranks and dark tones correspond to low ranks. Comparing Figure III.2.34(a) and (c), it can be observed that although the two strongest hits are clearly recognizable in both matrices, in the integer rank matrix the weaker hits are difficult to distinguish from the nonhits.

The more general fuzzy-ordering-based FRO detectors[8,53,54] are capable of making a reliable distinction between hits and nonhits due to the clustering property of fuzzy ranks.[4,6] Extending the MRSR detector to the FRO detector is a straightforward fuzzy extension:

$$H_k : \max_k \sum_{\ell=1}^{L} \tilde{r}_{k,\ell} \qquad\qquad \text{(III.2.75)}$$

Figure III.2.34(d) shows the effect of fuzzy ranking when applied to the decoded matrix. The consideration of sample spread results in a clear identification of hits and nonhits indicated by bright and dark gray tones, respectively.

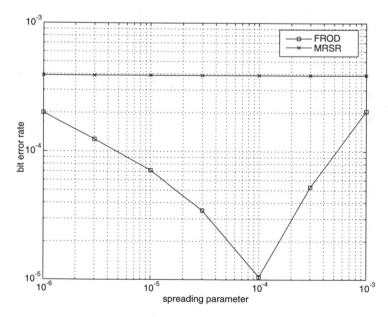

FIGURE III.2.35 Bit error rate of FRO detector vs. the fuzzy spreading parameter (λ) for $M = 2, Q = 16, L = 5$, and SNR $= 25$ dB. The bit error rate of the MRSR is shown for reference. (From Flaig, A., Barner, K.E., and Arce, G.R., Fuzzy ranking: theory and applications, *Signal Proc. Special Issue on Fuzzy Processing*, 80, 1017–1036, June 2000. With permission.)

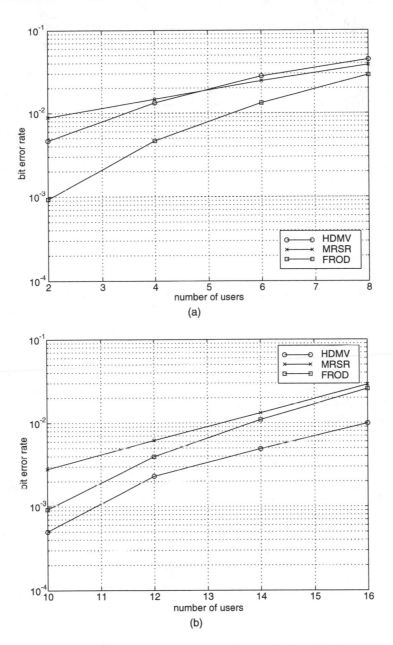

FIGURE III.2.36 Probability of bit error vs. the number of users for the HDMV, MRSR, and the FROD detectors for the $Q = 32$ and $L = 8$ case with (a) SNR = 10 dB and (b) SNR = 25 dB. (From Flaig, A., Barner, K.E., and Arce, G.R., Fuzzy ranking: theory and applications, *Signal Proc. Special Issue on Fuzzy Processing*, 80, 1017–1036, June 2000. With permission.)

To further illustrate the performance of the various detectors, consider the simplified Rayleigh fading channel.[58] In this model, the entries in the decoded matrix obey exponential distributions whose mean value reflects the presence or absence of a signal. More specifically, the samples in a correct row are distributed as $x \sim \lambda_1 e^{-\lambda_1 x}$, while the samples in incorrect rows are distributed as $x \sim p\lambda_1 e^{-\lambda_1 x} + (1 - p)\lambda_0 e^{-\lambda_0 x}$, where $1/\lambda_1 > 1/\lambda_0$. Note that $1/\lambda_1$, and $1/\lambda_0$ are the mean energies of samples that correspond to signal plus noise and noise only. The mixing parameter p is the proportion of samples generated by interfering users and can be found by invoking a randomization argument, which leads to $p = 1 - (1 - 2^{-log_2(Q)})^{(K-1)}$.

Figure III.2.35 depicts the bit error rate of the FRO detector for different values of the fuzzy spreading parameter. As expected, an appropriate amount of fuzzy spreading facilitates the separation of hits from nonhits, and thus reduces the probability of error of the FROD in comparison to its (zero spread) limiting case, the MRSR. Figure III.2.36 depicts the probability of bit error vs. the number of users for the HDMV, MRSR, and FROD detectors for the $Q = 32$ and $L = 8$ case with SNR $= 10\,$dB and SNR $= 25\,$dB. For the moderate SNR $= 10\,$dB case, the FRO detector clearly outperforms both the HDMV and the MRSR detectors. As the SNR is increased to the SNR $= 25\,$dB case, the HDMV detector achieves a slightly lower bit error rate than the FRO detector. This is not surprising, as it has been shown[57] that, for the simplified fading model, the HDMV detector is asymptotically optimal as the SNR $\to \infty$.

2.6   Conclusions

This chapter developed several broad classes of nonlinear filters from a Maximum Likelihood perspective. It is shown that for heavy-tailed Laplacian distributions, the ML development leads directly to the well-known median and weighted median filters. This development shows the fundamental importance of considering both spatial order and rank order of samples. The SR concept was formally established and then extended to the fuzzy case in which sample affinity, or spread (diversity), is considered. The concepts of sample affinity and affinity-based ordering lead to several classes of fuzzy filters. The fuzzy filters can, in general, be constructed as generalizations of existing filters that incorporate fuzzy measures or as new filtering paradigms based on fuzzy concepts. This chapter covers two broad classes of fuzzy filters: affine filters, which are generalizations of weighted sum filters that incorporate sample spread, and fuzzy weighted median filters, which generalize the standard class of weighted median filters by incorporating fuzzy samples.

The incorporation of affinity measures into filtering algorithms yields additional degrees of freedom that allow sample spread to be considered in the decision process, thus leading to improved performance. The performance gains realized by fuzzy methods were demonstrated for a wide range of problems, including robust frequency-selective filtering, inverse synthetic aperture radar image filtering, time-frequency plane filtering, multiresolution representations of signals, image smoothing, image zooming (interpolation), and multiuser detection. The wide applicability of fuzzy methods and their superior performance show that these concepts constitute important tools for modern signal processing applications.

References

1. Arce, G.R., Kim, Y.T., and Barner, K.E., Order-statistic filtering and smoothing of time series: Part 1, in *Order Statistics and their Applications* (Rao C.R. and Balakrishnan, N., Eds.), vol. 16 of *Handbook of Statistics*, pp. 525–554, Amsterdam, The Netherlands: Elsevier Science, 1998.
2. Barner K.E. and Arce, G.R., Order-statistic filtering and smoothing of time series: Part 2, in *Order Statistics and their Applications* (Rao C.R. and Balakrishnan, N., Eds.), vol. 16 of *Handbook of Statistics*, pp. 555–602, Amsterdam, The Netherlands: Elsevier Science, 1998.
3. Arce G.R. and Barner, K.E., Nonlinear signals and systems, in *Encyclopedia of Electrical and Electronics Engineering* (Webster, J.G., Ed.), pp. 612–630, New York: John Wiley & Sons, 1998.
4. Barner K.E. and Hardie, R.C., Spatial-rank order selection filters, in *Nonlinear Image Processing*. (Mitra S.K. and Sicuranza, G., Eds.), San Diego, CA: Academic Press, 2001.

5. Edgeworth, F.Y., A new method of reducing observations relating to several quantities, *Phil. Mag. (Fifth Series)*, vol. 24, 1887.

6. Barner, K.E., Flaig, A., and Arce, G.R., Fuzzy time-rank relations and order statistics, *IEEE Signal Proc. Lett.*, 5, 252–255, Oct. 1998.

7. Barner, K.E., Nie, Y., and An, W., Fuzzy ordering theory and its use in filter generalizations, *EURASIP J. Appl. Signal Proc. – Special Issue on Nonlinear Signal and Image Proc.*, Dec. 2001.

8. Flaig, A., Barner, K.E., and Arce, G.R., Fuzzy ranking: theory and applications, *Signal Proc. Special Issue on Fuzzy Processing*, 80, 1017–1036, June 2000.

9. Flaig, A., Arce, G.R., and Barner, K.E., Affine order statistic filters: a data-adaptive filtering framework for nonstationary signals, *IEEE Trans. Signal Proc.*, 46, 2101–2112, 1998.

10. Lee, Y.H. and Kassam, S.A., Generalized median filtering and related nonlinear filtering techniques, *IEEE Trans. Acoustics, Speech, and Signal Proc.*, 33, 672–683, June 1985.

11. Cha I. and Kassam, S.A., Channel equalization using adaptive complex radial basis function networks, *IEEE J. Select Areas of Communications*, 13, 964–975, Jan. 1995.

12. Cha I. and Kassam, S.A., RBFN restoration of nonlinearly degraded images, *IEEE Trans. Image Proc.*, 5, 122–131, June 1996.

13. Haykin, S., *Adaptive Filter Theory*. New Jersey: Prentice-Hall, 1991.

14. Arce, G.R., A generalized weighted median filter structure admitting negative weights, *IEEE Trans. Signal Proc.*, vol. 46, Dec. 1998.

15. Arce, G.R. and Paredes, J., Image enhancement with weighted medians, in *Nonlinear Image Proc.* (Mitra S. and Sicuranza, G., Eds.), pp. 27–67, Academic Press, 2000.

16. Arce, G.R. and Paredes, J.L., Recursive weighted median filters admitting negative weights and their optimization, *IEEE Trans. Signal Proc.*, 48, 768–779, Mar. 2000.

17. Yin, L., Yang, R., Gabbouj, M., and Neuvo, Y., Weighted median filters: a tutorial, *IEEE Trans. Circuits and Systems II: Analog and Digital Signal Proc.*, vol. 41, May 1996.

18. Barner, K.E., Colored $l-\ell$ filters with tap bias and their application in speech pitch detection, *IEEE Trans. Signal Proc.*, vol. 9, Sept. 2000.

19. Gandhi P., and Kassam, S.A., Design and performance of combination filters, *IEEE Trans. Signal Proc.*, 39, 1524–1540, July 1991.

20. Palmieri F. and Boncelet, C.G., Jr., Ll-filters-a new class of order statistic filters, *IEEE Trans. Acoustics, Speech, and Signal Proc.*, 37, 691–701, May 1989.

21. Shao M. and Nikias, C.L., Signal processing with fractional lower order moments: stable processes and their applications, *Proc. IEEE*, vol. 81, July 1993.

22. Strang and G., Nguyen, T.Q., *Wavelets and Filter Banks*. Wellesley-Cambridge Press, 1996.

23. Daubechies, I., *Ten Lectures on Wavelets*. Philadelphia: SIAM, 1992.

24. Mallat, S.G., A theory for mulitiresolution signal decomposition: the wavelet representation, *IEEE Trans. Pattern analysis and Machine Intelligence*, 11, 674–693, July 1989.

25. Asghar M. and Barner, K.E., Nonlinear multiresolution techniques with applications to scientific visualization in a haptic environment, *IEEE Trans. Visualization and Computer Graphics*, 7, 76–93, Mar. 2001.

26. Asghar, M.W., Nonlinear multiresolution techniques with applications to scientific visualization in a haptic environment, Master's thesis, University of Delaware, Newark, Delaware 19716, May 1999.

27. Zyweck A. and Bogner, R.E., High-resolution radar imagery of mirage III aircraft, *IEEE Trans. Antennas and Propagation*, 42, 1356–1360, Sept. 1994.

28. Cohen, L., Time-frequency distributions – a review, *Proc. IEEE*, 77, 941–981, July 1989.

29. Claasen, T. and Meclenbrauker, W., The Wigner distribution – a tool for time-frequency signal analysis – Part II: Discrete time signals, *Philips J. Research*, 35, 276–300, 1980.

30. Janssen, A., On the locus and spread of pseudo-density functions in the time-frequency plane, *Philips J. Research*, 37(3), 79–110, 1982.

31. Hlawatsch F., and Bourdeaux-Bartels, G., Linear and quadratic time-frequency signal representations, *IEEE Signal Proc. Mag.*, 9, 21–67, April 1992.

32. Hlawatsch F. and Flandrin, P., *The Wigner Distribution – Theory and Applications in Signal Processing*, ch. The interference structure of the Wigner distribution and related time-frequency signal representations. Elsevier Science, 1997.

33. Atlas, L., Fang, J., Loughlin, P., and Music, W., Resolution advantages of quadratic signal processing, in *Proc. SPIE*, 1566, 134–143, 1991.

34. Flandrin, P., Some features of time-frequency representations of multicomponent signals, in *Proc. Int. Conf. Acoustics, Speech, and Signal Proc. (ICASSP)*, pp. 41.B.4.1–41.B.4.4, 1984.

35. Arce, G.R. and Hasan, S.R., Elimination of interference terms of the discrete Wigner distribution using nonlinear filtering, *IEEE Trans. Signal Proc.*, 48, 2321–2331, Aug. 2000.

36. Qian, S. and Morris, J., Wigner distribution decomposition and cross-terms deleted representation, *Signal Proc.*, 27, 125–144, 1992.

37. Jones, D.L. and Baraniuk, R.G., A signal-dependent time-frequency representation: optimal kernel design, *IEEE Trans. Signal Proc.*, 41, 1589–1601, Apr. 1993.

38. Jones, D.L. and Baraniuk, R.G., An adaptive optimal-kernel time-frequency representation, *IEEE Trans. Signal Proc.*, 43, 2361–2371, Oct. 1995.

39. Tomii, K. and Kanehisa, M., Analysis of amino acids and mutation matrices for sequence comparison and structure prediction of proteins, *Protein Engineering*, vol. 9, Jan. 1996.

40. Cosic, I., Macromolecular bioactivity: Is it resonant interaction between macromolecules? – theory and applications, *IEEE Trans. Biomedical Engineering*, vol. 41, Dec. 1994.

41. Veljkovic, V., Cosic, I., Dimitrjevic, B., and Lalovic, D., Is it possible to analyze DNA and protein sequences by the methods of digital signal processing?, *IEEE Trans. Biomedical Engineering*, vol. 32, May 1985.

42. Bloch, K. and Arce, G.R., Time-frequency analysis of protein sequence data, in *Proc. IEEE–EURASIP Nonlinear Signal and Image Processing (NSIP) Workshop*, (Baltimore, MD), June 2001.

43. Fang, F. and Cosic, I., Prediction of active sites of fibroblast growth factors using continuous wavelet transforms and the resonant recognition model, in *Proc. Inaugural Con. Victorian Chapter of the IEEE EMBS*, 1999.

44. Arce, G.R. and Gallagher, N.C., Jr., State description of the root set of median filters, *IEEE Trans. Acoustics, Speech, and Signal Proc.*, vol. 30, Dec. 1982.

45. Arce, G.R., Gallagher, N.C., Jr., and Nodes, T.A., Median filters: theory and aplications, in *Adv. Computer Vision and Image Proc.* (Huang, T.S., Ed.), Vol. 2, JAI Press, 1986.

46. Bovik, A.C., Huang, T.S., and Munson, D.C., Jr., A generalization of median filtering using linear combinations of order statistics, *IEEE Trans. Acoustics, Speech, and Signal Proc.*, vol. 31, Dec. 1983.

47. Brownrigg, D.R.K., The weighted median filter, *Commun. Assoc. Comput. Mach.*, vol. 27, Aug. 1984.

48. Ko, S.-J. and Lee, Y.H., Center weighted median filters and their applications to image enhancement, *IEEE Trans. Circuits and Systems*, 38, 984–993, Sept. 1991.

49. Zeng, B., Optimal median-type filtering under structural constraints, *IEEE Trans. Image Proc.*, vol. 7, July 1995.

50. Coyle, E.J., Lin, J.-H., and Gabbouj, M., Optimal stack filtering and the estimation and structural approaches to image processing, *IEEE Trans. Acoustics, Speech, and Signal Proc.*, 37, 2037–2066, Dec. 1989.

51. Wendt, P., Coyle, E.J., and Gallagher, N.C. Jr., Stack filters, *IEEE Trans. Acoustics, Speech, and Signal Proc.*, 34, 898–911, Aug. 1986.

52. Flaig, A., Cooper, A.B., Arce, G.R., Tayong, H., and Cole-Rhodes, A., Fuzzy rank-order detectors for frequency hopping networks, in *Proc. 3rd Annual Fedlab Symp. Adv. Telecommunications Information Distribution Research Program (ATIRP)*, (College Park, MD), pp. 121–125, Feb. 1999.

53. Tayong, H., Cole-Rhodes, A., Cooper, B., Flaig, A., and Arce, G., A reduced complexity detector for fast frequency hopping, in *Proc. Annual Fedlab Symp. Adv. Telecommunications/Information Distribution Research Program (ATIRP)*, (College Park, MD), Mar. 2000.

54. Tayong, H., Beasley, A., Cole-Rhodes, A., Cooper, B., and Arce, G., Parametric diversity combining in fast frequency hopping, in *Proc. Annual Fedlab Symp. Adv. Telecommunications/Information Distribution Research Program (ATIRP)*, (College Park, MD), Mar. 2001.

55. Woinsky, M.N., Nonparametric detection using spectral data, *IEEE Trans. Information Theory*, 18, 110–118, Jan. 1972.

56. Goodman, D., Henry, P., and Prabhu, V., Frequency-hopped multilevel FSK for mobile radio, *Bell Systems Tech. J.*, 59, 1257–1275, Sept. 1980.

57. Viswanathan R. and Gupta, S.C., Performance comparison of likelihood, hard-limited, and linear combining receiver, *IEEE Trans. Communication Theory*, 31, 670–677, May 1983.

58. Viswanathan R. and Gupta, S.C., Nonparametric receiver for FH-MFSK mobile radio, *IEEE Trans. Communication Theory*, 33, 178–184, Feb. 1985.

3

Intelligent System Modeling of Bioacoustic Signals Using Advanced Signal Processing Techniques

Leontios J. Hadjileontiadis
Aristotle University of Thessaloniki

Yannis A. Tolias
Aristotle University of Thessaloniki

Stavros M. Panas
Aristotle University of Thessaloniki

0-8493-1121-7/03/$0.00+$1.50
© 2003 by CRC Press LLC

3.1   Introduction

Bioacoustics is defined as *the study of sounds produced by or affecting living organisms.*[1] When focusing on the sounds produced by the human body, a family of signals, namely the human bioacoustic signals or simply *Bioacoustic Signals* (BioS), could be formed. The BioS, being the audible outcome emitted from the human body, possess valuable diagnostic information regarding the functionality of the human organs involved in the sound production mechanisms. Proper extraction and use of this information can transform the art of BioS auscultation into a scientific discipline. This is achieved in part by the use of signal processing methods, which convert BioS from acoustic vibrations inside the human organism into graphs and parameters with diagnostic value. This diagnostic value is further revealed when advanced signal-processing techniques are employed in the analysis of BioS. In that way, novel diagnostic tools that objectively track the characteristics of the relevant pathology and assist the clinicians in everyday practice could be introduced.

This chapter is organized to provide a stepwise presentation of different approaches in the intelligent modeling and enhanced analysis of three types of BioS, i.e., *Lung Sounds* (LS), *Heart Sounds* (HS), and *Bowel Sounds* (BS), based on advanced signal-processing techniques. In particular, de-noising, detection, modeling, feature extraction, and classification procedures structured on *Wavelet Transform* (WT), *Fuzzy Logic* (FL), *Higher-Order Statistics/Spectra* (HOS), *Lower-Order Statistics* (LOS), and *Higher-Order Crossings* (HOC) are described in detail. The methods are grouped in accordance with the examined problem, set, each time, by the relevant diagnostic information embedded in the analyzed type of the bioacoustic signal. In that way, a comparison of the performance of different approaches to the same problem is feasible by the reader. Sufficient mathematical background information is provided when needed in order to assist comprehension of the theoretical basis of the methods. Moreover, examples of the application of the proposed methods on real data exemplify their efficient performance and help in the realization of their enhanced characteristics.

The techniques included in this chapter define new approaches in the BioS analysis and expand the frontiers of the 'mainstreaming' of their processing into new fields with significant scientific merit. The advantageous performance of the proposed methods turns them into appropriate candidates as diagnostic tools in clinical practice, facilitating their route from inception to practical fulfillment.

3.2 Lung, Heart, and Bowel Sounds: Definition, Categorization, and Main Characteristics

Before embarking on a thorough analysis of the selected BioS, i.e., LS, HS, and BS, it is pertinent to discuss first their definition, categorization, and main characteristics. Acquaintance with this information facilitates the reader to better correlate the mechanisms of the processing methods with the processed signal itself.

3.2.1 Lung Sounds

3.2.1.1 Definition and Historical Background

From the time of ancient Greeks and their doctrine of medical experimentation until at least the 1950s, the LS were considered as *the sounds originating from within the thorax* and they were justified mainly on the basis of their acoustic impression. For example, the writings of the Hippocratic School, in about 400 B.C., describe the chest (lung) sounds as splashing, crackling, wheezing, and bubbling sounds emanating from the chest.[2] An important contribution to the qualitative appreciation of lung and heart sounds was the invention of the stethoscope, invented by René Theophil Laënnec in 1816. Laënnec's gadget replaced the 'ear-upon-chest' detection procedure with the employment of a narrow paper tube, enhancing the emitted lung and heart sounds.[3] Unfortunately, it took over one century for the quantitative analysis to appear. Attempts for a quantitative approach date to 1930, but the first systematic, quantitative measurement of their characteristics, i.e., amplitude, pitch, duration, and timing in controls and in patients, is attributed to McKusick et al. in 1953;[4] the door to the acoustic studies in medicine was finally opened.

3.2.1.2 Categorization and Main Characteristics

The categorization of LS is summarized in Figure III.3.1. According to this figure, the LS are divided into two main categories, i.e., *Normal* LS (NLS) and *Abnormal* LS (ALS). The NLS are certain sounds heard over specific locations of the chest during breathing in healthy subjects. The

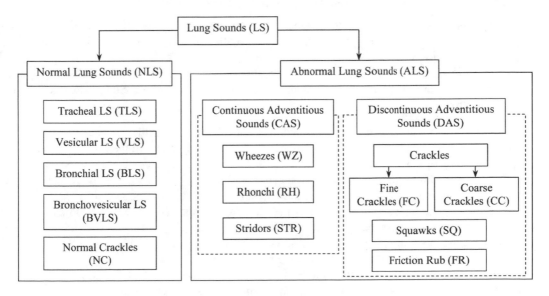

FIGURE III.3.1 Summary of the categorization of LS.

character of the NLS and the location at which they are heard defines them. Hence, the category of the NLS includes the following types:[5]

1. *Tracheal* LS (TLS), heard over the trachea having a high loudness
2. *Vesicular* LS (VLS), heard over dependent portions of the chest, not in immediate proximity to the central airways
3. *Bronchial* LS (BLS), heard in the immediate vicinity of central airways, but principally in the trachea and larynx
4. *Bronchovesicular* LS (BVLS), which refers to NLS with a character in between VLS and BLS, heard at intermediate locations between the lung and the large airways
5. *Normal Crackles* (NC), inspiratory LS heard over the anterior or the posterior lung bases

The ALS consist of LS of a BLS or BVLS nature that appear at typical locations (where VLS are the norm). The ALS are categorized between *Continuous Adventitious Sounds* (CAS) and *Discontinuous Adventitious Sounds* (DAS),[6] and include the following types:[7–9]

1. *Wheezes* (WZ), musical CAS that occur mainly in expiration, invariably associated with airway obstruction, either focal or general
2. *Rhonchi* (RH), low-pitched, sometimes musical CAS that occur predominantly in expiration, associated more with chronic bronchitis and bronchiectasis than with asthma
3. *Stridors* (STR), musical CAS that are caused by a partial obstruction in a central airway, usually at or near the larynx
4. *Crackles*, discrete, explosive, nonmusical DAS, further categorized between:
   a. *Fine Crackles* (FC), high-pitched exclusively inspiratory events that tend to occur in mid to late inspiration, repeat in similar patterns over subsequent breaths, and have a quality similar to the sound made by trips of Velcro being slowly pulled apart; they result from the explosive reopening of small airways that had closed during the previous expiration
   b. *Coarse Crackles* (CC), low-pitched sound events found in early inspiration and occasionally in expiration as well, arise from fluid in small airways, are of a 'popping' quality, and tend to be less reproducible than the FC from breath to breath
5. *Squawks* (SQ), short, inspiratory wheezes that appear to occur in allergic alveolitis and interstitial fibrosis, initiated with a crackle, caused by the explosive opening and fluttering of the unstable airway that causes the short wheeze
6. *Friction Rub* (FR), DAS localized to the area overlying the involved pleura and occur in inspiration and expiration when roughened pleural surfaces rub together, instead of gliding smoothly and silently

As the preceding paragraph has demonstrated, it is not difficult to provide evidence that the LS are directly related to the condition of the pulmonary function. The variety in the categorization of LS implies changes in the acoustic characteristics either of the source and/or the transmission path of the LS inside the lungs, due to the effect of a certain pulmonary pathology. It is likely that the time and frequency domain characteristics of the LS signals reflect these anatomical changes.[10]

In particular, the time domain pattern of the NLS (apart from the NC which follow the one of crackles) resembles a noise pattern bound by an envelope, which is a function of the flow rate.[10] Tracheal sounds have higher intensity and a wider frequency band (0–2 kHz) than the chest wall sounds (0–600 Hz) and contain acoustic energy at higher frequencies.[11] The CAS time domain pattern is a periodic wave that may be either sinusoidal or consist of more complex, repetitive sound structures.[10] In the case of WZ, the power spectrum contains several peaks ('polyphonic' WZ), or a single peak ('monophonic' WZ), usually in the frequency band of 200–800 Hz, indicating

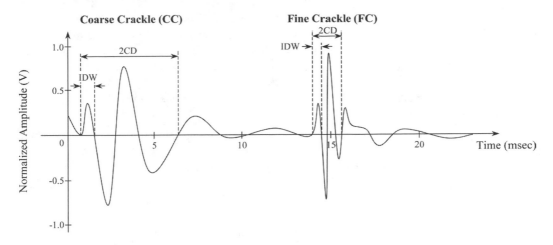

FIGURE III.3.2   Crackles morphology in a simulated LS signal. This morphology is met in LS from patients with interstitial fibrosis.[10] The two parameters (i.e., Initial Deflection Width (IWD) and Two-Cycle Duration (2CD)) provide a means for classification of crackles as Coarse Crackles (CC) and Fine Crackles (FC).

bronchial obstruction.[10] Crackles have an explosive time domain pattern, with a rapid onset and short duration.[10] Their time domain structural characteristics, i.e., a sharp, sudden deflection usually followed by a wave, provide a means for their categorization between FC and CC,[12] as is shown in Figure III.3.2.

   For an extensive description and a variety of examples regarding the LS structure and characteristics, the reader should refer to Reference 10.

3.2.2   Heart Sounds

3.2.2.1   Definition and Historical Background

Heart sounds are defined as *the repetitive 'lub-dub' sounds of the beating of the heart*.[13] Heart auscultation followed similar pathways with lung auscultation, due to the topological coexistence of the heart with lungs. Hippocrates (460–377 B.C.) was familiarized with heart auscultation and he may have used HS for diagnostic purposes.[14] Nevertheless, it took almost 2000 years for reevaluation of HS by William Harvey (1578–1657), and 300 years more, with the contribution of Laënnec's stethoscope (1816), for Dr. Joseph Skoda (1805–1881) first to describe the cardiac sounds and murmurs, by pinpointing their locations and defining the clinical auscultatory signs that have allowed the noninvasive diagnosis of cardiac pathology via auscultation.[14]

3.2.2.2   Categorization and Main Characteristics

The HS are named according to their sequence of occurrence and are produced at specific points in the cardiac cycle.[14] In particular, the HS are categorized as follows:

1. *First Heart Sound*, or S_1, occurs at the beginning of ventricular systole when ventricular volume is maximal and is considered as normal heart sound.
2. *Second Heart Sound*, or S_2, occurs at the end of ventricular systole and is also considered as normal heart sound.
3. *Third Heart Sound*, or S_3 or 'S_3 gallop', occurs just after the S_2, as a result of decreased ventricular compliance or increased ventricular diastolic volume and is a sign of congestive heart failure.

4. *Fourth Heart Sound*, or S_4 or 'S_4 gallop', occurs just before the S_1, as a result of decreased ventricular compliance or increased volume of filling and is a sign of ventricular stress.
5. *Murmurs* are sustained noises that are audible during the time periods of systole, diastole, or both, associated with backward regurgitation, forward flow through narrowed or deformed valve, a high rate of blood flow through normal or abnormal valves, vibration of loose structures within the heart (chordae tendineae), and continuous flow through A-V shunts. They are further categorized between:

   a. *Systolic Murmurs* (SM), sustained noises that are audible between S_1 and S_2, categorized as *Early* SM, which begin with S_1 and peak in the first third of systole; *Mid* SM (or 'ejection' murmur), which begin shortly after S_1, peak in midsystole, and do not quite extend to S_2; *Late* SM, which begin in the later one half of systole, peak in the later third of systole, and extend to S_2; and *Pansystolic Murmurs*, which begin with S_1 and end with S_2, thus heard continuously throughout systole.
   b. *Diastolic Murmurs* (DM), sustained noises that are audible between S_2 and the next S_1, categorized as *Early* DM, which begin with S_2 and peak in the first third of diastole; *Mid* DM, which begin after S_2 and peak in midsystole; *Late* DM, which begin in the latter one half of diastole, peak in the later third of diastole, and extend to S_1; and *Pansystolic Murmurs*, which begin with S_2 and extend throughout the diastolic period.

The acoustic content of HS is concentrated in low frequencies.[15] In fact, over 95% of the acoustic energy of S_1 and 99% of the one of S_2 is concentrated under 75 Hz.[16] The average spectrum decays exponentially from its peak at approximately 7 Hz and often contains one or more shallow and wide peaks.[17] The S_2 contains much more energy in the lower frequencies than S_1 does. In addition, the frequency content of most cardiac murmurs is also in the low range.[16] The two most common HS, i.e., S_1 and S_2, have much longer duration and thus a lower pitch than crackles and both of them are short-lived compared to NLS and CAS.[13] This contributes to the facilitation of their separation from the LS. Nevertheless, HS from patients with irregular cardiac rhythms (transient character) and/or loud murmurs tend to overlap more the LS and thus are less separable.

3.2.3 Bowel Sounds

3.2.3.1 Definition and Historical Background

Bowel sounds are defined as *the sounds heard when contractions of the lower intestines propel contents forward*.[18] Knowledge of BS has advanced little since Cannon's pioneering work in 1902,[19] which used the sounds as a way of studying the mechanical activity of the gastrointestinal tract. Clinical tradition assesses the BS in a 'passive' way, i.e., by tracing not their presence but their absence, since the latter is an indicator of intestinal obstruction or ileus (paralysis of the bowel).[18] This lack of interest in 'active' abdominal auscultation is due in part to its lack of support in scientific fact and definitely not due to its lack of diagnostic information. Bowel sound patterns in normal people have not been clearly defined, as only a small number of them have been studied.[20-24] In addition, the trivial signal-processing methods that have been involved[25-27] have also been a problem. The vague notion about the usefulness of the BS in clinical practice that has been cultivated all these years, along with the lack of a worldwide-accepted BS categorization has resulted in a reduced interest in their processing and evaluation, even nowadays.

3.2.3.2 Categorization and Main Characteristics

There is no reference of what can be considered normal bowel sound activity, and no physiological understanding of the significance of different types of BS exists. Attempts to describe the acoustic impression of normal BS, such as 'rushes,' 'gurgles,' etc., fail due to their increased subjectivity. Sporadic works in the literature try to analyze the time and frequency domain characteristics of BS, defining as normal BS those with a frequency content in the range of 100–1000 Hz, with durations within a range of 5–200 msec, and with widely varying amplitudes.[28] An analysis of the 'staccato pop,' which is one of the most common BS characteristics of the colon, proves that it has a frequency content between 500 and 700 Hz and a duration of 5–20 msec.[28]

Recent work based on the differences seen between the sound-to-sound intervals of BS from different pathologies, i.e., irritable bowel syndrome, Crohn's disease, and controls, introduces a time domain-based tool for relating BS to the associated bowel pathology.[29] In a similar vein, the approaches of BS presented in this chapter contribute to the provision of an extended, accurate, and objective alternative to the current BS processing and evaluation status.

3.3 An Overview of the Examined Issues

To help the reader become acquainted with the goals of the LS, HS, and BS assessment and to add continuity across the subsections of this chapter, a brief overview of the examined problems and issues is presented here.

3.3.1 Noise Reduction-Detection

In order to objectively analyze the BioS, a preprocessing procedure regarding the elimination of any noise effect in the recorded signal has to be undertaken. From the wide field of 'any noise effect' the examined problem is narrowed down to the following specific cases:

1. Elimination of HS effect in LS recordings
2. Elimination of the background noise from BS and DAS recordings and detection of DAS

The above cases are met in the routine recordings of LS, HS, and BS, and advanced signal-processing techniques are needed in order to efficiently circumvent these noise effects. Specific approaches, based on HOS, WT, and FL, are presented in the succeeding sections.

3.3.2 Modeling

Modeling of the system that produces the BioS contributes to the understanding of their production mechanisms. The way the pathology affects these mechanisms is reflected in the modeling parameters adopted. In this chapter, modeling of LS and impulsive BS production systems with techniques based on HOS and LOS methodologies is presented.

3.3.3 Feature Extraction-Classification

The characteristics of BioS provide a variety of structural features that are directly connected to the underlined pathology. The correlation of these features with parameters derived from advanced processing of BioS provides a means for their efficient classification and, consequently, for the classification of the associated pathology. In this chapter, the assessment of CAS with HOS, which provides new diagnostic features of CAS based on their nonlinear characteristics, along with the processing of DAS and BS with HOC- and LOS-based methodologies, for classification between different pathologies, are thoroughly described.

3.4 Reduction of Heart Sounds Noise in Lung Sounds Recordings

The problem addressed in this section is the reduction of the HS interference in the LS recordings. Heart beating produces an intrusive quasi-periodic interference that masks the clinical auscultative interpretation of lung sounds. This is a serious noise effect which creates major difficulties in LS analysis, due to the fact that the intensity of HS is three- to tenfold that of LS over the anterior chest and one- to threefold that of LS from the posterior bases.[30] This high relative intensity of the HS can easily saturate the analog amplifiers and/or the analog-to-digital (A/D) converter, leading to a truncated signal and artifacts, especially in the case of children and infants, where the heart rate is high and its HS are loud.[31,32] As was described in Section 3.2, the main frequency components of the HS are in the range of 20–150 Hz,[33,34] and overlap with the low-frequency components of breath sound spectrum in the range of 20–1600 Hz.[11] Two basic approaches for heart noise reduction were followed in the literature:

1. Highpass linear filtering, with a cutoff frequency varying from 50–150 Hz
2. Adaptive filtering

The first approach[11] effectively reduces the HS noise, but at the same time it degrades the respectively overlapped frequency region of breath sounds. The second approach, proposed by Iyer et al.,[33] uses the theory of adaptive filtering,[35] with a reference signal highly correlated to the noise component of the input signal, derived from a modified ECG signal. A modification to this method, proposed by Kompis and Russi,[36] combines the advantages of adaptive filtering with the convenience of using only a single microphone input, but has a moderate heart noise reduction of 24–49%.

Two signal processing methods for *Adaptive Noise Cancellation* (ANC) of HS interference in LS recordings, based on HOS and WT, respectively, are presented in the following subsections.

3.4.1 ANC-FOS: A Fourth-Order Statistics-Based ANC Technique

The proposed ANC is based on HOS, and more specifically on *Fourth-Order Statistics* (FOS). For the reader not familiarized with the definitions and the advantageous properties of HOS, a short outline is given below.

3.4.1.1 Brief Mathematical Background

Given a set of n real random variables $\{x_1, x_2, \ldots, x_n\}$, their *joint cumulants* of order $r = k_1 + k_2 + \cdots + k_n$ are defined as the coefficients in the Taylor expansion of the second characteristic function $\widetilde{\Psi}(\omega_1, \omega_2, \ldots, \omega_n) = \ln[E\{\exp(j(\omega_1 x_1 + \omega_2 x_2 + \cdots + \omega_n x_n))\}]$, about zero,[37-39] i.e.:

$$Cum[x_1^{k_1}, x_2^{k_{21}}, \ldots, x_n^{k_n}] = (-j)^r \frac{\partial^r \widetilde{\Psi}(\omega_1, \omega_2, \ldots, \omega_n)}{\partial \omega_1^{k_1} \partial \omega_2^{k_2} \ldots \partial \omega_n^{k_n}}\bigg|_{\omega_1 = \omega_2 = \cdots = \omega_n = 0} \tag{III.3.1}$$

If $\{X(k)\}, k = 0, \pm 1, \pm 2, \ldots$ is a real stationary random process and its moments up to order n exist, then its nth-order *cumulants*, i.e., its nth-order statistics, are given by the following equation:

$$c_n^x(\tau_1, \ldots, \tau_{n-1}) \equiv Cum[X(k), X(k + \tau_1), \ldots, X(k + \tau_{n-1})]. \tag{III.3.2}$$

In the case where $\{X(k)\}$ is Gaussian, its third- and fourth-order cumulants, $c_3^x(\tau_1, \tau_2)$, $c_4^x(\tau_1, \tau_2, \tau_3)$, are identically zero. For a stationary process $\{X_i(k)\}, i = 1, 2, \ldots, n$, the nth-order *cross-cumulant* sequence is defined as:

$$c_n^{x_1 \cdots x_n}(\tau_1, \ldots, \tau_{n-1}) \equiv Cum[X_1(k), X_2(k + \tau_1), \ldots, X_n(k + \tau_{n-1})]. \tag{III.3.3}$$

The relationship between fourth-order moments and cumulants of zero-mean signals[40] is given by:

$$Cum[x_1 x_2 x_3 x_4] = E\{x_1 x_2 x_3 x_4\} - E\{x_1 x_2\} E\{x_3 x_4\}$$
$$- E\{x_1 x_3\} E\{x_2 x_4\} - E\{x_1 x_4\} E\{x_2 x_3\}. \tag{III.3.4}$$

The *Second-Order Statistics* (SOS) of a signal are sufficient for the complete statistical description of a Gaussian process with a known mean. Unfortunately, phase relationships among the various frequencies of the signal are suppressed in the autocorrelation domain. Thus, in practical applications (e.g., detection, parameter estimation, and signal reconstruction problems) where extraction of signal's phase information, as well as information due to deviations from Gaussianity, reveal the real nature of the signal, SOS are not considered as a sufficient analysis tool. Techniques based on higher-order statistics preserve the phase relationships among the harmonic components of the signal, suppress Gaussian noise (or non-Gaussian noise having a symmetric probability density function of unknown spectral characteristics), and detect and characterize nonlinearities in time series.[40] For an analytical description of the HOS characteristics and properties, the reader should refer to Reference 40 and references therein.

3.4.1.2 Description of the ANC-FOS Algorithm Structure

The ANC-FOS algorithm is an adaptive de-noising tool that employs a reference signal $z(k)$ highly correlated with the HS noise. The ANC-FOS algorithm analyzes the incoming LS signal $x(k)$ to generate the reference signal $z(k)$. In order to generate the reference signal, the real location of the heart sound in the incoming LS signal must be detected. Therefore, the ANC-FOS scheme initially applies on $x(k)$ a peak detection algorithm, namely *LOcalized REference Estimation* (LOREE) algorithm,[41] which searches for the true locations of HS noise, based on amplitude, distance, and noise-reduction percentage criteria. Its output, $z(k)$, is a localized signal, with precise tracking of the first (S_1) and second (S_2) heartbeats, highly correlated with heart noise and with no extra recording requirement as Iyer's method does.[33]

The incoming signal $x(k)$ is the sum of LS $s(k)$, the interference HS $i(k)$, and the sensor noise $n_x(k)$, i.e:

$$x(k) = s(k) + i(k) + n_x(k). \tag{III.3.5}$$

It is assumed that $s(k)$ is a zero-mean, uncorrelated to the interference and the reference signal, while the noise $n_x(k)$ is a zero-mean, white or colored Gaussian random process independent of $s(k)$, $i(k)$, and $z(k)$. Furthermore, the reference signal $z(k)$ is a zero-mean, stationary non-Gaussian random process. It is also assumed that the interference and reference signals are related through the following linear-time-invariant transformation:[41]

$$i(k) = \sum_j g(j)z(k - j). \tag{III.3.6}$$

If $y(k)$ denotes the adaptive-filter output, i.e:

$$y(k) = \sum_{j=0}^{N-1} h(j)z(k - j), \tag{III.3.7}$$

where N denotes the number of taps and $\{h(n), n = 0, 1, \ldots, N - 1\}$ are the adaptive-filter coefficients, then the ANC system output, $\{e(k)\}$ is:

$$e(k) = x(k) - y(k) = s(k) + n_e(k), \tag{III.3.8}$$

where $\{n_e(k)\}$ is the ANC-system output noise, given by:

$$n_e(k) = n_x(k) + i(k) - \sum_{j=0}^{N-1} h(j)z(k - j). \tag{III.3.9}$$

As can be noticed from Equation III.3.9, the ANC-system output contains uncorrelated noise. This was the motivation for Kompis and Russi[36] to modify their least-mean-squares-based adaptation algorithm by introducing a low-pass filter (cutoff frequency of 250 Hz) in the error-signal path. In order to circumvent this problem more efficiently, the FOS are employed as follows.

Noticing that $s(k)$ and $z(k)$ are uncorrelated zero-mean processes and $s(k)$, $i(k)$, and $z(k)$ are independent of uncorrelated noises, the fourth-order cross-cumulants between $[x(k), z(k)]$ and $[y(k), z(k)]$ are given by:

$$C_{xzzz}(\tau_1, \tau_2, \tau_3) = C_{izzz}(\tau_1, \tau_2, \tau_3) = \sum_j g(j)C_{zzzz}(j + \tau_1, j + \tau_2, j + \tau_3), \qquad \text{(III.3.10)}$$

$$C_{yzzz}(\tau_1, \tau_2, \tau_3) = \sum_{j=0}^{N-1} h(j)C_{zzzz}(j + \tau_1, j + \tau_2, j + \tau_3). \qquad \text{(III.3.11)}$$

The objective is to minimize a cost function of the system-output error, with respect to the filter coefficients $h(n)$, so that $e(k) \approx s(k)$ and $n_e(k)$ is zero. In this case, Equation III.3.8 could be rewritten as:

$$x(k) - y(k) = s(k)\big|_{n_e(k)=0}. \qquad \text{(III.3.12)}$$

Since $s(k)$ and $z(k)$ are uncorrelated $C_{szzz}(\tau_1, \tau_2, \tau_3) = 0$. Thus, by taking the fourth-order cross-cumulants of both sides of Equation III.3.12, we have:

$$C_{xzzz}(\tau_1, \tau_2, \tau_3) - C_{yzzz}(\tau_1, \tau_2, \tau_3) = 0\big|_{n_e(k)=0}. \qquad \text{(III.3.13)}$$

From Equation III.3.13 it is observed that the error in the output signal is minimized if the following cost function is minimized:

$$\xi_3 = \sum_{\tau_1=0}^{M-1} \sum_{m,n=0}^{\tau_1} \left[C_{xzzz}(\tau_1, \tau_1 - m, \tau_1 - n) - \sum_{j=0}^{N-1} h(j)C_{zzzz}(j + \tau_1, j + \tau_1 - m, j + \tau_1 - n) \right]^2,$$
$$\text{(III.3.14)}$$

where M denotes the number of overdetermined equations, $m = \tau_1 - \tau_2$, $n = \tau_1 - \tau_3$, $0 \le \tau_1, \tau_2, \tau_3 \le M - 1$, $\tau_1 \ge \tau_2$, and $\tau_1 \ge \tau_3$. Equation III.3.14 could be rewritten in a matrix form as follows:

$$\xi_3 = \left(\mathbf{C}_{xzzz} - \mathbf{C}_{zzzz}\mathbf{H}_f\right)^T \left(\mathbf{C}_{xzzz} - \mathbf{C}_{zzzz}\mathbf{H}_f\right), \qquad \text{(III.3.15)}$$

where $\mathbf{C}_{xzzz}$ is an $(M(M+1)(2M+1)/6) \times 1$ column vector, $\mathbf{C}_{zzzz}$ is an $(M(M+1)(2M+1)/6) \times N$ matrix, and $\mathbf{H}_f$ denotes an $N \times 1$ ANC-filter coefficient vector, i.e.:

$$\mathbf{H}_f = [h(0), h(1), \ldots, h(N-1)]^T. \qquad \text{(III.3.16)}$$

The gradient of ξ_3 could be calculated as:

$$\nabla \equiv \frac{\partial \xi_3}{\partial \mathbf{H}_f} = 2 \cdot (\mathbf{C}_{zzzz}^T \mathbf{C}_{zzzz} \mathbf{H}_f - \mathbf{C}_{zzzz}^T \mathbf{C}_{xzzz}), \qquad \text{(III.3.17)}$$

so the filter-update equation becomes:

$$\mathbf{H}_f(k + 1) = \mathbf{H}_f(k) - \mu(k)\nabla(k), \qquad \text{(III.3.18)}$$

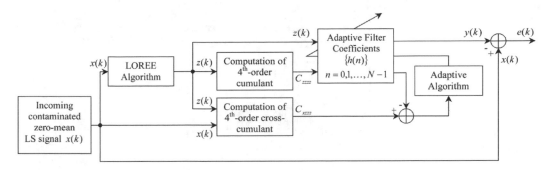

FIGURE III.3.3 Block diagram of the ANC-FOS scheme. An analytical description of the LOREE algorithm can be found in Reference 41.

where $\mu(k)$ is the step size, given by:

$$\mu(k) = \frac{\mu_f}{(a + tr\{\mathbf{C}_{zzzz}^T(k)\mathbf{C}_{zzzz}(k)\})}, \qquad \text{(III.3.19)}$$

where μ_f and a are constants[41] and tr denotes the matrix trace.

From Equation III.3.19 it can be noticed that the update equation consists only of the fourth-order cumulants of the incoming and reference signals and it is not affected by Gaussian uncorrelated noise. This observation reveals the enhancement achieved in the robustness of an ANC scheme when FOS are employed in its structure. A schematic representation of the ANC-FOS structure is depicted in Figure III.3.3.

3.4.1.3 Results from the Application of the ANC-FOS Scheme on Real Data

Figure III.3.4 illustrates an example of the results obtained when applying the ANC-FOS algorithm on contaminated LS, recorded from six healthy subjects. Specific implementation and experimental set-up issues are described in detail in Reference 41. In particular, Figure III.3.4(a) shows the input signal $x(k)$, i.e., contaminated LS recorded from a subject [HS (S_1 and S_2) are marked with small, black bars], Figure III.3.4(b) depicts the LOREE output $z(k)$, Figure III.3.4(c) illustrates the ANC-FOS output, and Figure III.3.4(d) presents the output from a highpass filter with a cutoff frequency of 75 Hz (HPF-75Hz) for the same input.

From the comparison of Figures III.3.4(a) and (b), it can be seen that the heart peak locations in input and reference signals are synchronized. Furthermore, there are many structural similarities between the HS signal in $x(k)$ and the estimated reference signal $z(k)$, indicating an efficient performance of the LOREE algorithm. From Figures III.3.4(a), (c), and (d), it can be noticed that the signal included between the HS remains unchanged in Figure III.3.4(c), but not in Figure III.3.4(d). Moreover, from Figures III.3.4(a) and (d), the joint loss of low frequencies of lung sound and heart sound reduction is obvious. These justify the adaptive performance of the ANC-FOS algorithm, since it has a localized effect on input signal at locations pointed out from the reference signal, whereas linear filtering deteriorates the whole signal due to the elimination of low frequencies.

The evaluation of the performance of the ANC-FOS scheme, besides the inspection and listening to the processed and unprocessed signals, could be achieved through a mean energy-based reduction criterion, namely *Local-HS Noise Reduction Percentage* (L-HSNRP), defined as follows:[41]

$$L\text{-}HSNRP(\%) = \left\{ \frac{E\{\mathbf{I}^2\} - E\{\mathbf{O}^2\}}{E\{\mathbf{I}^2\}} \right\}, \qquad \text{(III.3.20)}$$

where $\mathbf{I}$ is a windowed input signal, consisting of sections of the analyzed epoch, and $\mathbf{O}$ is the corresponding ANC-FOS algorithm output. The locations and the time duration of the windows

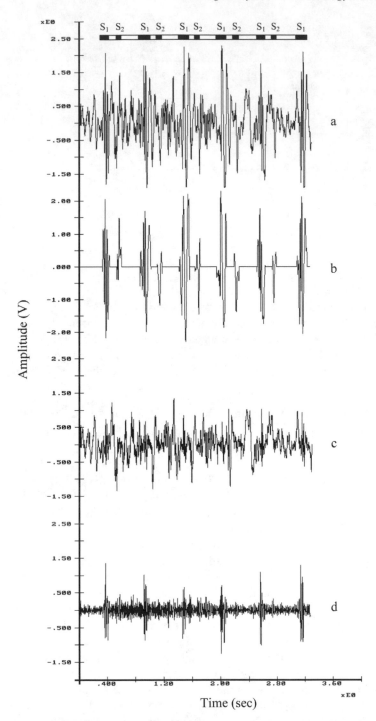

FIGURE III.3.4   (a) Time waveform of 3.27 sec raw contaminated LS recorded from a subject during normal breathing [input $x(k)$]. (b) Time waveform of the corresponding reference signal [LOREE output, $z(k)$]. (c) Time waveform of adaptively filtered LS (ANC-FOS output). (d) Time waveform after linearly filtering the input $x(k)$ (HPF-75 Hz output). The small, black lines in the input tracing indicate the presence of the HS noise.

applied on the input signal are pointed out by the starting points and the time duration of each corresponding nonzero section of the reference signal $z(k)$. In this way, the performance of the ANC-FOS algorithm is evaluated in the true locations of the heart-sound presence. For the input signal in Figure III.3.4(a), the corresponding L-HSNRP is 94.5%, indicating a significant reduction of the HS noise effect in the LS recordings. This is true for all subjects, since the L-HSNRP takes values over 90% for all subjects.[41]

3.4.1.4 Comparison with Other Works

Comparing the ANC-FOS algorithm with the adaptive schemes proposed by Iyer et al.[33] and Kompis and Russi,[36] it can be noticed that, unlike the latter, the ANC-FOS algorithm finds the true locations of S_1 and S_2, without assuming any similarity in the two heart sounds. In addition, neither Iyer's nor Kompis' methods take into account any nonlinear transformation that might be necessary to correctly account for nonlinearities in the system, as the ANC-FOS algorithm does due to the employment of FOS. Furthermore, although Kompis' method, unlike Iyer's, avoids the need of extra recording, it results in moderate noise reduction percentages compared to that of the ANC-FOS scheme. Finally, both Iyer's and Kompis' methods are sensitive to additive Gaussian noise, whereas the ANC-FOS algorithm is insensitive to Gaussian uncorrelated noise and independent of it, since it is based on FOS, which are zero for Gaussian processes.

3.4.2 ANC-WT: A Wavelet Transform-Based ANC Technique

Another approach of ANC is based on the WT, combining the multiresolution analysis with hard thresholding. For the reader not familiar with the definitions and the advantageous properties of WT, a short outline is given below.

3.4.2.1 Brief Mathematical Background

Wavelets are families of functions $\psi_{a,b}(t)$ generated from a single basic wavelet, namely the 'mother wavelet,' $\psi(t)$ by dilations and translations,[42] i.e.:

$$\psi_{a,b}(t) = \frac{1}{\sqrt{a}} \psi\left(\frac{t-b}{a}\right), a, b \in \Re, a > 0, \tag{III.3.21}$$

where a and b are the dilation (scale) and translation parameters, respectively. The continuous WT of a $1-D$ function $f(t) \in L^2(\Re)$, where $L^2(\Re)$ denotes the vector space of measurable, square-integrable one-dimensional functions $f(t)$, is defined in a Hilbert space, as the projection of the $f(t)$ onto the wavelet set $\psi_{a,b}(t)$ as follows:

$$CWTf(a, b) = \langle f, \psi_{a,b} \rangle = |a|^{-0.5} \int_{-\infty}^{\infty} f(t)\psi^*\left(\frac{t-b}{a}\right) dt, \tag{III.3.22}$$

where $\langle \cdot \rangle$ and $*$ denote inner product and complex conjugate, respectively.

The procedure of ANC employs the WT through the realization of the multiresolution analysis.[43] The latter consists of a sequence of embedded closed subspaces ... $V_2 \subset V_1 \subset V_0 \subset V_{-1} \subset V_{-2} \ldots$, which satisfies certain properties.[43,44] The family:

$$\phi_{j,k}(t) \equiv 2^{j/2}\phi(2^j t - k), j, k \in Z \tag{III.3.23}$$

forms an orthonormal basis for V_{-j}, while the family:

$$\psi_{j,k}(t) \equiv 2^{-j/2}\psi(2^{-j} t - k), j, k \in Z \tag{III.3.24}$$

forms an orthonormal basis for O_j, where O_j is the orthogonal complement of V_j in V_{j-1} defined as:

$$V_{j-1} = V_j \oplus O_j, j \in Z, \tag{III.3.25}$$

with $V_j \perp O_j$, where $\oplus$ denotes the direct sum. With the use of Equation III.3.25, the V_0 space can be decomposed in the following way:

$$V_0 = O_1 \oplus \cdots \oplus O_{J-1} \oplus O_J \oplus V_J, \tag{III.3.26}$$

by simply iterating the decomposition J times. The subspace sequences $\{V_j\}_{\{j \in Z\}}$ and $\{O_j\}_{\{j \in Z\}}$ of $L^2(\mathfrak{R})$ are called approximation and detail spaces, respectively.[44] The orthogonal projection of the input signal onto $O_1, \ldots O_{J-1}$, O_J and V_J, using the orthonormal basis of Equations III.3.23 and III.3.24, decomposes the input signal into a very coarse resolution that exists in V_j and added details that exist in the $O_i, i = 1, \ldots, J$ spaces. By Equation III.3.26, the sum of the coarse version and all the added details yields back the original signal.[44] This decomposition of the signal into approximation and detail space is called *multiresolution approximation*,[44] and it can be realized through a pair of FIR filters H, G (and their *adjoints* H^*, G^*), which are lowpass and highpass, respectively, defining a *MultiResolution Decomposition-MultiResolution Reconstruction* (MRD-MRR) scheme, where $f \in V_j$ is equivalent to $f(2 \cdot) \in V_{j+1}$.

For an analytical description and details regarding the applications of the wavelet transform and the multiresolution analysis, the reader should refer to References 42–44.

3.4.2.2 Description of the ANC-WT Algorithm Structure

The ANC-WT algorithm is an adaptation of the Wavelet Transform-based STationary-NonSTationary (WTST-NST) filter, proposed by Hadjileontiadis and Panas for crackle detection,[45] in the case of HS noise reduction.[46] The application of WTST-NST filter in the DAS analysis will be described in detail in Section 3.6.1.

The main concept adopted by the ANC-WT algorithm is the fact that explosive peaks in the time domain (heart sound peaks) have large signal over many wavelet transform scales. On the contrary, 'noisy' background (LS) dies out swiftly with increasing scale. The definition of 'noise' is not always clear. For example, in this case, the HS peaks are temporary, considered as the 'desired' signal in order to be isolated from the LS (background noise); after their isolation, the residual, i.e., the de-noised LS, is now the desired signal and the isolated HS peaks are the noise. Hence, it is better to view an N-sample signal as being noisy or incoherent relative to a basis of waveforms if it does not correlate well with the waveforms of the basis.[47] From this notion, the separation of HS from LS becomes a matter of breath sounds coherent structure extraction.

The ANC-WT scheme employs an iterative MRD-MRR procedure to form different levels of noise separation. Specifically, at k iteration, the MRD is applied on $f(\lambda)$ (note that for $k = 1$, $f(\lambda) = X(\lambda), \lambda = 1, \ldots, N$, where $X(\lambda)$ is the normalized input signal) at m adjacent resolution scales ($m = 1, \ldots, M$, where $M = 2 \log_2 N$), using previously defined libraries of orthonormal bases.[42] The resulted WT coefficients at j scale are compared with a hard threshold, defined as:

$$T_j^k = \sigma_j^k \cdot F_{adj}, \tag{III.3.27}$$

where σ_j^k is the standard deviation of WT at k iteration and j scale, and F_{adj} is an adjusting multiplicative factor, used to sustain the threshold at high value, at different scales. After the thresholding, the WT coefficients are divided into large ($> T_j^k$) and small ($\leq T_j^k$) ones, $WT^{kC}(\lambda)$ and $WT^{KR}(\lambda)$, respectively. By applying MRR(m scales) to $WT^{kC}(\lambda)$ and $WT^{kR}(\lambda)$, coefficients, the $f(\lambda)$ is decomposed into $C_k(\lambda)$ (coherent part of the signal) and $R_k(\lambda)$ (less coherent part of the signal), respectively. The iterative procedure continues by setting $f(\lambda) = R_k(\lambda)$, and stops after the following STopping Criterion (STC) is satisfied:[46]

$$STC = |E\{R_{k-1}^2(\lambda)\} - E\{R_k^2(\lambda)\}| < \varepsilon, 1 \gg \varepsilon > 0. \tag{III.3.28}$$

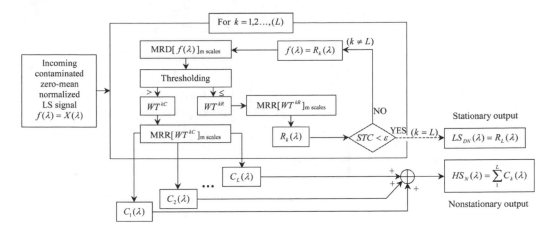

FIGURE III.3.5 Block diagram of the ANC-WT scheme.

After the last iteration (L), the coherent part of the signal that corresponds to the HS noise is composed superimposing the coherent parts derived at each iteration k, i.e.:

$$HS_N(\lambda) = \sum_{k=1}^{L} C_k(\lambda), \qquad (\text{III.3.29})$$

while the remains (de-noised LS):

$$LS_{DN}(\lambda) = R_L(\lambda) \qquad (\text{III.3.30})$$

qualify as 'noise,' since it cannot be well-represented by any sequence of waveforms of the basis. In fact, the $LS_{DN}(\lambda)$ is the desired signal and not the noise, since the ANC-WT focuses on separating the HS bursts from the LS only at locations of their presence, keeping unchanged the rest of the input signal. In this way, the ANC-WT scheme acts as an adaptive de-noising tool that peels the recorded contaminated LS in layers, reveals the noise coherent structure (HS), and efficiently separates it from the desired signal (LS), resulting in an adaptively de-noised LS signal. A schematic representation of the ANC-WT algorithm is shown in Figure III.3.5.

3.4.2.3 Results from the Application of the ANC-WT Scheme on Real Data

Figure III.3.6 illustrates an example of the results obtained when applying the ANC-WT algorithm on the same contaminated LS used for the evaluation of ANC-FOS algorithm. Specific implementation and experimental set-up issues are described in detail in References 41 and 46. In particular, Figure III.3.6(a) shows the input signal $X(k)$ i.e., contaminated LS recorded from a subject [HS (S_1 and S_2) are marked with small, black bars]; Figure III.3.6(b) depicts the corresponding de-noised LS, $LS_{DN}(k)$; Figure III.3.6(c) illustrates the corresponding estimated HS, $HS_N(k)$; and Figure III.3.6(d) presents the corresponding output from a highpass filter with a cutoff frequency of 75 Hz (HPF-75Hz).

From the comparison of Figures III.3.6(a) and (c), it can be seen that the heart peak locations in input $X(k)$ and in ANC-WT output $HS_N(k)$ are synchronized, *without employing any reference signal*. Furthermore, there are many structural similarities between the HS signal in $X(k)$ and the ANC-WT output $HS_N(k)$, indicating an efficient estimation of the true location and morphology of the HS noise in the contaminated LS by the ANC-WT algorithm. In addition, from Figures III.3.6(a), (b), and (d), it can be noticed that the signal included between the HS remains unchanged in Figure III.3.6(b), but not in Figure III.3.6(d). Moreover, from Figures III.3.6(a) and (d), the joint loss of low frequencies of lung sound and heart sound reduction is obvious. These justify the adaptive performance of the

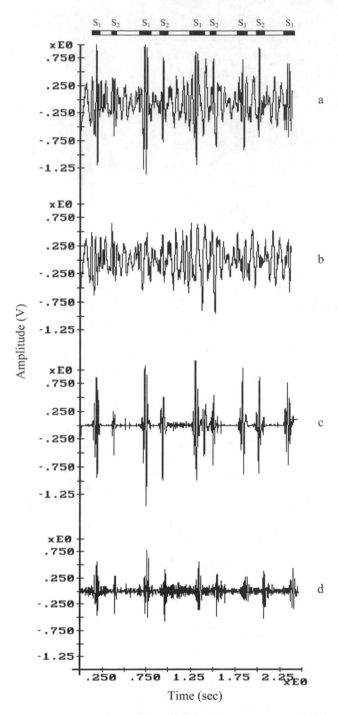

FIGURE III.3.6   (a) Time waveform of 2.457-sec raw contaminated LS recorded from a subject during normal breathing [input $X(k)$]. (b) Time waveform of the corresponding stationary output of the ANC-WT algorithm [$LS_{DN}(k)$]. (c) Time waveform of the corresponding nonstationary output of the ANC-WT algorithm [$HS_N(k)$]. (d) Time waveform after linearly filtering the input $X(k)$ (HPF-75 Hz output). The small, black lines in the input tracing indicate the presence of the HS noise.

ANC-WT algorithm, since it has a localized effect on input signal without requiring any reference signal, whereas linear filtering deteriorates the whole signal due to the elimination of low frequencies.

The evaluation of the performance of the ANC-WT scheme, besides the inspection and listening to the processed and unprocessed signals, could be achieved through the L-HSNRP measure, defined in Equation III.3.20,[41] where **I** is a windowed input signal, consisting of sections of the analyzed epoch at the true locations of heart sound presence, and **O** is the corresponding ANC-WT algorithm output. For the input signal in Figure III.3.6(a), the corresponding L-HSNRP is 84.1%, indicating a significant reduction of the HS noise effect in the LS recordings. This is true for all subjects, since the L-HSNRP takes values over 80% for all subjects.[46]

3.4.2.4 Comparison with Other Works

Comparing the ANC-WT algorithm[46] with the adaptive schemes proposed by Iyer et al.[33] and Kompis and Russi,[36] and the ANC-FOS[41] described in Section 3.4.1, it can be noticed that the ANC-WT scheme possesses all the advantages of the ANC-FOS algorithm over Iyer's and Kompis' methods, such as finding the true locations of S_1 and S_2 without assuming any similarity in the two heart sounds, avoiding the need of extra recording, and resulting in high-noise reduction. In addition, the ANC-WT scheme performs better than the ANC-FOS one when the additive noise has an impulsive character (e.g., friction sound due to movement, impulsive ambient noise, etc). On the other hand, the ANC-FOS scheme performs better than the ANC-WT one when the input LS signal is contaminated by additive Gaussian uncorrelated noise.[41,46] Although the ANC-WT algorithm results in slightly smaller L-HSNRP values than the ANC-FOS algorithm, its 'reference-free' structure equalizes this loss, characterizing both of them as attractive noise reduction schemes.

3.5 Elimination of the Background Noise in Bowel Sounds Recordings

The problem examined in this section is the elimination of background noise in the BS recordings. This is a major issue to be considered prior to the diagnostic analysis of BS, since the presence of noise severely influences the clinical auscultative interpretation of BS. The elimination of noise results in de-noised BS, which provide a more reliable and accurate characterization of the underlying gastrointestinal malfunction. The noise sources in the case of the BS recordings include instrumentation noise introduced during the recording process, sounds from the stomach, as well as cardiac and respiratory sounds, which occur mostly in the case of infants. In order to yield a successful BS classification, an effective reduction of noise from the contaminated BS signal is required, since the presence of the noise introduces pseudoperiodicity, masks the relevant signal, and modifies the energy distribution in the spectrum of BS. This is clear from Figure III.3.7, where the spectrum of de-noised BS along with the one of the corresponding background noise is plotted.

From this figure it is evident that there is a strong overlap in the frequency content of the de-noised BS and the background noise, indicating the necessity for an intelligent manipulation of the recorded BS signal in order to successfully eliminate the noise interference.

From the available studies related to BS analysis, only two are focused on the BS de-noising process in order to extract their original structure before any further diagnostic evaluation.[26,48] Unfortunately, the method used for noise reduction in Reference 26 was based only on assumptions and general descriptions of the noise characteristics, forming a static, rather than a dynamic, noise reduction scheme. In addition, the noise is manually extracted, after subjective characterization and localization of the noise presence in the histogram of the recorded BS. The second de-noising method of BS, used by Mansy and Sandler,[48] is based on adaptive filtering[35] and eliminates the HS noise from BS recorded from rats. Although it results in satisfactory enhancement of BS, it still requires careful construction of a noise reference signal and needs an empirical set-up of its adaptation parameters.

The use of the ANC-WT scheme as an adaptive noise canceler in BS recordings is presented in the following subsection.

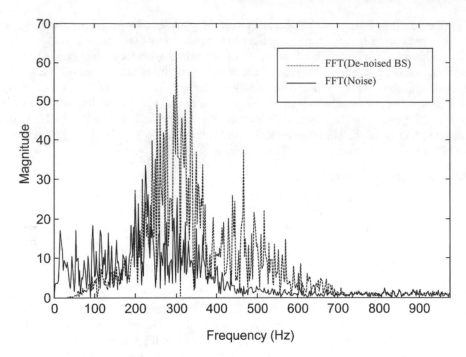

FIGURE III.3.7   The fast Fourier transform (FFT) of one case of de-noised BS (dashed line) and of the corresponding interference (solid line), as recorded from a patient.

3.5.1   The ANC-WT Scheme as a Noise Eliminator in BS Recordings

The methodology described in Section 3.4.2 offers an efficient tool for the elimination of the background noise in BS recordings. The only changes needed to the ANC-WT scheme of Figure III.3.5 refer to the input signal, which is now the contaminated BS instead of the LS recordings; thus $X(\lambda)$ in Figure III.3.5 represents the BS, and to the two outputs (nonstationary and stationary ones) in Figure III.3.5, which are now $BS_{DN}(\lambda)$ and $BG_N(\lambda)$, corresponding to the de-noised BS and the background noise, respectively.

3.5.1.1   Results from the Application of the ANC-WT Scheme on Real Data

Figure III.3.8 illustrates an example of the results obtained when applying the ANC-WT algorithm on moderately and severely contaminated BS, recorded from patients with gastrointestinal pathology.[49] Specific implementation and experimental set-up issues have been described in detail.[49] In particular, Figure III.3.8(a) shows the input signal $X_1(k)$, i.e., moderate contaminated BS recorded from a patient with irritable bowel syndrome (the desired BS are marked with small, black arrows); Figure III.3.8(b) depicts the corresponding de-noised BS, $BS_{DN}^1(k)$ (nonstationary ANC-WT output); Figure III.3.8(c) illustrates the corresponding background noise, $BG_N^1(k)$ (stationary ANC-WT output); Figure III.3.8(d) presents the input signal $X_2(k)$, i.e., severe contaminated BS recorded from the same patient (the desired BS are marked again with small, black arrows); Figure III.3.8(e) depicts the corresponding de-noised BS, $BS_{DN}^2(k)$ (nonstationary ANC-WT output); and Figure III.3.8(f) illustrates the corresponding background noise, $BG_N^2(k)$ (stationary ANC-WT output).

Comparing Figures III.3.8(a) and (b), it can be seen that the locations of the BS peaks in the input and in the de-noised output of the ANC-WT scheme are synchronized, without employing any reference signal. The same holds for Figures III.3.8(d) and (e), where, in Figure III.3.8(d), unlike in Figure III.3.8(a), the background noise and the two BS included in that section, apart from similar

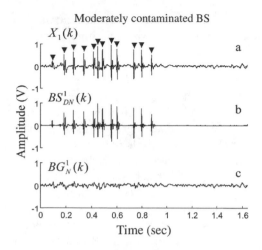
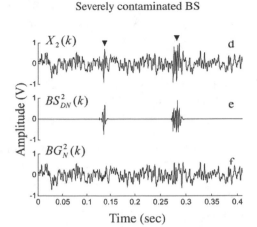

FIGURE III.3.8 Experimental results from the analysis of moderately $X_1(k)$, and severely $X_2(k)$ contaminated BS, recorded from a patient with irritable bowel syndrome. In each case, (a, d) correspond to the input signal of the ANC-WT scheme $X_1(k)$ and $X_2(k)$, respectively; (b, e) correspond to the de-noised BS $BS_{DN}^1(k)$ and $BS_{DN}^2(k)$, respectively; and (c, f) correspond to the background noise $BG_N^1(k)$ and $BG_N^2(k)$, respectively. The arrowheads in the input tracing indicate the events of interest that correspond to BS.

dynamic range, also have similar frequency content, as is shown in Figure III.3.7. Furthermore, comparing Figures III.3.8(a) and (c), and Figures III.3.8(d) and (f), it can be noticed that the signal between the BS remains unchanged, indicating the adaptive performance of the ANC-WT scheme, despite the level of contamination of the BS.

The efficiency of the ANC-WT scheme, besides the inspection and listening to the processed and unprocessed signals, can be measured by calculating the following quantitative evaluators,[45,49] which express the ability of the algorithm to detect the desired signal, i.e., BS:

$$\textbf{Rate of Detectability (\%): } D_R = \left(1 - \frac{N_R - N_E}{N_R}\right) \cdot 100, \tag{III.3.31}$$

$$\textbf{Total Performance (\%): } TD_R = mean(D_R), \tag{III.3.32}$$

$$\textbf{Quality Factor: } 0 \le S_q = \left(\frac{area|X(k)| - area|BS_{DN}(k)|}{area|X(k)|}\right) \le 1, k = 1, \ldots, N, \tag{III.3.33}$$

where $|\cdot|$ denotes absolute value, N_R the number of visually recognized BS by a physician, N_E the number of estimated BS using the ANC-WT scheme, $X(k)$ the normalized contaminated BS, and $BS_{DN}(k)$ the de-noised BS. The first two quantitative evaluators (D_R and TD_R) describe the ability of the ANC-WT scheme to find the correct number of BS, at the correct position in the raw data, and to separate them from the noise. The third quantitative evaluator (S_q) gives a measure of the quality of the separation of the BS from the noise, regarding structural extraction and the preservation of the morphological characteristics of the original BS. According to Hadjileontiadis and Panas,[45] the optimum range of the S_q values that avoids over/underestimation of the de-noised BS is 0.4–0.75. Values of S_q very close to zero correspond to overestimation of BS, while values of S_q very close to unity correspond to underestimation of BS.

For the analyzed BS shown in Figure III.3.8, D_R was equal to 100% for both $X_1(k)$ and $X_2(k)$. This was also true for all analyzed BS,[49] resulting in $TD_R = 100\%$ and mean $S_q = 0.6 \pm 0.1$, which indicate that the ANC-WT scheme has a localized effect on the input signal, only at true locations of the BS presence, without requiring any reference signal, and it deteriorates neither the structure nor the morphology of the BS as a result of the noise reduction procedure, even under extreme noise contamination, as is proved from the noise robustness test of the ANC-WT algorithm described in detail in Reference 49.

3.5.1.2 Comparison with Other Works

Comparing the ANC-WT algorithm[49] with the adaptive scheme proposed by Mansy and Sandler,[48] it can be noticed that, unlike the ANC-WT scheme, Mansy's and Sandler's method is a 'nonreference-free' scheme, since, apart from the BS recordings, it requires a reference signal for its performance. In addition, their scheme constructs a heart sound template since it deals only with heart sound interference, with averaged heart sounds from previous successive heartbeats, to estimate the characteristics of the heart noise. This is not required by the ANC-WT scheme in order to describe the characteristics of the aggregated interference. Moreover, Mansy's and Sandler's scheme is only tested on rats for the construction of the heart noise template, using their structural characteristics. On the contrary, the ANC-WT scheme is implemented in BS analysis derived from humans, with or without any gastrointestinal pathology. Finally, its implementation, unlike Mansy's and Sandler's scheme, does not require empirical definition of a set of parameters which is always prone to subjective human judgment.

3.6 Intelligent Separation of Discontinuous Adventitious Lung Sounds from Background Noise

As was mentioned in the description of the LS characteristics in Section 3.2.1, the discontinuous adventitious lung sounds, or DAS, behave as a nonstationary explosive noise superimposed on breath sounds. The DAS are only heard in pathological cases, indicating an underlying pathological malfunction. This means that their separation from the background sound, i.e., the vesicular lung sounds or VLS, could reveal significant information since the structure of the DAS isolates their diagnostic character.

In order to achieve automated separation, the nonstationarity of DAS must be taken into account. Thus, neither highpass filtering, which destroys the waveforms, nor level slicing, which cannot overcome the small amplitude of FC, is adequate for this task. Application of time-expanded waveform analysis in crackle time domain analysis[12,50] results in separation; it is, however, time consuming, with large inter-observer variability. Nonlinear processing, proposed by Ono et al.[51] and modified by Hadjileontiadis and Panas,[52] obtains more accurate results, but requires empirical definition of the set of parameters of its *STationary-NonSTationary* (ST-NST) filter.

In the next three subsections, three methods for intelligent separation of three types of DAS, namely fine crackles (or FC), coarse crackles (or CC), and squawks (or SQ), from VLS are presented, based on WT and fuzzy logic (or FL).

3.6.1 The ANC-WT (or WTST-NST) Scheme as an Efficient DAS Detector

The first method for automated separation of DAS from VLS described in this section is based on WT. In fact, it is a form of the ANC-WT scheme, adapted for this task. Since the problem addressed here deals with the separation of the nonstationary part of the LS from the stationary one, the ANC-WT scheme is renamed *Wavelet Transform-based ST-NST* (WTST-NST) filter, originally introduced by Hadjileontiadis and Panas.[45] The only changes needed to the ANC-WT scheme of Figure III.3.5 refer to the input signal, which is now the breath sounds (DAS and VLS) and thus $X(\lambda)$ in Figure III.3.5 represents the breath sounds, and to the two outputs (nonstationary and stationary ones) in Figure III.3.5, which are now $DAS(\lambda)$ and $VLS(\lambda)$, corresponding to the separated DAS and the background VLS, respectively.

3.6.1.1 Results from the Application of the WTST-NST Filter on Real Data

Figure III.3.9 illustrates an example of the results obtained when applying the WTST-NST filter on breath sounds containing FC, CC, and SQ, recorded from patients with pulmonary fibrosis, chronic bronchitis, and interstitial fibrosis, respectively.[45] Specific implementation and experimental set-up issues are described in detail in Reference 45. In particular, Figure III.3.9(a) shows the input signal $X_1(k)$, i.e., breath sounds with FC (the desired FC are marked with small, black arrows); Figure III.3.9(b) depicts the corresponding separated FC, $DAS_{FC}(k)$ (nonstationary WTST-NST filter output); Figure III.3.9(c) illustrates the corresponding background VLS, $VLS_{FC}(k)$ (stationary WTST-NST filter output); Figure III.3.9(d) shows the input signal $X_2(k)$, i.e., breath sounds with CC

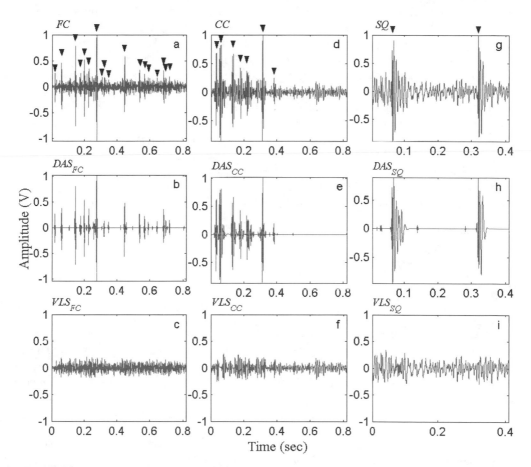

FIGURE III.3.9 (a) A time section of 0.8192 sec of FC recorded from a patient with pulmonary fibrosis, considered as an input to the WTST-NST filter. (b) The nonstationary output of the WTST-NST filter $DAS_{FC}(k)$. (c) The stationary output of the WTST-NST filter $VLS_{FC}(k)$. (d) A time section of 0.8192 sec of CC recorded from a patient with chronic bronchitis, considered as an input to the WTST-NST filter. (e) The nonstationary output of the WTST-NST filter $DAS_{CC}(k)$. (f) The stationary output of the WTST-NST filter $VLS_{CC}(k)$. (g) A time section of 0.4096 sec of SQ recorded from a patient with interstitial fibrosis, considered as an input to the WTST-NST filter. (h) The nonstationary output of the WTST-NST filter $DAS_{SQ}(k)$. (i) The stationary output of the WTST-NST filter $VLS_{SQ}(k)$. The arrowheads in the input tracing indicate events of interest which have been visually identified and correspond to DAS.

(the desired CC are marked with small, black arrows); Figure III.3.9(e) depicts the corresponding separated CC, $DAS_{CC}(k)$ (nonstationary WTST-NST filter output); Figure III.3.9(f) illustrates the corresponding background VLS, $VLS_{CC}(k)$ (stationary WTST-NST filter output); Figure III.3.9(g) shows the input signal $X_3(k)$, i.e., breath sounds with SQ (the desired SQ are marked with small, black arrows); Figure III.3.9(h) depicts the corresponding separated SQ, $DAS_{SQ}(k)$ (nonstationary WTST-NST filter output); and Figure III.3.9(i) illustrates the corresponding background VLS, $VLS_{SQ}(k)$ (stationary WTST-NST filter output).

From Figure III.3.9(a), the explosive, nonstationary character of FC, along with their short time duration are evident, since they are superimposed on stationary VLS with lower amplitude. Comparing the nonstationary part of Figure III.3.9(a) with Figure III.3.9(b), it is clear that, in Figure III.3.9(b), all structural components of the DAS are easily recognized, while their various morphologies and locations are clearly distinguished. Furthermore, the stationary part of Figure III.3.9(a), i.e., the pure VLS, is faithfully reproduced in Figure III.3.9(c), retaining its original amplitude level and structure. Although the sequence of CC depicted in Figure III.3.9(d) differs from previously described FC, either in structure and shape or in their reproducible character, the WTST-NST filter still performs well, as can be seen from the comparison of Figures III.3.9(d) and (e), regarding the CC identification, and of Figures III.3.9(d) and (f), regarding the reproducibility of the pure VLS of Figure III.3.9(d) in the stationary output of the WTST-NST filter $VLS_{CC}(k)$. The two SQ included in Figure III.3.9(g) possess the two main characteristics of a squawk, i.e., an underlying fine crackle (at the locations pointed out by the arrowheads) followed by a short wheeze, with an almost exponential decay. These characteristics are clearly identified in the nonstationary output of the WTST-NST filter, $DAS_{SQ}(k)$, depicted in Figure III.3.9(h). Again, the pure VLS of Figure III.3.9(g) are accurately reconstructed in the stationary output of the WTST-NST filter $VLS_{SQ}(k)$, shown in Figure III.3.9(i).

The efficiency of the WTST-NST filter, apart from visual comparisons between the marked DAS and those picked up by the WTST-NST filter, can be measured by calculating the same quantitative evaluators defined in Equations III.3.31 and III.3.32, i.e., D_R and TD_R, respectively (with N_R now denoting the number of visually recognized DAS by a physician, and N_E now denoting the number of estimated DAS using the WTST-NST filter), along with the following adapted version of the third one (Equation III.3.33):

$$\textbf{Quality Factor}: 0 \le S_q = \left(\frac{area|X(k)| - area|DAS(k)|}{area|X(k)|} \right) \le 1, k = 1, \ldots, N, \qquad \text{(III.3.34)}$$

where $X(k)$ the normalized breath sounds (DAS and VLS) and $DAS(k)$ the estimated DAS form the nonstationary output of the WTST-NST filter.[45] For the analyzed DAS shown in Figure III.3.9, D_R was equal to 100%, for all DAS (FC, CC, and SQ). This was also true for almost all analyzed DAS,[45] resulting in TD_R ranging from 97.5–100% and mean S_q near 0.55,[45] which indicate that the WTST-NST filter has a localized effect on the input signal, only at true locations of the DAS presence, without requiring any reference signal, and it deteriorates neither the structure nor the morphology of the DAS as a result of the separation procedure.

After applying the WTST-NST filter to the input signal of crackles, a more efficient estimation of the two main features of crackles, i.e., IDW and 2CD (see also Figure III.3.2), can be obtained by time domain analysis of the accurately separated crackles. Such an analysis is performed in Reference 45, from which it is derived that the estimated IDW and 2CD values, either of FC or of CC, are almost identical to those proposed by Cohen,[53] indicating again the accuracy of the WTST-NST filter in separating FC and CC from VLS.

3.6.1.2　Comparison with Other Works

Comparing the results of the WTST-NST algorithm[45] with those derived from the application of the nonlinear scheme (ST-NST) proposed by Ono et al.[51] at the same breath sounds (analytically shown in Table II in Reference 45), it is clear that the WTST-NST filter performs better than the ST-NST filter, in all cases of each type of DAS. This is due to the empirical definition of the parameters employed in the performance of the ST-NST filter, depending on the characteristics of the input signal. The WTST-NST filter may be applied to all types of DAS requiring neither empirical definition nor adaptive updating of its parameters according to the characteristics of the input signal. In addition, the WTST-NST filter, unlike the ST-NST filter, separates the whole DAS, without leaving the later parts, i.e., the part of DAS that follows the initial peak, as can be seen from the Figure III.3.9.

Regarding the performance of the WTST-NST filter with that of the rest of the separation techniques mentioned in Section 3.6, the WTST-NST filter clearly overcomes the disadvantages of highpass filtering, since it results in nondestroyed DAS and VLS; unlike a level slicer, it detects all DAS, even those with small amplitude; and unlike time-expanded waveform analysis,[12,50] it results in accurate, fast, and objective results, regardless of DAS type and without employing any human intervention.

3.6.2 GFST-NST: A Fuzzy Rule-Based Real-Time DAS Detector

Although the WTST-NST filter described in the previous subsection obtains the most accurate separation results among all other methods reported in the recent literature, it cannot serve as the optimum solution when the real-time analysis of the lung sounds is the primary aim. In this subsection, an enhanced *real-time* DAS detector based on fuzzy logic is described. It is based on a generalized version of the *Fuzzy logic-based ST-NST* (FST-NST) filter proposed by Tolias et al.,[54] namely *Generalized FST-NST* (GFS-NST) filter, that uses one *Fuzzy Inference System* (FIS) dedicated to the estimation of the stationary part of the signal to provide a reference to the second FIS that estimates the nonstationary part.[55]

For the nonfamiliarized reader, a short outline regarding the general FIS structure is given below.

3.6.2.1 Brief Description of FIS

An FIS is composed of five functional blocks: a *rule base* that contains a number of fuzzy if-then rules,[56,57] a *database* that defines the membership functions of the fuzzy sets used by the fuzzy rules, a *decision-making subsystem* that performs the inference operations on the rules, a *fuzzification* interface that transforms crisp measurement to degrees of membership to different fuzzy sets, and, finally, a *defuzzification* interface that transforms the fuzzy results into a crisp output (e.g., a control signal, a predicted value, etc). A fuzzy inference system performs the following processing steps on the given inputs:

1. *Fuzzification*: Compare the input variables with the membership functions that constitute the database on the premise part to obtain the membership values of each linguistic label.
2. *Determination of the firing strength of each rule*: Combine the membership values on the premise part.
3. *Generation of the consequent of each rule*: Dependent on the firing strength.
4. *Defuzzification*: Aggregation of the consequents to produce a crisp output.

Depending on the type of fuzzy reasoning and the fuzzy if-then rules that are used, there are three types of inference systems. The overall output in the first type is the weighted average of each rule's crisp output induced by the rule's firing strength and output membership functions. The output membership functions must be nondecreasing functions in the universe of discourse. In the second type, by applying the maximum operator to the fuzzy outputs, the overall fuzzy output is derived and the final crisp output is calculated using an appropriate defuzzification method (area, bisector of area, center of mass, etc.). Finally, the third type of fuzzy inference systems uses the Sugeno type approach,[58] and the fuzzy output for each rule is a linear combination of input variables with an additional constant term. The final output is the weighted average of each rule's output.

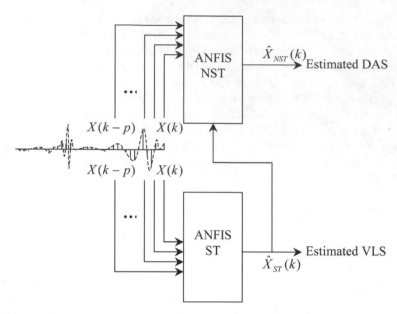

FIGURE III.3.10 Block diagram of the GFST-NST scheme. ANFIS NST denotes the adaptive network-based FIS for the estimation of the nonstationary part of the signal (DAS), while ANFIS ST denotes the adaptive network-based FIS for the estimation of the stationary part of the signal (VLS).

3.6.2.2 The GFST-NST Filter Structure

The GFST-NST is based on two *Adaptive Network-based FIS* (ANFIS) proposed by Jang.[59] Two ANFIS-based fuzzy inference systems are employed for the estimation of the stationary and the nonstationary parts of the input signal, ANFIS-ST and ANFIS-NST, respectively. The GFST-NST filter operates in two steps. Initially, the normalized zero-mean recorded lung sounds ($X(k)$) and their p delays ($X(k-1), \ldots, X(k-p)$) are fed into the ANFIS-ST part of the GFST-NST filter. Since the ANFIS-ST part is pre-trained (using the stationary output of the WTST-NST filter[55]) to focus on the stationary part of the signal, its output ($\hat{X}_{ST}(k)$) is the estimated stationary part of the recorded lung sounds, i.e., the VLS. In the following step, the output of the ANFIS-ST part ($\hat{X}_{ST}(k)$), the input ($X(k)$), and its p delays ($X(k-1), \ldots, X(k-p)$) are fed into the ANFIS-NST part. Similarly, since the ANFIS-NST part is pre-trained (using the nonstationary output of the WTST-NST filter[55]) to focus on the nonstationary part of the signal, its output ($X_{NST(k)}$) results in the estimated nonstationary part of the recorded lung sounds, i.e., the DAS.

From this description, it is evident that the GFST-NST filter forms a very simplified procedure that it is based on dedicated pre-trained FIS, which require only the recorded input and a small number of its delays (usually 1 or 2 are enough) to estimate the stationary and the nonstationary part of the recorded breath sounds. Consequently, the GFST-NST filter can be easily implemented using only a small number of multiplications and additions and it can provide a *real-time* separation of DAS from VS.[55] The structure of the GFST-NST filter is illustrated in Figure III.3.10.

3.6.2.3 Results from the Application of the GFST-NST Filter on Real Data

Figure III.3.11 illustrates an example of the results obtained when applying the GFST-NST filter on the same breath sounds used in Reference 45, which apart from VLS also contain FC, CC, and SQ recorded from patients with pulmonary fibrosis, chronic bronchitis, and interstitial fibrosis, respectively. Specific training, implementation, and experimental set-up issues are described in detail in Reference 55. In particular, Figure III.3.11(a) shows the input signal $X_1(k)$, i.e., breath sounds with FC (the desired FC are marked with small, black arrows); Figure III.3.11(b) depicts the

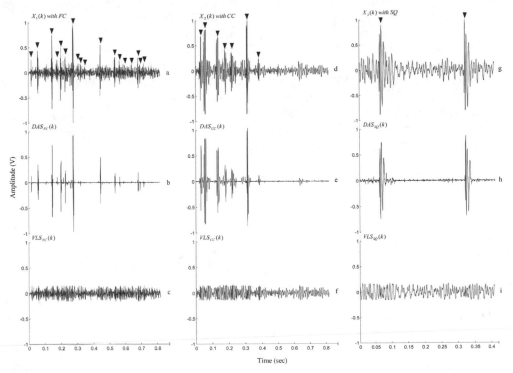

FIGURE III.3.11 (a) A time section of 0.8192 sec of FC recorded from a patient with pulmonary fibrosis, considered as an input to the GFST-NST filter. (b) The nonstationary output of the GFST-NST filter $DAS_{FC}(k)$. (c) The stationary output of the GFST-NST filter $VLS_{FC}(k)$. (d) A time section of 0.8192 sec of CC recorded from a patient with chronic bronchitis, considered as an input to the GFST-NST filter. (e) The nonstationary output of the GFST-NST filter $DAS_{CC}(k)$. (f) The stationary output of the GFST-NST filter $VLS_{CC}(k)$. (g) A time section of 0.4096 sec of SQ recorded from a patient with interstitial fibrosis, considered as an input to the GFST-NST filter. (h) The nonstationary output of the GFST-NST filter $DAS_{SQ}(k)$. (i) The stationary output of the GFST-NST filter $VLS_{SQ}(k)$. The arrowheads in the input tracing indicate events of interest which have been visually identified and correspond to DAS.

corresponding separated FC, $DAS_{FC}(k)$ (nonstationary GFST-NST filter output); Figure III.3.11(c) illustrates the corresponding background VLS, $VLS_{FC}(k)$ (stationary GFST-NST filter output); Figure III.3.11(d) shows the input signal $X_2(k)$, i.e., breath sounds with CC (the desired CC are marked with small, black arrows); Figure III.3.11(e) depicts the corresponding separated CC, $DAS_{CC}(k)$ (nonstationary GFST-NST filter output); Figure III.3.11(f) illustrates the corresponding background VLS, $VLS_{CC}(k)$ (stationary GFST-NST filter output); Figure III.3.11(g) shows the input signal $X_3(k)$, i.e., breath sounds with SQ (the desired SQ are marked with small, black arrows); Figure III.3.11(h) depicts the corresponding separated SQ, $DAS_{SQ}(k)$ (nonstationary GFST-NST filter output); and Figure III.3.11(i) illustrates the corresponding background VLS, $VLS_{SQ}(k)$ (stationary GFST-NST filter output).

Comparing the nonstationary part of Figures III.3.11(a), (d), and (g) with Figures III.3.11(b), (e), and (h), respectively, it is clear that all structural components of DAS are easily recognized, for all cases (FC, CC, and SQ), while their various morphologies and locations are clearly distinguished. Furthermore, the stationary part of Figures III.3.11(a), (d), and (g), i.e., the pure VLS, is faithfully reproduced in Figures III.3.11(c), (f), and (i), respectively, retaining its original amplitude level and structure. For the case of SQ, their two main characteristics, i.e., an underlying fine crackle (at the locations pointed out by the arrowheads) followed by a short wheeze, with an almost exponential decay, are identified in the nonstationary output of the GFST-NST filter, $DAS_{SQ}(k)$, depicted in Figure III.3.11(h). Again, the pure VLS of Figure III.3.11(g) are accurately reconstructed in the stationary output of the GFST-NST filter $VLS_{SQ}(k)$, shown in Figure III.3.11(i).

The performance of the GFST-NST filter, apart from visual comparisons between the marked DAS and those picked up by the GFST-NST filter, can be evaluated by calculating the same quantitative evaluators defined in Equations III.3.31 and III.3.32, i.e., D_R and TD_R, respectively (with N_R now denoting the number of visually recognized DAS by a physician, and N_E now denoting the number of estimated DAS using the GFST-NST filter), along with the S_q defined in Equation III.3.34 and the *Root Mean Squares Error* (RMSE), defined as:

$$RMSE = \sqrt{\frac{\sum_k (X_{WTST-NST}(k) - X_{GFST-NST}(k))^2}{k - 1}}, \qquad \text{(III.3.35)}$$

where $X_{WTST-NST}(k)$ denotes the results of the WTST-NST filter[45] on a signal and $X_{GFST-NST}(k)$ the estimated version of that signal using the GFST-NST filter. For the analyzed DAS shown in Figure III.3.11, D_R was equal to 100% for all DAS (FC, CC, and SQ). This was also true for almost all analyzed DAS,[55] resulting in TD_R ranging from 97.5–100%, a mean S_q near 0.6,[45] and a mean *RMSE* of 0.05, which indicate that the GFST-NST filter has a localized effect on the input signal, only at true locations of the DAS presence, without requiring any reference signal, and it deteriorates neither the structure nor the morphology of the DAS as a result of the separation procedure. Since it requires a small number of multiplications, and considering that the time required for additions and look-up table operations is negligible to the one for the multiplications, *the GFST-NST filter could be easily implemented in a real-time context*.[55] The comparison of the GFST-NST filter with the WTST-NST one is included in a succeeding subsection, after the description of another FL-based approached, described in the following subsection.

3.6.3   OLS-FF: An Orthogonal Least Squares-Based Fuzzy Filter as a Real-Time DAS Detector

An alternative approach in developing fuzzy models for real-time separation of DAS from VLS is described in this subsection. In this approach, the model generation process is based on the *Orthogonal Least Squares* (OLS) concept,[60] providing structure and parameter identification as well as performing input selection. In particular, the parameters of the resulting models are calculated in a forward manner and, in contrast to ANFIS, *no training algorithm is required*. Therefore, the method does not suffer from drawbacks inherent in gradient descent learning, such as tapping to local minima and extensive training time. This approach uses two FIS that operate in parallel and form the *OLS-based Fuzzy Filter* (OLS-FF).[61]

3.6.3.1   The OLS-FF Structure

The OLS-FF is based on the FIS proposed by Takagi, Sugeno, and Kang, i.e., TSK FIS,[58] but in an extended new form, namely *Extended TSK FIS* (ETSK FIS).[61] It employs two FIS for the estimation of the stationary and the nonstationary parts of the input signal, i.e., ETSK FIS-ST and ETSK FIS-NST, respectively. The two FIS operate in parallel and are both fed with the premise part input vector $\bar{\mathbf{z}}$, which includes the normalized zero-mean recorded breath sounds, $X(k)$, and some of its delays. The consequent part input vectors, consisting of $X(k)$ and some of its delays, are different for each FIS and are extracted during a modeling process described in detail in Reference 61, which is carried out independently for each FIS and signal category. As a result, the output of the ETSK FIS-ST part ($\hat{X}_{ST}(k)$) is an estimation of the VLS, while the output of the ETSK FIS-NST

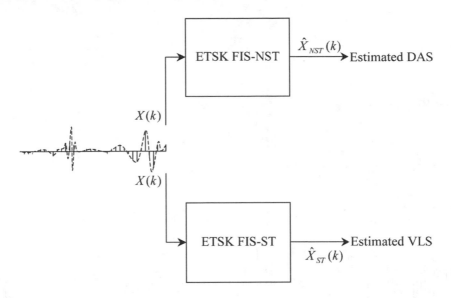

FIGURE III.3.12 Block diagram of the OLS-FF. ETSK FIS-NST denotes the ETSK FIS for the estimation of the nonstationary part of the signal (DAS), and ETSK FIS-ST denotes the ETSK FIS for the estimation of the stationary part of the signal (VLS), after the model generation process.

part $(\hat{X}_{NST}(k))$ is the estimated DAS. The main assumption that is employed in the OLS-FF is that the latter is capable, after the model generation process, of separating the DAS from the VLS using specifically built models, which estimate the stationary and the nonstationary parts of the signal. A block diagram of the OLS-FF is shown in Figure III.3.12.

3.6.3.2 Results from the Application of the OLS-FF on Real Data

Figure III.3.13 illustrates an example of the results obtained when applying the OLS-FF on the same breath sounds used in References 45 and 55, which, apart from VLS, also contain FC, CC, and SQ recorded from patients with pulmonary fibrosis, chronic bronchitis, and interstitial fibrosis, respectively. Specific model generation, training, implementation, and experimental set-up issues are described in detail in Reference 61. In particular, Figure III.3.13(a) shows the input signal $X_1(k)$, i.e., breath sounds with FC (the desired FC are marked with small, black arrows); Figure III.3.13(b) depicts the corresponding separated FC, $DAS_{FC}(k)$ (nonstationary OLS-FF output); Figure III.3.13(c) illustrates the corresponding background VLS, $VLS_{FC}(k)$ (stationary OLS-FF output); Figure III.3.13(d) shows the input signal $X_2(k)$, i.e., breath sounds with CC (the desired CC are marked with small, black arrows); Figure III.3.13(e) depicts the corresponding separated CC, $DAS_{CC}(k)$ (nonstationary OLS-FF output); Figure III.3.13(f) illustrates the corresponding background VLS, $VLS_{CC}(k)$ (stationary OLS-FF output); Figure III.3.13(g) shows the input signal $X_3(k)$, i.e., breath sounds with SQ (the desired SQ are marked with small, black arrows); Figure III.3.13(h) depicts the corresponding separated SQ, $DAS_{SQ}(k)$ (nonstationary OLS-FF output); and Figure III.3.13(i) illustrates the corresponding background VLS, $VLS_{SQ}(k)$ (stationary OLS-FF output).

Comparing the nonstationary part of Figures III.3.13(a), (d), and (g) with Figures III.3.13(b), (e), and (h), respectively, it is clear that all structural components of the DAS are easily recognized, for all cases (FC, CC, and SQ), while their various morphologies and locations are clearly distinguished. Furthermore, the stationary part of Figures III.3.13(a), (d), and (g), i.e., the pure VLS, is faithfully reproduced in Figures III.3.13(c), (f), and (i), respectively, retaining its original amplitude level and structure. For the case of SQ, their two main characteristics, i.e., an underlying fine crackle (at the locations pointed out by the arrowhead) followed by a short wheeze, with an almost exponential decay,

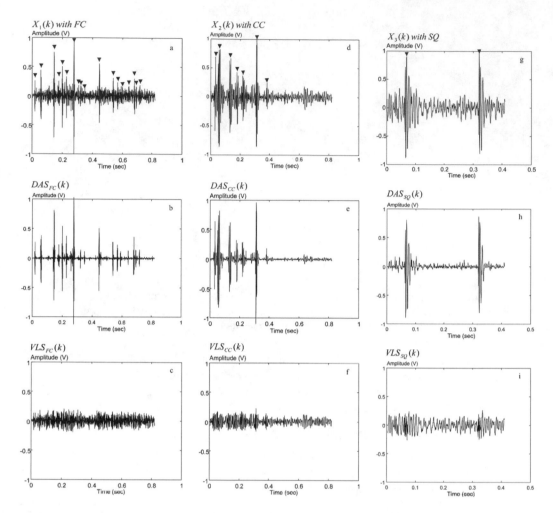

FIGURE III.3.13   (a) A time section of 0.8192 sec of FC recorded from a patient with pulmonary fibrosis, considered as an input to the OLS-FF. (b) The nonstationary output of the OLS-FF $DAS_{FC}(k)$. (c) The stationary output of the OLS-FF $VLS_{FC}(k)$. (d) A time section of 0.8192 sec of CC recorded from a patient with chronic bronchitis, considered as an input to the OLS-FF. (e) The nonstationary output of the OLS-FF $DAS_{CC}(k)$. (f) The stationary output of the OLS-FF $VLS_{CC}(k)$. (g) A time section of 0.4096 sec of SQ recorded from a patient with interstitial fibrosis, considered as an input to the OLS-FF. (h) The nonstationary output of the OLS-FF $DAS_{SQ}(k)$. (i) The stationary output of the OLS-FF $VLS_{SQ}(k)$. The arrowheads in the input tracing indicate events of interest which have been visually identified and correspond to DAS.

are identified in the nonstationary output of the OLS-FF, $DAS_{SQ}(k)$, depicted in Figure III.3.13(h). The pure VLS of Figure III.3.13(g) are accurately reconstructed in the stationary output of the OLS-FF, $VLS_{SQ}(k)$, shown in Figure III.3.13(i). Nevertheless, a small section of the initial part of the SQ is identified in the stationary output $VLS_{SQ}(k)$.

The performance of the OLS-FF, apart from visual comparisons between the marked DAS and those picked up by the OLS-FF, can be evaluated by calculating the same quantitative evaluators defined in Equations III.3.31 and III.3.32, i.e., D_R and TD_R, respectively (with N_R now denoting the number of visually recognized DAS by a physician, and N_E now denoting the number of estimated

DAS using the OLS-FF), along with the S_q defined in Equation III.3.34, and the *RMSE* defined in Equation III.3.35, adapted for the OLS-FF as:

$$RMSE = \sqrt{\frac{\sum_k (X_{WTST-NST}(k) - X_{OLS-FF}(k))^2}{K - 1}} \qquad \text{(III.3.36)}$$

where $X_{WTST-NST}(k)$ denotes the results of the WTST-NST filter[45] on a signal, and $X_{OLS-FF}(k)$ the estimated version of that signal using the OLS-FF. For the analyzed DAS shown in Figure III.3.13, D_R was equal to 94.45% for FC, and 100% for CC and SQ, respectively. This was also true for almost all analyzed DAS,[55] resulting in TD_R ranging from 80–96.4%, a mean S_q near 0.64,[61] and a mean *RMSE* of 0.07, which indicate that the OLS-FF has a localized effect on the input signal, only at true locations of the DAS presence, without requiring any reference signal, and it deteriorates neither the structure nor the morphology of the DAS as a result of the separation procedure. The computational time per output sample depends practically on the number of multiplications, since the time required for additions and look-up table operations is negligible. For the OLS-FF, the number of multiplications per input sample varies from 44–79.[61] A usual sampling period used in LS recordings is of 0.4 msec; hence, it is obvious that *the OLS-FF could be easily implemented in a real-time context* using either an ordinary PC or dedicated hardware.[61]

3.6.3.3   Comparison of the OLS-FF with the GFST-NST and WTST-NST Filters

As a conclusion to the DAS detection problem, a comparison of the three algorithms described in this section is adopted. In general, the OLS-FF improves the ANFIS structure employed in Reference 55, introducing a more flexible architecture. In this way, it activates the optimum employed fuzzy rules, uses a lower number of rules, and comprises the most appropriate inputs for each of them. This is a clear advantage over the previous work,[55] since ANFIS uses a complete rule base and consequent parts with fixed number of terms. In addition, unlike the GFST-NST filter, the OLS-FF does not require a training phase for the estimation of the optimum model parameters, since they are calculated on the spot. Table III.3.1 presents a comparison among the three DAS detectors, namely the WTST-NST and GFST-NST filters and the OLS-FF, with respect to the mean values of the quantitative evaluators defined in Equations III.3.32, III.3.34–III.3.36. Although the OLS-FF exhibits an inferior performance compared to the WTST-NST filter,[45] it satisfies the real-time implementation issue. However, Table III.3.1 shows that the GFST-NST filter[55] performs better than the OLS-FF, by 1% up to 4% with respect to $\overline{TD}_R$, for the cases of CC and SQ, and by 14% for the case of FC, as a trade-off between the structural simplicity of the OLS-FF. On the contrary, the OLS-FF requires a significantly smaller (by 62%) computational load than the GFST-NST filter, improving the procedure of clinical screening of DAS under a real-time implementation context.[61]

TABLE III.3.1   Performance Indexes for the WTST-NST, GFTS-NST, and OLS-Algorithms

Mean ± Standard Deviation	Fine Crackles			Coarse Crackles			Squawks		
	WTST-NST	GFST-NST	OLS-FF	WTST-NST	GFST-NST	OLS-FF	WTST-NST	GFST-NST	OLS-FF
$\overline{TD}_R$	100	100	79.55	97.5	97.5	96.36	100	100	96.00
$\pm S_{TD_R}$	0	0	16.66	5.59	5.59	12.01	0	0	8.94
$\overline{S}_q$	0.523	0.56	0.62	0.4176	0.71	0.67	0.5838	0.59	0.61
$\pm S_{S_q}$	0.0609	0.06	0.16	0.0812	0.12	0.18	0.0502	0.09	0.11
$RMSE_{ST}$	–	0.0463	0.0487	–	0.0647	0.0679	–	0.0669	0.0666
$\pm S_{RMSE_{ST}}$	–	0.011	0.0107	–	0.0173	0.0166	–	0.0516	0.0109
$RMSE_{NST}$	–	0.04885	0.0484	–	0.0678	0.071	–	0.0666	0.0663
$\pm S_{RMSE_{NST}}$	–	0.01	0.009	–	0.0183	0.0214	–	0.0519	0.0125

Note: — denotes not defined for WTST-NST.

3.7 Intelligent Modeling of Lung and Bowel Sounds Source and Transmission Path Using Higher- and Lower-Order Statistics

In this subsection, the problem of efficient modeling of the system related to the BioS source and transmission path is addressed. This is a very important issue in the analysis of BioS, since intelligent modeling could lead to an objective description of the changes a disease imposes to the production and/or transmission path of the BioS. In particular, the use of Higher-Order Statistics/Spectra, *Alpha-Stable Distributions*, and Lower-Order Statistics, for modeling LS and impulsive DAS and BS, respectively, is described in the following subsections.

3.7.1 AR-TOS: Autoregressive, Third-Order, Statistics-Based Modeling of Lung Sounds

Lung sounds originating inside the airways of the lung are modeled as the input to an all-pole filter, which describes the transmission of lung sounds through the parenchyma and chest wall structure.[62] The output of this filter is considered to be the LS recorded at the chest wall. As already mentioned in previous subsections, the recorded LS also contain heart sound interference, the reduction of which has been thoroughly addressed in Section 3.4. Muscle and skin noise, along with instrumentation noise, are modeled as an additive Gaussian noise.

Using the aforementioned model, given a signal sequence of LS at the chest wall, an *AutoRegressive* (AR) analysis based on *Third-Order Statistics* (TOS), namely AR-TOS, can be applied to compute the model parameters. Therefore, the source and transmission filter characteristics can be separately estimated,[63] as is described in the succeeding subsection.

3.7.1.1 Modeling of Transmission of the LS

The AR-TOS prediction filter performs autoregressive prediction based on third-order statistics and is described by the following equation:

$$y_n + \sum_{i=1}^{p} a_i y_{n-i} = w_n, a_0 = 1,$$
(III.3.37)

where y_n represents a pth-order AR process of N samples $(n = 0, 1, \ldots, N - 1)$, a_i are the coefficients of the AR-TOS model, and w_n is an independent and identically distributed (i.i.d.), non-Gaussian, third-order stationary, zero-mean process, with $E\{w_n^3\} = \beta \neq 0$ and y_n independent of w_l for $n < l$. Since w_n is third-order stationary, y_n is also third-order stationary, assuming it is a stable AR model. The cumulant-based 'normal' equations[64] for the model of Equation III.3.37 are given by:

$$\sum_{i=0}^{p} a_i R(\tau_i - i, \tau_2) = 0, \tau_1 = 1, \ldots, p, \ \& \ \tau_2 = -p, \ldots, 0,$$
(III.3.38)

where $R(\tau_1, \tau_2)$ is the third-order cumulant sequence of the AR process, described in Section 3.4.2.1. In practice, we use sample estimates of the cumulants. Equation III.3.38 yields consistent estimates of the AR parameters $\hat{a}_i$, maintaining the orthogonality of the prediction error sequence to an instrumental process derived from the data.[65] The calculation of the order p of the AR-TOS model is reduced to a rank determination problem and it is calculated according to Giannakis' and Mendel's method, described in Reference 66. Estimations of the transfer function $H(\omega)$ of the system and the normalized parametric third-order spectrum or *Parametric Bispectrum* $C_3^y(\omega_1, \omega_2)$ can be derived using the following equations, respectively:[40]

$$\hat{H}(\omega) = \frac{1}{1 + \sum\limits_{i=1}^{p} \hat{a}_i e^{-j\omega i}}, \quad |\omega| \leq \pi, \tag{III.3.39}$$

$$\frac{1}{\hat{\beta}} C_3^y(\omega_1, \omega_2) = \hat{H}(\omega_1)\hat{H}(\omega_2)\hat{H}^*(\omega_1 + \omega_2), \quad |\omega_1| \leq \pi, |\omega_2| \leq \pi, |\omega_1 + \omega_2| \leq \pi, \tag{III.3.40}$$

where $*$ denotes the complex conjugate part. The profound motivations behind the use of third-order statistics in the estimation of a_i were discussed in Section 3.4.1.1. We remind here the suppression of Gaussian noise, since third-order statistics of Gaussian signals are identically zero. Hence, when the analyzed waveform consists of a non-Gaussian signal in additive symmetric noise (e.g., Gaussian), the parameter estimation of the original signal using third-order statistics takes place in a high *signal-to-noise ratio* (SNR) domain, and the whole parametric presentation of the process is more accurate and reliable.[40,64]

3.7.1.2 Source Modeling of LS

The model used for the LS originating inside the airways considers the lung sound source as the output from an additive combination of three kinds of noise sequences.[11] The first sequence (periodic impulse) describes the *CAS sources*, since they have characteristic distinct pitches associated with them, and they are produced by periodic oscillations of the air and airway walls (see also Section 3.2.1.2).[7,13] The second sequence (random intermittent impulses) describes the *crackle sources*, since they are produced by sudden opening/closing of airways or bubbling of air through extraneous liquids in the airways, both phenomena associated with sudden intermittent bursts of sounds energy (see also Section 3.2.1.2).[7,13] Finally, the third sequence (white non-Gaussian noise) describes the *breath sound sources*, since they are produced by turbulent flow in a large range of airway dimensions (see also Section 3.2.1.2).[7,13] The estimation of the AR-TOS model input (lung sound source) can be derived from the prediction error of Equation III.3.37, by means of inverse filtering.[63,67]

3.7.1.3 Results from the Application of the AR-TOS Model on Real Data

Representative results of AR-TOS modeling of LS (FC and WZ) are shown in Figure III.3.14. Detailed discussion on experimental and implementation issues along with additional experimental results can be found in Reference 63. Figures III.3.14(a)–(d) show the results of AR-TOS modeling of an expiratory asthmatic segment containing wheezes (CAS-WZ). The estimated source [Figures III.3.14(b1), (b2)] is a periodic train of impulses (with period $T \approx 0.004$ sec). The order p of the AR-TOS model used is found equal to 10.[63] The estimated filter response $H^{WZ}(\omega)$ [Figure III.3.14(c)] has a lowpass filter characteristic, with a frequency band range around 0–550 Hz, and two distinct resonance peaks at 250 Hz and 500 Hz. According to the estimated bispectrum $C_3^{WZ}(\omega_1, \omega_2)$ [Figure III.3.14(d)], the power of the signal is located in the resonance frequencies, which result is consistent with the production mechanism of wheezes, i.e., distinct resonance in the transmission path (airway walls).[7,13]

In the case of modeling the inspiratory segment with FC shown in Figure III.3.14(e), the estimated source waveform [Figure III.3.14(f)] contains impulsive bursts, corresponding to fine crackles, and white non-Gaussian noise. This could be explained by the associated with fine crackles phenomenon of explosive reopening of small airways that had closed during the previous expiration. The order p of the AR-TOS model used is found equal to 2.[63] The estimated filter response $H^{FC}(\omega)$ [Figure III.3.14(g)] has a bandpass filter characteristic, with a frequency band range around 350–700 Hz, with central frequency at 530 Hz. According to the estimated bispectrum $C_3^{FC}(\omega_1, \omega_2)$ [Figure III.3.14(h)], the power of the signal is shifted to higher frequencies. These results are consistent with the understanding of pulmonary fibrosis, since the associated abnormal airway closure that precedes the 'crackling' reopening is due to increased lung stiffness, which probably causes the transmission of higher frequencies.[7,13]

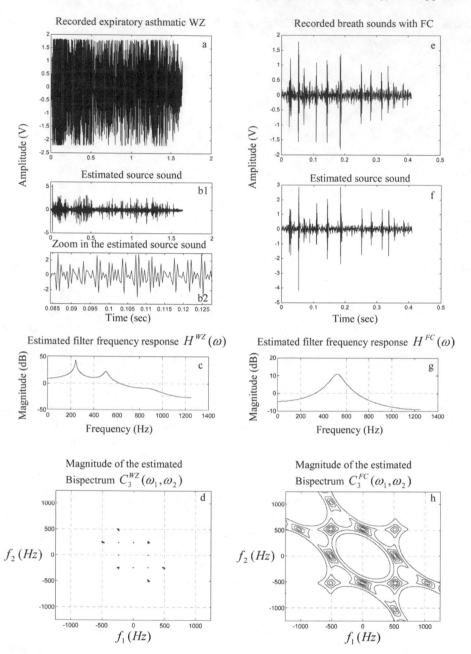

FIGURE III.3.14 (a) A time section of 1.63 sec of asthmatic wheezes (expiration) subjected to AR-TOS modeling process. (b1) The estimated LS source. (b2) A zoomed version of the estimated LS source. (c) The estimated filter frequency response $H^{WZ}(\omega)$. (d) The estimated magnitude of the parametric bispectrum $C_3^{WZ}(\omega_1, \omega_2)$ of the recorded sounds. (e) A time section of 0.4096 sec of breath sounds containing fine crackles (inspiration). (f) The estimated lung sound source. (g) The estimated filter frequency response $H^{FC}(\omega)$. (h) The estimated magnitude of the parametric bispectrum $C_3^{FC}(\omega_1, \omega_2)$ of the recorded sounds.

From the derived results it is evident that the AR-HOS modeling of lung sounds characterizes their source and transmission efficiently, providing useful diagnostic information to the clinicians regarding the influence of the pathology either to the source and/or to the transmission path of the LS.

3.7.2 Modeling of Impulsive Lung Sounds and Bowel Sounds Using *Alpha*-Stable Distributions and Lower-Order Statistics

As described in Section 3.2, among the different kinds of BioS recorded from the human body, there are some that exhibit more sharp peaks or occasional bursts of outlying observations than one would expect from normally distributed signals, such as the discontinuous adventitious sounds and some types of impulsive bowel sounds. In addition, these BioS occur with a different *degree* of impulsiveness. Due to their inherent impulsiveness, these bioacoustic signals have a density function that decays in the tails less rapidly than the Gaussian density function.[68] Since modeling based on *alpha-stable distribution* is appropriate for enhanced description of impulsive processes,[68] its use could provide a useful processing tool for the analysis of impulsive BioS. Under the non-Gaussian assumption in the context of analyzing impulsive processes, among the various distribution models that were suggested in the past, the *alpha*-stable distribution is the only one that is motivated by the *generalized central limit theorem*. This theorem states that *the limit distribution of the sum of random variables with possibly infinite variances is stable distribution.*[69] Stable distributions are defined by the stability property, which says that *a random variable, X, is stable if and only if the sum of any two independent random variables with the same distribution as X also has the same distribution.*[69]

Following the aforementioned approach, this subsection is focused on modeling of impulsive LS (i.e., DAS) and BS with explosive character, by means of *alpha-stable distribution* and *Lower-Order Statistics* (LOS). In this way, an innovating perception in the analysis of impulsive BS is introduced, suggesting a new field in their processing for diagnostic feature extraction. For the reader not familiar with *alpha*-stable distributions and LOS, a short outline of the definitions and mathematical background needed is given below.

3.7.2.1 *Alpha*-Stable and Symmetric *Alpha*-Stable Distributions

A univariate distribution function $F(x)$ is called *alpha*-stable if its characteristic function can be expressed in the following form:[68]

$$\varphi(t) = \exp\{jat - \gamma|t|^{\alpha}[1 + j\beta\,sign(t)\omega(t, \alpha)]\}, \qquad (III.3.41)$$

where:

$$\omega(t, \alpha) = \begin{cases} \tan\frac{\alpha\pi}{2} & \text{for } \alpha \neq 1 \\ \frac{2}{\pi}\log|t| & \text{for } \alpha = 1; \end{cases} \qquad (III.3.42)$$

$sign(t) = 1, 0, -1$ for $t > 0, t = 0$, and $t < 0$, respectively; and $-\infty < a < \infty, \gamma > 0$, $0 < \alpha \leq 2, -1 \leq \beta \leq 1$.

Thus, the following four parameters can completely determine the *alpha*-stable distribution:

1. The location parameter a, which is the symmetry axis
2. The scale parameter γ, also called the dispersion, which, in analogy to the variance of the Gaussian distribution, is a measure of the deviation around the mean
3. The symmetry parameter β, which is the index of skewness
4. The characteristic exponent *alpha*, α, which is a measure of the thickness of the tails of the distribution (a small value of α implies considerable probability mass in the tails of the distribution, while a large value of α implies considerable probability mass in the central location of the distribution)

The special cases $\alpha = 2$ and $\alpha = 1$ with $\beta = 0$ correspond to the Gaussian distribution and the Cauchy distribution, respectively. When $\beta = 0$, the distribution is symmetric about the center a, and it is called *Symmetric α-Stable*, or *SαS*, and its characteristic function is of the form:

$$\varphi(t) = \exp\{jat - \gamma|t|^{\alpha}\}. \qquad (III.3.43)$$

3.7.2.2 Fractional Lower-Order Moments and the Covariation Coefficient

Although the second-order moment of a $S\alpha S$ random variable X with $0 < \alpha < 2$ does not exist, all moments of order less than α do exist and are called *Fractional Low-Order Moments* (FLOM), defined as follows:[68]

$$E(|X|^p) = C(p, \alpha)\gamma^{p/\alpha}, 0 < p < \alpha, a = 0, \tag{III.3.44}$$

where:

$$C(p, \alpha) = \frac{2^{p+1}\Gamma(\frac{p+1}{2})\Gamma(-p/\alpha)}{\alpha\sqrt{\pi}\ \Gamma(-p/2)}, \tag{III.3.45}$$

with $E(\cdot)$ and $\Gamma(\cdot)$ denoting the expectation value and the usual gamma function, respectively.

The *covariation coefficient* of jointly $S\alpha S$ random variables X and Y with $\alpha > 1$ is given by the following equation:[68]

$$\lambda_{XY} = \frac{E(XY^{\langle p-1 \rangle})}{E(|Y|^p)}, \tag{III.3.46}$$

for any $1 \leq p < \alpha$. Despite the fact that it plays the role of correlation coefficient of second-order random variables, it can become unbounded. For a thorough description of alpha-stable distributions, and LOS, the reader should refer to Reference 68.

3.7.2.3 Analysis Tools for $S\alpha S$ Distributions- and LOS-Based Modeling of LS and BS

Since non-Gaussian stable distributions do not have finite variance, a test based on the convergence of the variance of the population distribution can lead to the adoption of the Gaussian or non-Gaussian assumption. Specifically, if the population distribution $F(x)$ has finite variance, then the *Running Sample Variance* (RSV) S_n^2 given by:

$$S_n^2 = \frac{1}{n}\sum_{k=1}^{N}(X_k - \overline{X}_n)^2 \qquad 1 \leq n \leq N, \tag{III.3.47}$$

where $\overline{X}_n = \frac{1}{n}\sum_{k=1}^{n} X_k$, and $X_k, k = 1, \ldots, N$, are samples from the distribution, *should converge to a finite value.*[70] In such a case, the Gaussian assumption is adopted; otherwise, we adopt the non-Gaussian one.

Since the three parameters α, a, and γ determine the $S\alpha S$ distribution, their estimation from the realizations of a symmetric stable random variable could reflect the differences among various types of impulsive BioS. Although the estimation of these parameters is generally severely hampered by the lack of known closed-form density functions (for all but a few members of the stable family), there are some numerical methods that perform reliable parameter estimation, such as the *fractile*,[71] *regression*,[72] $\log|S\alpha S|$,[68] and *negative-order moment* methods.[68]

In the same vein with the second- or third-order statistics-based AR modeling, a Pth order AR $S\alpha S$ process is described by:

$$X(n) = \sum_{i=1}^{P} a_i X(n - i) + U(n), \tag{III.3.48}$$

where $\{U(n)\}$ is a sequence of i.i.d. $S\alpha S$ random variables of characteristic exponent α and dispersion γ_u. Furthermore, the stationary process $X(n)$ is $S\alpha S$ and $X(n)$ and $U(n + j)$ are independent for any $j > 0$. From the several methods proposed for the estimation of $a_i, i = 1, \ldots, P$ coefficients of Equation III.3.48 from observations of the output $X(n)$, the *least squares* (LS) method[73] and the method which combines FLOMs with the Yule-Walker (YW) equation[68] (YW-FLOMs) provide reliable estimates.

3.7.2.4 Results from the Application of the $S\alpha S$ Distribution- and LOS-Based Modeling on Real Data

Representative results of the $S\alpha S$ distribution- and LOS-based modeling of an impulsive burst of squawks (SQ), recorded from a patient with allergic alveolitis, and *Explosive Bowel Sounds* (EBS), recorded from a patient with irritable bowel syndrome, are shown in Figure III.3.15. Detailed discussion on experimental and implementation issues, along with additional experimental results, can be found in Reference 74. For better isolation of the impulsive character of the analyzed signals, the WTST-NST filter[45,49] is initially used to eliminate the background noise from the recordings. The contaminated and de-noised waveforms are shown in Figures III.3.15(a) and (b) and III.3.15(f) and (g) for the SQ and the EBS, respectively.

For adopting the non-Gaussian assumption, the convergence of the RSV (S_n^2) is tested. The testing results depicted in Figures III.3.15(c) and (h) for the SQ and the EBS waveforms, respectively, show a nonconverging RSV, justifying the adoption of the non-Gaussianity assumption. The AR modeling of SQ and EBS with both the LS[73] and YW-FLOMs[68] methods shows that their estimates for $a_1, i = 1, \ldots, P$ (for all cases[74]) are almost identical. Using inverse filtering, an estimation of the input $U(n)$ (for all cases[74]) is found. This is depicted in Figures III.3.15(d) and (i) for the SQ and the EBS waveforms, respectively. Their visual inspection confirms the explosive character of the source sound of impulsive BioS, since they are produced by sudden changes in the acoustic properties of lung and bowel. In addition, Figures III.3.15(e) and (k) show the estimated $a_i, i = 1, \ldots, P$ for all cases of SQ and EBS categories,[74] respectively, calculated by the YW-FLOMs method, indicating consistent estimation.

From the estimated parameters of the $S\alpha S$ distribution of the analyzed impulsive BioS using the $\log|S\alpha S|$ method[68] (since the fractile method[71] works only for $\alpha \geq 0.6$, and the regression method[72] needs empirical parameter setting), it is derived that when the contaminated (without de-noising) impulsive BS are analyzed, the values of α are large enough (near 1.5); while when the AR-estimated inputs of the de-noised data are analyzed, these values decrease dramatically. This fact is due to the increase of impulsiveness of the analyzed signal, since the vesicular sound and the background noise are waived after de-noising. In fact, the impulsive source sound, initiated inside the lung or the bowel, reaches the surface with different signal characteristics, indicating a shifting toward the Gaussian distribution, due to the superimposed Gaussian noise. Generally, the values of dispersion γ are low enough, indicating small deviation around the mean, while the location parameter a has mean values around zero, indicating low mean and median for $1 < \alpha \leq 2$ and $0 < \alpha < 1$, respectively.[74]

Using Equation III.3.46, the covariation coefficient λ_{SQ-FC} is calculated for the cases of the sound sources of SQ and FC. It is found[74] that an almost 50% correlation exists between the SQ and FC, confirming the accepted theory that SQ are produced by the explosive opening, due to a FC, and decaying fluttering of an unstable airway.[7,13] This proves that $S\alpha S$ distribution- and LOS-based modeling of impulsive BioS provide a measure of their impulsiveness, and, at the same time, reveal the underlying relationships between the associated production mechanisms and pathology.

3.8 Feature Extraction of Lung, Heart, and Bowel Sounds Using Higher-Order Statistics/Spectra

The properties of the Higher-Order Statistics/Spectra (HOS), briefly described in Sections 3.4.1.1 and 3.7.1.1, provide a fruitful field of analysis and feature extraction of BioS. In this subsection, HOS-based features that are derived from the HOS-based analysis of musical LS, HS, and BS are described.

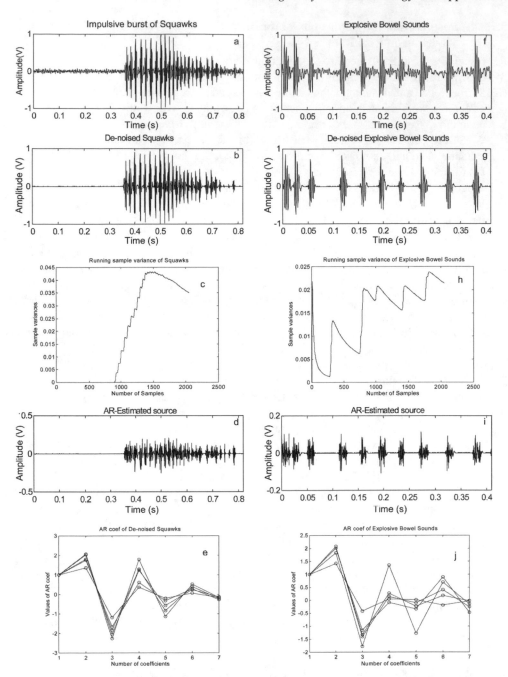

FIGURE III.3.15   Results from the analysis of impulsive BioS by means of $S\alpha S$ distributions analysis tools. (a) The original recorded impulsive burst of squawks (SQ).[74] (b) The de-noised output from the WTST-NST filter.[45] (c) The results from the converging variance test of de-noised impulsive SQ. (d) The AR-estimated sound source. (e) The values of the estimated AR coefficients (5 cases/category,[74] $P = 6$). (f) The original recorded explosive bowel sounds (EBS). (g) The de-noised output from the WTST-NST filter.[49] (h) The results from the converging variance test of de-noised EBS. (i) The AR-estimated sound source. (j) The values of the estimated AR coefficients (5 cases/category,[74] $P = 6$).

3.8.1 HOS-Based Parameters that Serve as Feature Extractors

In the succeeding subsections, the HOS-based parameters used for feature extraction are described.

3.8.1.1 The Bispectrum

The bispectrum $C_3^x(\omega_1, \omega_2)$ of an N-sample process $\{X(k)\}$ is defined as:[40]

$$C_3^x(\omega_1, \omega_2) = E\{X(\omega_1)X(\omega_2)X^*(\omega_1 + \omega_2)\}, |\omega_1| \le \pi, |\omega_2| \le \pi, |\omega_1 + \omega_2| \le \pi, \quad \text{(III.3.49)}$$

where $X(\omega_i)$, $i = 1, 2$ is the complex Fourier coefficient of the process $\{X(k)\}$; $E\{\cdot\}$ and * denotes the expectation value and the complex conjugate, respectively.

3.8.1.2 The Bicoherence and Cross-Bichorence Index

The *normalized bispectrum* or *bicoherency* (Bic) is defined as:[40]

$$Bic(\omega_1, \omega_2) = \frac{C_3^x(\omega_1, \omega_2)}{[C_2^x(\omega_1)C_2^x(\omega_2)C_2^x(\omega_1 + \omega_2)]^{1/2}} \quad \text{(III.3.50)}$$

where $C_2^x(\omega_i)$, $i = 1, 2$ is the power spectrum. From Equation III.3.50, it is apparent that the bicoherency is a function that combines two completely different entities, namely, the bispectrum, $C_3^x(\omega_1, \omega_2)$, and the power spectrum, $C_2^x(\omega)$, of the process. The magnitude of the bicoherency:

$$BI(\omega_1, \omega_2) = |Bic(\omega_1, \omega_2)| \quad \text{(III.3.51)}$$

is called the *Bicoherence Index* (BI) and is not bounded above by 1.0.[75] Under the Gaussian assumption, the expected value of the bicoherency is zero. The $BI(\omega_1, \omega_2)$ quantifies the presence of *quadratic phase coupling* (QPC) between any two frequency components in the process due to their nonlinear interactions. Two frequency components are said to be quadratically phase coupled (with the $BI(\omega_1, \omega_2)$ close to or greater than 1.0) when *there exists a third frequency component whose frequency and phase are the sum of the frequencies and phases of the first two components*.[40] The *cross-bicoherency* refers to two N-sample processes $\{X(k)\}$, $\{Y(k)\}$, and is defined through the cross-bispectrum, $C_3^{xxy}(\omega_1, \omega_2)$ as:

$$CBic(\omega_1, \omega_2) = \frac{C_3^{xxy}(\omega_1, \omega_2)}{[C_2^x(\omega_1)C_2^x(\omega_2)C_2^y(\omega_1 + \omega_2)]^{1/2}} \quad \text{(III.3.52)}$$

The magnitude of the cross-bicoherency:

$$CBI(\omega_1, \omega_2) = |CBic(\omega_1, \omega_2)| \quad \text{(III.3.53)}$$

is referred to as *Cross-Bicoherence Index* (CBI), and quantifies the presence of QPC between any two frequency components in the two processes due to their nonlinear interactions.

3.8.1.3 Significance Level and Noncentrality Parameter

The estimation of the $BI(\omega_1, \omega_2)$ by Equation III.3.51 initially employs the estimation of the bispectrum and the power spectrum of the process. For enhanced fidelity in the bifrequency domain, the parametric bispectrum of Equation III.3.40 (combined with the parametric estimation of the power spectrum[76]) is adopted. In that way, even adjacent frequency pairs with QFC could be located. The *Significance Level* (SL) of the bicoherence index is estimated assuming that the mean bicoherence power:

$$S = \sum |bic(\omega_1, \omega_2)|^2, \quad \text{(III.3.54)}$$

(where the summation is performed over the nonredundant region of the bispectrum,[40] is chi-square distributed with p_{df} degrees of freedom and a *non-centrality parameter* (NCP) λ.[77] The NCP is a measure of skewness (i.e., the zero-lag third-order statistics[40]) of the distribution of S from a central chi-square distribution. The SL is the probability of false alarm, i.e., the probability that the data may actually have zero skewness when accepting the hypothesis that the data have nonzero skewness. If SL is large, then the hypothesis of zero skewness cannot be easily rejected, i.e., the estimations of bicoherence indexes would have low validity.

3.8.1.4   Distance Between Sample and Theoretical Inter-Quartile Range

Under the assumption of linearity, the bispectrum $C_3^x(\omega_1, \omega_2)$ is constant for all ω_1 and ω_2. Let:

$$\Lambda(\omega_1, \omega_2) \equiv \frac{1}{\sqrt{N^{1-2k}}} C_3^x(\omega_1, \omega_2), \qquad \text{(III.3.55)}$$

where k is a resolution parameter which lies between 0.5 and 1.0 (common value $k = 0.51$). The $\Lambda(\omega_1, \omega_2)$ are chi-square distributed with noncentrality parameter λ.[77] In order to accept the nonlinear hypothesis of the process, the *Sample Inter-Quartile Range* (SI-QR) (R) of the $\Lambda(\omega_1, \omega_2)$'s is estimated and compared to the *Theoretical Inter-Quartile Range* (TI-QR) (R') of a chi-squared distribution with two degrees of freedom and a noncentrality parameter λ. For a nonlinear process, the distance between the estimated SI-QR and TI-QR values is expected to be high enough.[77]

3.8.2   HOS-Based Features of Musical Lung Sounds

The lung sounds that are characterized as musical are the continuous adventitious sounds (CAS), i.e., wheezes (WZ), rhonchi (RH), and stridors (STR), described in Section 3.2.1.2. The musical character of these LS is reflected in the power spectrum through distinct frequency peaks.[11] Since the second-order statistics (autocorrelation) suppress any phase information, they fail to detect and characterize the nonlinear interactions of the distinct harmonics. Although spectral-based methodologies have been widely applied to frequency domain analysis of musical lung sounds (mainly wheezes),[11,78,79] they do not take into account the nonlinearity and non-Gaussianity of the analyzed processes. Since HOS preserve the phase character of the signals,[40] they could be used as a useful tool for the detection of nonlinearity and deviation from Gaussianity of the signals. In this way, the process of musical lung sounds is expanded to more accurate estimations of their true character. The application of the HOS-based parameters, described in Section 3.8.1, to musical lung sounds, for estimating their nonlinearities, follows.

3.8.2.1   Results from the Application of the HOS-Based Parameters on Real Data of Musical Lung Sounds

Figure III.3.16 illustrates two examples of the parametric bispectrum analysis of an (a) inspiratory stridor (ISTR) and (d) expiratory stridor (ESTR), recorded from an infant with croup.[80] A detailed discussion on experimental and implementation issues, along with additional experimental results can be found.[80]

The power spectrum of the analyzed ISTR is depicted in Figure III.3.16(b), from where three peaks (410.2 Hz, 820.4 Hz, and 1230.6 Hz) are easily recognized. The first two peaks (410.2 Hz and 820.4 Hz) are also evident in the power spectrum of the analyzed ESTR, shown in Figure III.3.16(e). The parametric bispectrum (Equation III.3.40), with an order of $p = 12$, of the ISTR and ESTR is shown in Figure III.3.16(c) and (e), respectively. From these figures, a sharp peak in the bifrequency domain, located at $(f_1, f_2) = (410.2\,\text{Hz}, 410.2\,\text{Hz})$, is evident both in the ISTR and ESTR cases. This sharp peak in the bispectrum domain indicates a strong QPC among the frequencies of the related frequency pair (f_1, f_2). In this case, $f_1 = f_2 = f_0 = 410.2\,\text{Hz}$, resulting in *quadratic phase self-coupling*.[40] This fact is also justified from the power spectrum of Figures III.3.16(b)

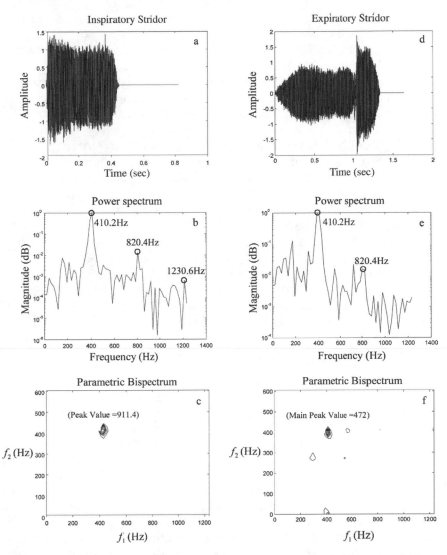

FIGURE III.3.16   Quadratic phase coupling among frequencies of an (a) inspiratory and (d) expiratory stridor, recorded from an infant with croup.[80] (b), (e) The estimated power spectrum for each case, respectively. (c), (f) The magnitude of the estimated parametric bispectrum (AR order $p = 12$) for each case, respectively.

and (e), where the three main peaks with the largest magnitude for the ISTR, and the two ones for the ESTR, are equal to $(f_0, 2f_0,$ and $3f_0)$, and $(f_0, 2f_0)$, respectively. Since the relevant BI has values greater than 1.0 (mean value equal to 1.25[80]), it is deduced that a strong QPC among these frequency harmonics is involved, resulting from a nonlinear production mechanism of stridors, and retained both in inspiratory and expiratory phases of the breathing cycle.

From the nonlinear analysis of the musical lung sounds, with respect to the HOS-based parameters mentioned in Section 3.8.1, new features emerge. In particular, based on the results presented in Reference 80, the high values (greater than or near 1) of $BI(\omega_1, \omega_2)$ indicate the existence of strong QPC. Furthermore, the values of SL always smaller than 0.05 justify the validity of the bicoherency estimates, while the values of S always high indicate the strong non-Gaussianity of the analyzed processes. Finally, the high values of skewness (mainly in inspiration), as measured by the NCP (λ) show a strong deviation from normality.

When *Monophonic Wheezes* (MWZ) are subjected to nonlinear analysis, the revealed frequency pair of QPC belongs in the low-frequency band. The fundamental frequency is evident in the inspiratory phase of breath cycle, while its first harmonic dominates in the expiratory phase. This fact leads to the conclusion that the monophonic wheeze consists of a single note or a single tonality, established by the octaves of the fundamental frequency, due to a nonlinear mechanism.

In the case of analyzed *polyphonic wheezes*, the revealed frequency pair of QPC belongs in a higher frequency band than that of monophonic wheezes. In the inspiratory phase, pairs of high frequencies perform QPC, while in the expiratory phase, pairs with their submultiples perform QPC. The harmonics that are involved result in a polyphonic chord, since apart from the fundamental frequency (f_0), the octave ($2f_0$) and the fifth of the chord ($3f_0$), perform QPC. This result is consistent with the accepted theory that polyphonic wheezes are made up of several dissonant notes starting and ending simultaneously, like a chord.[81] As is well known, stridor intensity is the only thing that distinguishes a stridor from a monophonic wheeze.[81] From the results of nonlinear analysis,[80] it can be seen that $(f_1, f_2)|_{inspiratory\ MWZ} = (1/5)(f_1, f_2)|_{inspiratory\ ST}$, justifying their relationship, but the degree of nonlinearity is much larger in stridor than in monophonic wheeze. From the values of the $CBI(\omega_1, \omega_2)$,[80] i.e., $|CBic^{MWZ-ST}(78.12, 78.12)| = 0.43\ |CBic^{MWZ-ST}(390.5, 390.5)| = 1.21$ estimated by Equation III.3.53, a common nonlinear production mechanism of MW and ST is implied, which becomes more evident in the high-frequency QPC pair.

3.8.3 HOS-Based Features of Heart Sounds

Spectral analysis and parametric modeling have been widely applied to HS analysis.[82–85] Nevertheless, methods of HS analysis based on second-order statistics ignore the nonlinearity and non-Gaussianity of the analyzed processes, since they suppress any phase information.[40] This information is recovered in the HOS domain; thus, the HOS-based parameters described in Section 3.8.1 provide new features for HS identification and characterization of the associated heart pathology.

3.8.3.1 Results from the Bispectrum-Based Analysis of Real Data of HS

An example of the analysis[86] of HS, recorded from healthy subjects and patients with different cardiac pathologies, with one HOS-based parameter, i.e., bispectrum [see Equation III.3.49], is illustrated in Figure III.3.17. Discussion on experimental and implementation issues, along with additional experimental results, can be found in Reference 86. In particular, Figure III.3.17(a1) illustrates the first (S_1) and second (S_2) HS recorded from a normal subject, while from the latter, it is clear that the two HS have low-frequency components ($< 200\ Hz$) and exhibit distinct frequency peaks at ($f_1 \cong 70\ Hz$, $f_2 \cong 70\ Hz$).

Sometimes, the acoustic impression of similar HS, i.e., heart murmurs in patients with mitral regurgitation and heart murmurs in those with tricuspid regurgitation, confuses the physician in their interpretation.[87] Figures III.3.17(b1) and (c1) present such a case. More specifically, Figures III.3.17(b1) illustrates a time section of 0.512 sec of a murmur recorded from a patient with mitral regurgitation, while Figure III.3.17(b2) depicts the corresponding bispectrum. Furthermore, Figure III.3.17(c1) shows a time section of 0.512 sec of a murmur recorded from a patient with tricuspid regurgitation, while Figure III.3.17(c2) depicts the corresponding bispectrum.

From the comparison of the two estimated bispectra depicted in Figures III.3.17(b2) and (c2), it is apparent that, despite the similarity which might exist in the HS auscultation (note the similarity in the time domain), the structural differences in the bispectrum domain are apparent, providing a discrimination feature. In addition, the severity of a cardiac dysfunction is reflected through the changes in the bifrequency domain when different stages of the cardiac pathology are examined. This case is shown in Figures III.3.17(d1) and (e1), where a time section of 0.512 sec of a murmur

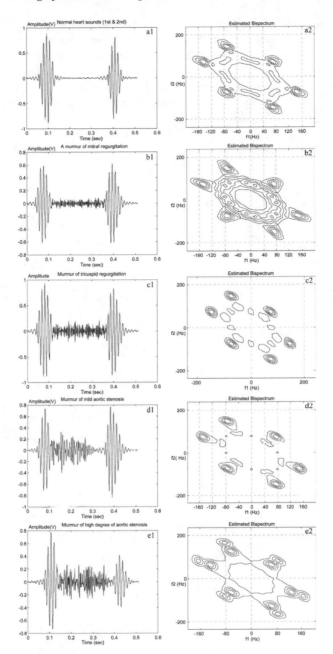

FIGURE III.3.17   Examples from the bispectral analysis[86] of HS: (a1) a time section of 0.512 sec of HS recorded from healthy subjects (normal HS). (a2) the corresponding estimated bispectrum; (b1) a time section of 0.512 sec of a murmur recorded from a patient with mitral regurgitation; (b2) the corresponding estimated bispectrum; (c1) a time section of 0.512 sec of a murmur recorded from a patient with tricuspid regurgitation; (c2) the corresponding estimated bispectrum; (d1) a time section of 0.512 sec of a murmur recorded from a patient with mild aortic stenosis; (d2) the corresponding estimated bispectrum. (e1) a time section of 0.512 sec of a murmur recorded from a patient with a high degree of aortic stenosis; (e2) the corresponding estimated bispectrum.

recorded from a patient with mild aortic stenosis, and a time section of 0.512 sec of a murmur recorded from a patient with a high degree of aortic stenosis are illustrated, respectively. The corresponding estimated bispectra are depicted in Figures III.3.17(d2) and (e2), respectively. From the comparison in the bispectrum domain, it is clear that, in the case of severe aortic stenosis [Figure III.3.17(e2)], new peaks in the bifrequency domain are generated, compared to the bispectrum of mild aortic stenosis [Figure III.3.17(d2)]. In this way, the bispectrum serves as a diagnostic feature that distinguishes among different stages of aortic stenosis.

The results of bispectral analysis of HS presented above, along with those included in,[86] indicate that *an efficient discrimination of cardiac pathologies could be established in the bispectrum domain, which can be used for further elicitation of the diagnostic character of heart sounds.*

3.8.4   HOS-Based Features of Bowel Sounds

The HOS-based parameters described in Section 3.8.1 are also used in the case of bowel sounds analysis for feature extraction.[88] An example from these results is described below.

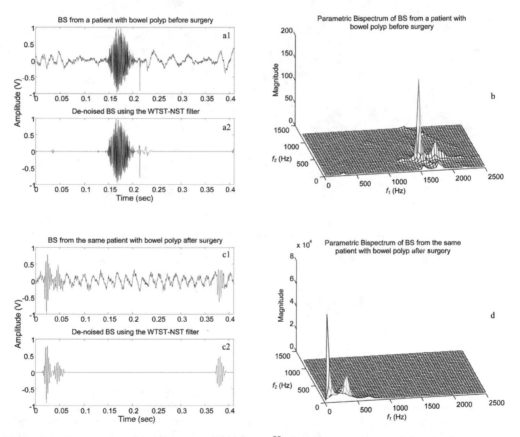

FIGURE III.3.18   Examples from the bispectral analysis[88] of BS recorded from a patient with bowel polyp: (a1) the BS recorded from the patient *before* the surgery; (a2) the de-noised (using the WTST-NST filter[49]) BS depicted in (a1); (b) corresponding to the de-noised BS estimated parametric bispectrum with $p = 12$; (c1) the BS recorded from the patient *after* the polypectomy; (c2) the de-noised (using the WTST-NST filter[49]) BS depicted in (c1); (d) corresponding to the de-noised BS estimated parametric bispectrum with $p = 12$.

3.8.4.1 Results from the Application of the HOS-Based Parameters on Real Data of Bowel Sounds

An example of the bispectral analysis[88] of BS is illustrated in Figure III.3.18. Discussion on experimental and implementation issues, along with additional experimental results, can be found in Reference 88. The case described in Figure III.3.18 shows the ability of the bispectrum to differentiate the BS recorded from a patient with a bowel polyp before and after the polypectomy, hence to reflect the anatomical changes of the bowel. In particular, Figures III.3.18(a1) and (a2) show the recorded and de-noised (using the WTST-NST filter[49]) BS, respectively, recorded from the patient *before* the surgery. Figure III.3.18(b) depicts the corresponding estimated parametric bispectrum with $p = 12$ (see also Equation III.3.40). In addition, Figures III.3.18(c1) and (c2) show the recorded and de-noised (using the WTST-NST filter[49]) BS, respectively, recorded from the patient *after* the polypectomy, while Figure III.3.18(d) illustrates the corresponding estimated parametric bispectrum with $p = 12$.

From the comparison of Figures III.3.18(c) and (d), it is clear the extraction of the polyp produces a shift of the QPC frequency pairs, seen in the bispectrum domain, to the low-frequency band. From the analysis results described in Reference 88, the QPC frequency pairs after the polypectomy coincide with those corresponding to BS from controls (healthy subjects). This indicates that the polypectomy was successfully performed, with the success reflected in the HOS-based parameters.

Results from the analysis[88] with the HOS-based parameters of additional BS indicate that for BS recorded from controls, the median estimated QPC harmonic pair is located in the area of 175 Hz with mean BI greater than 1.0 exhibiting a strong deviation from normality and linearity; for BS recorded from patients with diverticular disease, the median estimated QPC harmonic pair is located in the area of 254 Hz with mean BI near 0.0 (0.1), indicating almost an absence of QPC, but with increased non-Gaussianity and decreased nonlinearity; for BS recorded from patients with ulcerative colitis, the median estimated QPC harmonic pair is located in the area of 351 Hz with mean BI greater than 1.0, exhibiting a strong deviation from normality and linearity; and for BS recorded from patients with irritable bowel syndrome, the median estimated QPC harmonic pair is located in the area of 273 Hz with mean BI below 0.5 (0.3), indicating a weak QPC, but with increased non-Gaussianity and decreased nonlinearity.[88] In all cases, it is found that SL $\ll 0.0001$,[88] justifying the correctness of the adopted characterization of the processes. These results show that HOS provide a new perspective in the computerized analysis of BS, defining an enhanced field of new features which reinforce the diagnostic character of BS.

3.9 Classification of Lung and Bowel Sounds Using *Alpha*-Stable Distributions and Higher-Order Crossings

In this section, the classification of lung and bowel sounds based on the *alpha*-stable distributions parameters, discussed in Section 3.7.2, and *Higher-Order Crossings* (HOC), is presented.

3.9.1 Classification of Crackles and Artifacts Using $S\alpha S$ Distribution-Based Parameters

In Section 3.6, three methods for separating the discontinuous adventitious sounds (DAS) from the vesicular lung sounds (VLS), considered as background noise, were presented. These methods provide a nonstationary and stationary part that correspond to DAS and VLS, respectively. Nevertheless, they do not provide with a classification of the sounds included in the nonstationary output.

Moreover, in some cases, the input signal contains *artifacts* (A), which might be present in the nonstationary output. In order to overcome this problem and result in a successful classification of the signals included in the nonstationary output, the *alpha* parameter, described in Section 3.7, is employed. This is due to the direct relationship of the *alpha* values and the degree of impulsiveness of the analyzed process. In particular, considering the non-Gaussianity and the impulsiveness of crackles and artifacts, we could model them as $S\alpha S$ distributed processes, in the way it is described Section 3.7.2. By estimating the characteristic exponent α for each category (crackles, background noise, and A) using the $\log|S\alpha S|$ method,[68] a classification criterion could be established, based on the estimated *alpha* values.

3.9.1.1 Results from the Classification of Crackles and Artifacts Using the *Alpha* Parameter

The experimental results[89] depicted in Figure III.3.19 prove that the values of the characteristic exponent α could be used as a classification index for classifying coarse crackles (CC), fine crackles (FC), background noise (BN), and A, based on their differences in the degree of impulsiveness. Figure III.3.19(a) depicts a section of recorded LS that contain CC, FC, BN, and A. Details on experimental and implementation issues can be found in Reference 89. The arrowheads in Figure III.3.19(a) indicate the position of the nonstationary events of interest, marked and categorized by a physician. Nevertheless, this categorization includes ten cases of marked but not categorized events, marked as 'A?'.

Figures III.3.19(b), (c), and (d) show the estimated *alpha* values from the $S\alpha S$ distribution-based modeling of the de-noised with the WTST-NST filter[45] CC, FC, and A, respectively. From these figures it can be noticed that the FC follow approximately the Cauchy distribution (mean $\alpha = 0.92$, which is close to $\alpha = 1.0$ that holds in the case of the Cauchy distribution), while the CC deviate both from Cauchy and Gaussian distributions (mean $\alpha = 1.33$. The artifacts have high impulsiveness (mean $\alpha = 0.0218$); thus, events that could not be classified by the doctor (marked as A?) are clearly classified by the proposed method, as is shown in Figure III.3.19(e). The events that resemble the VLS are seen as background noise and are modeled as Gaussian processes (values of $\alpha > 1.8$ or very close to 2.0, values that hold in the case of the Gaussian distribution).

From the above analysis it is evident that the *alpha* parameter could be used as an efficient classification criterion for the categorization of impulsive events; hence, it could serve as an extension to the DAS detectors described in Section 3.6, forming an integrated automated detection and classification tool for DAS diagnostic analysis.

3.9.2 Classification of Crackles, Bronchial, and Bowel Sounds Using Higher-Order Crossings

In this subsection, a HOC-based classification of lung and bowel sounds is described. For the nonfamiliarized reader, the subsection begins with a brief mathematical description of HOC-based parameters.

3.9.2.1 Definition

The zero-crossing counts resulting from the application of a bank of filters to a time series are called *higher-order crossings* (HOC).[90] The latter constitute a domain by itself, which 'sits' between the time and spectral domains. Parameters and procedures defined in this domain could be used for the classification analysis described in this subsection. A short description follows.

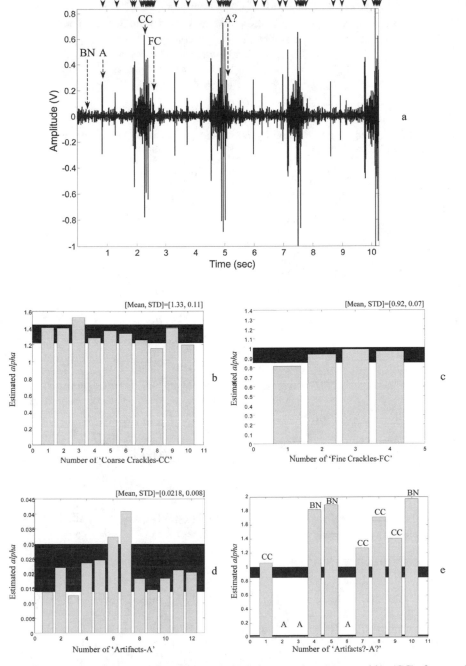

FIGURE III.3.19 (a) A section 10.244 sec of recorded LS that contain coarse crackles (CC), fine crackles (FC), background noise (BN), and artifacts (A).[89] The arrowheads in the input tracing indicate the position of the nonstationary events of interest, marked and categorized by a physician. Unclearly categorized events are marked as 'A?'. The five arrows indicate an example of each category. (b), (c), and (d) show the estimated *alpha* values, along with their mean and standard deviation from $S\alpha S$ the distribution-based modeling of the de-noised with the WTST-NST filter[45] CC, FC, and A, respectively. (e) Results from the classification of ten uncategorized events (A?), using the mean *alpha* values found in (b)–(d).

3.9.2.2 Zero-Crossings and kth-Order Zero-Crossings

Let $Z_1, Z_2, \ldots, Z_N$ be a zero-mean stationary real-valued time series. The zero-crossing count in discrete time is defined as the number of symbol changes in the corresponding clipped binary stationary time series $X_1, X_2, \ldots, X_N$, defined by the nonlinear transformation:[90]

$$X_t = \begin{cases} 1 & \text{if } Z_t \geq 0 \\ 0 & \text{if } Z_t < 0 \end{cases}. \tag{III.3.56}$$

The number of *zero-crossings*, denoted by D, is defined in terms of $\{X_t\}$ as follows:

$$D = \sum_{t=2}^{N} [X_t - X_{t-1}]^2. \tag{III.3.57}$$

The number of kth-order zero-crossings (in general HOC) D_k, is defined as follows[90]:

$$D_k = NZC\{\nabla^k Z_t\}, \tag{III.3.58}$$

where $NZC\{\cdot\}$ denotes the calculation of the number of zero-crossings; the kth-order difference operator is given by:

$$\nabla^k Z_t = \sum_{j=0}^{k} \binom{k}{j} (-1)^j Z_{t-j}. \tag{III.3.59}$$

3.9.2.3 Distance from White Gaussian Noise or White Noise Test

In the discrimination procedure between two N-sample signals, the measurement of their distance from *White Gaussian Noise* (WGN) could be employed, for which the expected simple HOC are given by the following equation:[90]

$$E[D_k] = (N-1) \left[\frac{1}{2} + \frac{1}{\pi} \sin^{-1} \left(\frac{k-1}{k} \right) \right] \pm 1.96(N-1)^{1/2} \left\{ \frac{1}{4} - \left[\frac{1}{\pi} \sin^{-1} \left(\frac{k-1}{k} \right) \right]^2 \right\}^{1/2}. \tag{III.3.60}$$

This measurement is also referred to as *White Noise Test* (WNT). For a detailed discussion on HOC and their applications, the reader should refer to Reference 90.

3.9.2.4 Classification of Crackles Using HOC

The diversity of lung sounds characteristics yields the grounds for developing efficient discrimination tools in the HOC domain, exploring the LS features by processing their time domain oscillations. The projection of the LS recordings in the HOC domain reveals a new field in their discrimination analysis. In this subsection, the results from the application of two HOC-based discrimination indexes, i.e., the WNT and the *HOC-Scatter Plot* (HOC-SP), which is a scattergram that depicts a pair of simple HOC (D_j, D_k),[90] on crackles classification are described.

The WNT is employed for measuring the distance of the two categories of crackles (i.e., FC and CC) from the WGN. In this way, the two categories are indirectly compared to each other. The ith order of HOC that results in the maximum distance between the two categories, defined as the maximum distance from the lower curve of one category to the upper curve of the other at the ith order, provides with a linear discrimination in the HOC domain. In particular, by plotting D_i vs. $D_{i\pm m}$, where m usually equals to 1, 2, or 3 (for $i \geq 5$), the resultant HOC-SP reveals two *linearly* separated classes. Due to the linear character of the separation, a simple decision rule of the form $w_1 D_i + w_2 D_{i-1} + w_3 = 0$, where $w_i (i = 1, 2, 3)$ are the coefficients of the line equation found

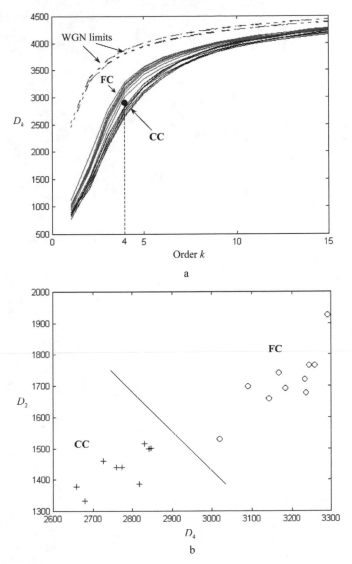

FIGURE III.3.20   Discrimination of coarse from fine crackles using HOC-based analysis. (a) The white noise test. The fourth HOC order provides the maximum 'opened eye.' (b) The HOC-SP (D_4, D_2) provides linear discrimination between CC and FC.

(e.g., by a simple single-layer perceptron neural network[92]) could be used for decision upon class membership.

Results from the HOC-based analysis of crackles,[91] initially de-noised with the WTST-NST filter,[45] are illustrated in Figure III.3.20. Discussion on experimental and implementation issues, along with additional experimental results, can be found in Reference 91. From Figure III.3.20(a) it is clear that the order, which provides the maximum distance between FC and CC, the so-called 'opened-eye,' equals 4.0. This distance is reflected in the corresponding HOC-SP (D_4, D_2) of Figure III.3.20(b), where FC and CC are concentrated in two distant, linearly separated classes.

From the above results, it is apparent that the HOC-based indexes result in fast, easy, and efficient discrimination of crackles, providing with a simple decision rule for categorizing FC and CC.

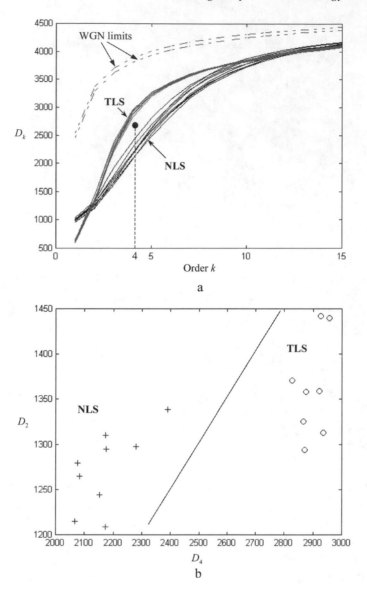

FIGURE III.3.21   Discrimination of normal lung sounds (NLS) from abnormal tracheobronchial lung sounds (TLS), using HOC-based analysis. (a) The white noise test. The fourth HOC order provides the maximum 'opened eye.' (b) The HOC-SP (D_4, D_2) provides linear discrimination between NLS and TLS.

3.9.2.5   Classification of Normal and Abnormal LS Using HOC

The analysis described in the previous subsection has been applied[91] to another pair of LS with similar acoustic impression (i.e., the normal LS (NLS)) and the abnormal bronchial LS heard over the trachea [tracheobronchial LS (TLS)]. Results from the HOC-based analysis of these categories of LS[91] are shown in Figure III.3.21. Discussion on experimental and implementation issues along with additional experimental results can be found in Reference 91. From Figure III.3.21(a) it is clear that the order, which provides the maximum 'opened eye' between NLS and TLS equals 4.0. This distance is reflected in the corresponding HOC-SP (D_4, D_2) of Figure III.3.21(b), where NLS and TLS are concentrated in two distant, linearly separated classes, as in the case of crackles.

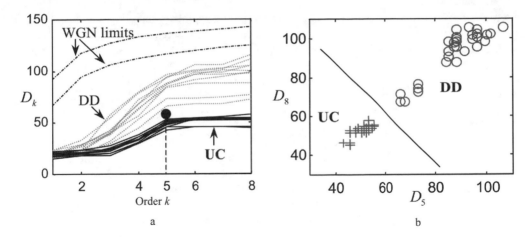

FIGURE III.3.22 Discrimination of bowel sounds recorded from patients with ulcerative colitis (UC), from the ones recorded from patients with diverticular disease (DD), using HOC-based analysis. (a) The white noise test. The fifth HOC order provides the maximum 'opened eye.' (b) The HOC-SP (D_5, D_8) provides linear discrimination between BS from UC and DD.

From the comparison of Figures III.3.20 and III.3.21, it is clear that the HOC-based analysis results in more distant classes when NLS and TLS rather than CC and FC are analyzed. Nevertheless, in all cases, a linear separation is feasible.

3.9.2.6 Classification of Bowel Sounds Using HOC

By applying the HOC-based analysis followed in LS to bowel sounds, similar results are obtained.[93] The issue in this case is to correctly classify de-noised BS, recorded from patients with different pathologies of the large bowel, such as ulcerative colitis (UC), diverticular disease (DD), etc., in order to increase the efficacy of BS analysis. Results from the HOC-based analysis of previously de-noised with the WTST-NST filter[49] BS, recorded from patients with UC and DD,[93] are shown in Figure III.3.22. Discussion on experimental and implementation issues, along with additional experimental results, can be found in Reference 93. From Figure III.3.22(a) it is clear that the order, which provides the maximum 'opened eye' between UC and DD, equals 5.0. This distance is reflected the corresponding HOC-SP (D_5, D_8) of Figure III.3.21(b), where UC and DD are concentrated in two linearly separated classes, as in the case of lung sounds.

From the HOC-based analysis of LS and BS, described in the preceding three subsections, it is apparent that the HOC-based indexes (i.e., the WNT and HOC-SP) act as efficient classifiers and group the examined cases into linearly separated classes, providing an easy decision rule for successful categorization of BioS, and, consequently, identification of the associated pathologies.

3.10 Epilogue

The methods discussed in this chapter adopt the same endeavor: exploitation and surfacing of the valuable diagnostic information that exists in BioS. Under this common aim, the different approaches of de-noising, detection, modeling, feature extraction, and classification, based on artificial intelligence and advanced signal processing tools described in this chapter, introduce new pathways of BioS analysis and give rise to their objective evaluation. Apart from this, they reveal the call for a common interest in interdisciplinary undertakings, both by biomedical engineers and clinicians. This should not go unnoticed; although it is unknown when (or even whether) traditional clinicians will cope with alternative approaches in everyday clinical practice, the results shown here are the outcome of joint efforts and scientific resources from both engineers and clinical medical doctors. It is well understood that transition from medical doctors' profession symbol, i.e., stethoscope, to an electronic device that would include the approaches described here in an integrated form, has a long way to go. This chapter contributes to that direction, offering the reader a new way to embark on a noninvasive and efficient evaluation of bioacoustic signals.

References

1. ⟨URL:http://www.dictionary.com/cgi-bin/dict.pl?term= bioacoustics⟩, *The American Heritage Dictionary of the English Language*, 4th ed., Houghton Mifflin Company, 2000.
2. McKusick, V.S., *Cardiovascular Sound in Health and Disease*, Williams & Wilkins, Baltimore, 1958, 3.
3. Rapoport, J., Laënnec and the discovery of auscultation, *Israel J. Med.*, 22, 597–601, 1986.
4. McKusick, V.S., *Cardiovascular Sound in Health and Disease*, Williams & Wilkins, Baltimore, 1958, 13.
5. Kraman, S.S., Vesicular (normal) lung sounds: How are they made, where do they come from and what do they mean? *Semin. Respir. Med.*, 6, 183–191, 1985.
6. Robertson, A.J., Rales, ronchi, and Laënnec, *Lancet*, 1, 417–423, 1957.
7. Kraman, S.S., *Lung Sounds: An Introduction to the Interpretation of Auscultatory Findings*, American College of Chest Physicians, Northbrook, IL, 1983, 14–21.
8. Murphy, R.L.H., Discontinuous adventitious lung sounds, *Semin. Respir. Med.*, 6, 210–219, 1985.
9. Cugell, D.W., Lung sound nomenclature, *Am. Rev. Respir. Dis.*, 136, 1016, 1987.
10. Gavriely, N. and Cugell, D.W., *Breath Sounds Methodology*, CRC Press, Boca Raton, FL, 1994, chap. 2.
11. Gavriely, N., Palti, Y., and Alroy, G., Spectral characteristics of normal breath sounds, *J. Appl. Physiol.*, 50, 307–314, 1981.
12. Murphy, R.L.H., Holford, S.K., and Knowler, W.C., Visual lung-sound characterization by time-expanded wave-form analysis, *New Engl. J. Med.*, 296, 968–971, 1978.
13. Gavriely, N. and Cugell, D.W., *Breath Sounds Methodology*, CRC Press, Boca Raton, FL, 1994, 8.
14. Erickson, B., *Heart Sounds and Murmurs: A Practical Guide*, 3rd ed., Mosby-Year Book, Inc., 1997, 2.
15. Cabot, R.C. and Dodge, H.F., Frequency characteristics of heart and lung sounds, *JAMA*, 84, 1793–1795, 1925.
16. Gavriely, N. and Cugell, D.W., *Breath Sounds Methodology*, CRC Press, Boca Raton, FL, 1994, 116.
17. Yoganathan, A.P., Gupta, R., Miller, J.W., Udwardia, F. E., Corcoran, W.H., Sarma, R., Johnson, J.L., and Bing, R.J., Use of the fast Fourier transform for frequency analysis of the first heart sound in normal man, *Med. Biol. Eng.*, January, 69–81, 1976.
18. ⟨URL: http://www.graylab.ac.uk/cgi-bin/omd?bowel + sounds⟩, *On-line Medical Dictionary*, Academic Medical Publishing & CancerWEB, 1997.
19. Cannon, W.B., Auscultation of the rhythmic sounds produced by the stomach and intestine, *Am. J. Physiol.*, 13, 339–353, 1905.
20. Horn, G.E. and Mynors, J.M., Recording the bowel sounds, *Med. Biol. Eng.*, 4, 205–208, 1966.
21. Watson, W.C. and Knox, E.C., Phonoenterography: the recording and analysis of BS, *Gut*, 8, 88–94, 1967.
22. Dalle, D., Devroede, G., Thibault, R., and Perrault, J., Computer analysis of BS, *Comp. Biol. Med.*, 4, 247–256, 1975.

23. Politzer, J.P., Devroede, G., Vasseur, C., Gerand, J., and Thibault, R., The genesis of bowels sounds: influence of viscus and gastrointestinal content, *Gastroenterology*, 71, 282–285, 1976.

24. Weidringer, J., Sonmoggy, S., Landauer, B., Zander, W., Lehner, F., and Blumel, G., BS recording for gastrointestinal motility, in *Motility of the Digestive Tract*, Weinbeck M, Ed., Raven Press, New York, 1982, 273–278.

25. Michael, S. and Redfern, M., Computerised Phonoenterography: The clinical investigation of a new system, *J. Clin. Gastroenterol.*, 18(2), 139–144, 1994.

26. Yoshino, H., Abe, Y., Yoshino, T., and Ohsato, K., Clinical application of spectral analysis of BS in intestinal obstruction, *Dis. Col. Rect.*, 33(9), 753–757, 1990.

27. Sandler, R.H., Mansy, H.A., Kumar, S., Pandya, P., and Reddy, N., Computerised analysis of bowel sounds in human subjects with mechanical bowel obstruction vs. ileus, *Gastroenterology*, 110(4), A752, 1993.

28. Bray, D., Reilly, R.B., Haskin, L., and McCormack, B., Assessing motility through abdominal sound monitoring, in *Proc. 19th Int. Conf. IEEE/EMBS*, Myklebust, B., and Myklebust, J., Eds., IEEE Press, Chicago, IL, 1997, 2398–2400.

29. Craine, B.L., Silpa, M.L., and O'Toole, C.J., Enterotachogram analysis to distinguish irritable bowel syndrome from Crohn's disease, *Dig. Dis. and Sciences*, 46(9), 1974–1979, 2001.

30. Iyer, V.K., Ramamoorthy, P.A., and Ploysongsang, Y., Quantification of heart sounds interference with lung sounds, *J. Biomed. Eng.*, 11, 164–165, 1989.

31. McLellan, N.J. and Barnett, T.G., Cardiorespiratory monitoring in infancy with acoustic detector, *Lancet*, Dec., 1397–1398, 1983.

32. Tal, A., Sanchez, I., and Pasterkamp, H., Respisonography in infants with acute bronchiolitis, *Am. J. Dis. Child.*, 145, 1405–1410, 1991.

33. Iyer, V.K., Ramamoorthy, P.A., Fan, H., and Ploysongsang, Y., Reduction of heart sounds from lung sounds by adaptive filtering, *IEEE Trans. Biomed. Eng.*, 33(12), 1141–1148, 1986.

34. Ploysongsang, Y., Iyer, V.K., and Ramamoorthy, P.A., Characteristics of normal lung sounds after adaptive filtering, *Am. Rev. Respir. Dis.*, 139, 951–956, 1989.

35. Widrow, B., Glover, J.R., McCool, J., Kaunitz, M.J., Williams, C.S., Hearn, R.H., Zeidler, J.R., Dong, E., and Goodlin, R. C., Adaptive noise canceling: principles and applications, *Proc. IEEE*, 63(12), 1692–1716, 1975.

36. Kompis, M. and Russi, E., Adaptive heart-noise reduction of lung sounds recorded by a single microphone, in *Proc. 14th Int. Conf. IEEE/EMBS*, IEEE Press, Paris, 1992, 2, 691–692.

37. Shiryaev, A.N., Some problems in the spectral theory of higher-order moments I, *Theory Probl. Appl.*, 5, 265–284, 1960.

38. Brillinger, D.R., An introduction to polyspectra, *Ann. Math. Statist.*, 36, 1351–1374, 1965.

39. Rosenblatt, M., Cumulants and cumulant spectra, in *Handbook of Statistics*, Brillinger, D.R. and Krishnaiah P.R., Eds., Elsevier Science Publishers B.V., Amsterdam, 1983, Volume 3, 369–382.

40. Nikias, C.L. and Petropulu, A.P., *Higher-Order Spectra Analysis: A Nonlinear Signal Processing Framework*, 1st ed., Nikias, C. L., Ed., Prentice-Hall, Englewood Cliffs, NJ, 1993, chaps. 1–2.

41. Hadjileontiadis, L.J. and Panas, S.M., Adaptive reduction of heart sounds from lung sounds using fourth-order statistics, *IEEE Trans. Biomed. Eng.*, 44(7), 642–648, 1997.

42. Daubechies, I., Orthonormal bases of compactly supported wavelets, *Commun. Pure Appl. Math.*, 41, 909–996, 1988.

43. Veterli, M. and Kovačević, J., *Wavelets and Subband Coding*, Prentice-Hall, Englewood Cliffs, NJ, 1995, 201–298.

44. Mallat, G., A theory for multiresolution signal decomposition: the wavelet representation, *IEEE Trans. Patt. Anal. Machine Intell.*, 11(7), 674–693, 1989.

45. Hadjileontiadis, L.J. and Panas, S.M., Separation of discontinuous adventitious sounds from vesicular sounds using a wavelet-based filter, *IEEE Trans. Biomed. Eng.*, 44(12), 1269–1281, 1997.

46. Hadjileontiadis, L.J. and Panas, S.M., A wavelet-based reduction of heart sounds noise from lung sounds, *Int. J. Med. Inf.*, 52, 183–190, 1998.

47. Coifman, R. and Wickerhauser, M.V., Adapted waveform 'de-noising' for medical signals and images, *IEEE Eng. Med. Biol. Soc.*, 14(5), 578–586, 1995.

48. Mansy, H.A. and Sandler, R.H., Bowel-sound signal enhancement using adaptive filtering, *IEEE Eng. Med. Biol.*, 16(6), 105–117, 1997.

49. Hadjileontiadis, L.J., Liatsos, C.N., Mavrogiannis, C. C., Rokkas, T.A., and Panas, S.M., Enhancement of bowel sounds by wavelet-based filtering, *IEEE Trans. Biomed. Eng.*, 47(7), 876–886, 2000.

50. Loudon, R. and Murphy, R.L., Jr., Lung sounds, *Am. Rev. Respir. Dis.*, 130, 663–673, 1984.

51. Ono, M., Arakawa, K., Mori, M., Sugimoto, T., and Harashima, H., Separation of fine crackles from vesicular sounds by a nonlinear digital filter, *IEEE Trans. Biomed. Eng.*, 36(2), 286–291, 1989.

52. Hadjileontiadis, L.J. and Panas, S.M., Nonlinear separation of crackles and squawks from vesicular sounds using third-order statistics, in *Proc. 18th Int. Conf. IEEE/EMBS*, IEEE Press, Amsterdam, 1996, 5, 2217–2219.

53. Cohen, A., Signal processing methods for upper airway and pulmonary dysfunction diagnosis, *IEEE Eng. Med. Biol. Mag.*, 3, 72–75, 1990.

54. Tolias, Y.A., Hadjileontiadis, L.J., and Panas, S.M., A fuzzy rule-based system for real-time separation of crackles from vesicular sounds, in *Proc. 19th Int. Conf. IEEE/EMBS*, IEEE Press, Chicago, IL, 1997, 3, 1115–1118.

55. Tolias, Y.A., Hadjileontiadis, L.J., and Panas, S.M., Real-time separation of discontinuous adventitious sounds from vesicular sounds using a fuzzy rule-based filter, *IEEE Trans. Biomed. Eng.*, 2(3), 204–215, 1998.

56. Zadeh, L.A., Fuzzy sets, *Inf. Contr.*, 8, 338–353, 1965.

57. Zadeh, L.A., Outline of a new approach to the analysis of complex systems and decision processes, *IEEE Trans. Syst., Man, Cyber.*, 3, 28–44, 1973.

58. Takagi, T. and Sugeno, M., Fuzzy identification of systems and its application to modeling and control, *IEEE Trans. Syst., Man, Cyber.*, 15, 116–132, 1985.

59. Jang, J.S.R., ANFIS: Adaptive network-based fuzzy inference system, *IEEE Trans. Syst., Man, Cyber.*, 23, 665–685, 1993.

60. Chen, S., Cowan, C.F.N., and Grant, P.M., Orthogonal least squares learning algorithm for radial basis functions networks, *IEEE Trans. Neural Networks*, 2, 302–309, 1991.

61. Mastorocostas, P.A., Tolias, Y.A., Theocharis, J.B., Hadjileontiadis, L.J., and Panas, S.M., An orthogonal least squares-based fuzzy filter for real-time analysis of lung sounds, *IEEE Trans. Biomed. Eng.*, 47(9), 1165–1176, 2000.

62. Iyer, V.K., Ramamoorthy, P.A., and Ploysongsang, Y., Autoregressive modeling of lung sounds: Characterization of source and transmission, *IEEE Trans. Biomed. Eng.*, 36(11), 1133–1137, 1989.

63. Hadjileontiadis, L.J. and Panas, S.M., Autoregressive modeling of lung sounds using higher-order statistics: Estimation of source and transmission, in *Proc. IEEE Signal Processing Workshop on Higher-Order Statistics '97 (SPW-HOS '97)*, Petropulu, A., Ed., IEEE Signal Processing Society, Banff, 1997, 4–8.

64. Swami, A., Mendel, J.M., and Nikias, C.L., *Higher-Order Spectral Analysis Toolbox*, 3rd ed., The Mathworks, Inc., Natick, 1998, chap. 1.

65. Swami, A., System Identification Using Cumulants, Ph.D. dissertation, University of South California, 1988, 107–108.

66. Giannakis, G.B. and Mendel, J.M., Cumulant-based order determination of non-Gaussian ARMA models, *IEEE Trans. Acoust., Speech, and Sign. Proc.*, 38(8), 1411–1423, 1990.

67. Hadjileontiadis, L.J., Analysis and Processing of Lung Sounds Using Higher-Order Statistics-Spectra and Wavelet Transform, Ph.D. dissertation, Aristotle University of Thessaloniki, 1997, 139–175.

68. Nikias, C.L. and Shao Min, *Signal Processing with Alpha- Stable Distributions and Applications*, 1st ed., Wiley & Sons, Inc., New York, 1995, chaps. 1–7.

69. Feller, W., *An Introduction to Probability Theory and Its Applications*, Wiley & Sons, Inc., New York, 1966, volume II.

70. Granger, W. and Orr, D., Infinite variance and research strategy in time series analysis, *J. Am. Statist. Soc.*, 67, 275–285, 1972.

71. McCulloch, J.H., Simple consistent estimators of stable distribution parameters, *Commun. Statist. Simula.*, 15(4), 1109–1136, 1986.

72. Koutrouvelis, I.A., An iterative procedure for the estimation of the parameters of stable laws, *Commun. Statist. Simula.*, 10(1), 17–28, 1981.

73. Kanter, M. and Steiger, W.L., Regression and autoregression with infinite variance, *Adv. Appl. Prob.*, 6, 768–783, 1974.

74. Hadjileontiadis, L.J. and Panas, S.M., On modeling impulsive bioacoustic signals with symmetric α-Stable distributions: Application in discontinuous adventitious lung sounds and explosive bowel sounds, in *Proc. 20th Int. Conf. IEEE/EMBS*, IEEE Press, Hong Kong, 1998, 20(1), 13–16.

75. Elgar, S. and Guza, R.T., Statistics of bicoherence, *IEEE Trans. ASSP*, 36(10), 1667–1668, 1988.

76. Marple, S.L., Jr., *Digital Spectral Analysis with Applications*, Prentice-Hall, Inc., New Jersey, 1987, chap. 8.

77. Hinich, M.J., Testing for Gaussianity and linearity of a stationary time series, *J. Time Series Analysis*, 3, 169–176, 1982.

78. Chowdhury, S.K. and Majumder, A.K., Digital spectrum analysis of respiratory sound, *IEEE Trans. Biomed. Eng.*, 28(11), 784–788, 1981.

79. Fenton, T.R., Pasterkamp, H., Tal, A., and Chernick, V., Automated spectral characterization of wheezing in asthmatic children, *IEEE Trans. Biomed. Eng.*, 32(1), 50–55, 1985.

80. Hadjileontiadis, L.J. and Panas, S.M., Nonlinear analysis of musical lung sounds using the bicoherence index, in *Proc. 19th Int. Conf. IEEE/EMBS*, IEEE Press, Chicago, IL, 1997, 3, 1126–1129.

81. Lehrer, S., *Understanding Lung Sounds*, 2nd ed., W.B. Saunders Co., Philadelphia, 1993, chap. 5.

82. Rosen, R.M., Shakkottai, B.S., Parthasarathy, P., Turner, A.F., Blankenhorn, D.H., and Roschke, E.J., Phonoangiography by autocorrelation, *Circulation*, 55(4), 626–633, 1977.

83. Stein, P.D., Sabbah, H.N., Lakier, J.B., Kemp, S.R., and Magilligan, D.J., Frequency spectra of the first heart sound and of the aortic component of the second heart sounds in patients with degenerated porcine bioprosthetic valves, *Am. J. Cardiol.*, 53, 557–561, 1984.

84. Beyer, R., Levkovitz, S., Braun, S., and Polti, Y., Heart sounds processing by average and variance calculation—Physiologic basic and clinical implications, *IEEE Trans. Biomed. Eng.*, 31(9), 591–596, 1984.

85. Akay, M., Semmlow, J., Welkowitz, W., Bauer, M., and Kostis, J., Detection of coronary occlusions using AR modeling of diastolic heart sounds, *IEEE Trans. Biomed. Eng.*, 37, 366–373, 1990.

86. Hadjileontiadis, L.J. and Panas, S.M., Discrimination of heart sounds using higher-order statistics, in *Proc. 19th Int. Conf. IEEE/EMBS*, IEEE Press, Chicago, IL, 1997, 3, 1138–1141.

87. Tilkian, A.G. and Conover, M.B., *Understanding Heart Sounds and Murmurs with an Introduction to Lung Sounds*, 3rd ed., W.B. Saunders Co., Philadelphia, 1993, chaps. 1–3.

88. Liatsos, C.N., Hadjileontiadis, L.J., Mavrogiannis, C.C., Rokkas, T.A., and Panas, S.M., On revealing new diagnostic features of bowel sounds using higher-order statistics, *Digestion*, 59(3), 694, 1998.

89. Hadjileontiadis, L.J., Giannakidis, A.J., and Panas, S.M., α-Stable modeling: A novel tool for classifying crackles and artifacts, in *Proc. 25th Int. Lung Sounds Assoc. Conf. ILSA 2000*, Pasterkamp, H., Ed., Chicago, 2000.

90. Kedem, B., *Time Series Analysis by Higher-Order Crossings*, IEEE Press, New York, 1994, chaps. 1, 4, and 8.

91. Hadjileontiadis, L.J., Saragiotis, C.D., and Panas, S. M., Discrimination of lung sounds using higher-order crossing, in *Proc. 24th Int. Lung Sounds Assoc. Conf. ILSA 1999*, Wichert, P., Ed., Marburg, 1999.

92. Haykin, S., *Neural Networks*, 2nd ed., Prentice-Hall, Inc., New Jersey, 1999, chap. 3.

93. Hadjileontiadis, L.J., Kontakos, T.P., Liatsos, C.N., Mavrogiannis, C.C., Rokkas, T.A., and Panas, S.M., Enhancement of the diagnostic character of bowel sounds using higher-order crossings, in *Proc. 1st Joined Int. Conf. IEEE BMES/EMBS*, IEEE Press, Atlanta, 1999, 2, 1027.

4

Automatic Recognition of Multichannel EEG Signals Using a Committee of Artificial Neural Networks

Björn O. Peters
John von Neumann Institute for Computing

Gert Pfurtscheller
Institute of Biomedical Engineering

Henrik Flyvbjerg
Risø National Laboratory

Abstract.   We present a multichannel classification method that uses a "committee" of artificial neural networks to recognize intentions to move left or right index finger, or right foot in the electroencephalograms (EEGs) of three subjects. This classification method *automatically* finds dominant EEG frequency bands and spatial regions on the skull relevant for the classification task, hence for the man–machine interactions that can be based on it. Correct recognition was achieved in 85–98% of trials not seen previously by the committee, on the basis of single EEGs of 1-sec duration. Frequency filtering did not enhance the classification success, i.e., the information relevant for the classification task is encoded in a very wide frequency band. Classification was optimal during the actual movement, but possible also several seconds before and after. Three seconds before the actual movement, all subjects showed a first peak in the classification success rate.

4.1   Introduction

Human EEG activity is altered before, during, and after sensory-motor processing and other mental activities. This has long been known. The active parts of the cortex produce electrophysiological signals in specific regions on the skull and specific frequency bands, which can be measured with the EEG.[1–3] In the present article, we present in some detail a committee of neural networks which can learn to interpret quickly and with precision such multichannel EEG data recorded during a motor task. This makes a Brain–Computer Interface possible. Our results were published with a brief description of methods in Reference 4. Here, we first discuss what others have achieved in the same vein, and then present the methods we employ to achieve unprecedented precision of recognition. We give sufficient details, we hope, to enable others to proceed from our description as from a recipe, *and* with a good understanding of it. Not that we urge others to copy us. On the contrary, we should like our results to be superceded by a near-perfect recognition success rate. In the search for what that requires, our architecture might be a suitably flexible starting point.

Our own starting point is that during rest, there is large EEG activity in the alpha frequency band (8–12 Hz). Several seconds before, e.g., a finger movement, this activity is diminished over the contralateral side of the skull. The alpha band activity recovers only several seconds after the movement. On the central part of the skull, alpha band activity increases during foot movement.[5] These phenomena are called event-related desynchronization (ERD) in the case of attenuation, and event-related synchronization (ERS) for an increase of EEG activity in a specific frequency domain.[6,7]

Before, during, and after movements of fingers and feet, ERD and ERS can be observed in different frequency bands and spatial regions, but vary significantly between different subjects.[3,8–10] The phenomena of ERD and ERS during distinct mental tasks lead to the question of whether there is a way to *recognize* sensory-motor activity with EEG. It has been known for some time that this is, indeed, to some extent possible. EEGs produced during a very limited set of mental tasks can be classified, hence recognized, according to tasks. Off-line analysis of multichannel EEG of the kind discussed here has been performed by Keirn and Aunon;[11] by Wolpaw and Pfurtscheller and co-workers (see References) since 1992; and more recently by Anderson, Devalapalli, and Stolz.[12]

Once a limited set of mental tasks can be recognized reliably, the classifier doing this can be used to control a device by having each task correspond to a command. This concept is referred to as a *brain–computer interface* (BCI).[11] The authors classified five different mental tasks pairwise by analyzing EEG recordings from six judiciously placed electrodes. The "nature" of the five tasks were "idling," counting visually, writing a letter, multiplying two large numbers, and mentally rotating a three-dimensional object. The classification success rates for the pairwise discrimination tasks averaged to 80–90%. Those results were, however, obtained by training and testing the classifier on the same data (so-called in-sample classification results).

In a different approach, Wolpaw et al.[13] let four subjects teach themselves to *control* electrical potentials from two bipolar electrodes located on the left and right sides of the skull. The potentials were used to control the movement of a cursor on a screen. Subjects had to move the cursor into predefined areas. After 20–30 training sessions, subjects could move the cursor to given corners of the screen with 41–70% success rates. For comparison, a random cursor movement would have a 25% success rate.

Anderson et al.[12] reconsidered the EEG data already studied by Keirn and Aunon; 0.5-sec time windows of individual recordings were modeled with autoregressive (AR) models of order 6 (see Section 4.3.1.3). An artificial neural network (ANN, see Section 4.3.2) with one hidden layer was trained on the resulting coefficients of the AR model to recognize which of five classes a given set of coefficients belonged to. They achieved 31–54% correct classification, depending on the subject. This result should be compared with the 20% correct classification that would be achieved by a "classifier" that assigns trials at random to five classes.

By averaging the ANN output for 20 consecutive, overlapping time windows, corresponding to 5 sec of real recording time, the classification success went up to 33–71%. By studying only the three tasks (rest, multiplication, and writing), and after averaging over 3 sec of EEG recordings, 89% correct classification was achieved.[14] One notes that the duration of the time interval shown to a classifier is crucial: The longer the interval, the higher the success rate of classification. In technical applications, e.g., using the brain–computer interface as a means for communication, the response time of the classifier preferably should not exceed 0.5 sec, since it is desirable that commands can be given at least once per second.

EEG activity in narrow frequency bands and a small number of well-chosen channels have been used to classify *intentions* of movement of left and right index finger, right foot, and tongue.[15–17] By training a number of different classifiers, including ANNs, on the time-dependent power spectra obtained by fitting an AR model to overlapping time segments of the EEG of the individual channels, 85% correct recognition was achieved when considering just two intended movements, left and right index finger,[15] and 70% correct recognition when considering three[17] or four[16] intended movements.

Modern EEG recording devices make it possible to study EEG activity in 50–256 channels in parallel, and with arbitrary sampling rates. Since classifiers typically work well only when the dimensionality of the classifier's input space is not too large, it is essential to identify a limited number of *features* of the EEG signals for the classifiers to work with. Such features can be spatio-temporally resolved ERD and ERS patterns, the coefficients obtained by fitting AR models (Section 4.3.1.3) to segments of EEG data in individual channels, the coefficients of multivariate AR models, or, very recently, spatial patterns of EEG activity in broad frequency bands.[18]

In general, the search for optimal features — i.e., optimal frequency bands and an optimal combination of channels in which to look for specific changes in EEG activity — is tedious.[19] We present here a method that *automatically* finds relevant frequency domains and channels. For the analysis presented here, we used data similar to that recorded and studied previously by Pfurtscheller et al.[16] Data were recorded on three subjects before, during, and after movements of right foot, left index finger, right index finger, and tongue. We omitted here the data for tongue movement, since those trials were contaminated by electromyographic artifacts from the tongue movement itself. In order to find the best electrodes and best frequency bands for an optimal brain-state classifier, we developed the procedure described below.

In the following section, we present the EEG experiment. In Section 4.3, we present the preprocessing scheme, our method of feature extraction, and a description of our classification scheme. In Section 4.4, we discuss first our choice of filter settings, then we give the complete time course of classification of three types of movement, for all three subjects, for each type of movement, and for various frequency bands and spatial regions on the skull separately. In the last section, we compare our classification success rates, over 90%, to results obtained by other groups, and discuss briefly our results.

4.2 Experimental Setup

4.2.1 Recording Scheme

The test subjects' EEG potentials were measured on 56 silver/silver-chloride electrodes with a reference electrode on the nose-tip. The electrodes were positioned on a rectangular grid with an approximate spacing between neighboring electrodes of 2.5 cm. Four electrodes corresponded to the international 10–20 system[20] (Figure III.4.1). The sampling rate (f_s) was 128 Hz, giving $56 \times 128 = 7168$ values per second to be stored. Data was recorded for at least 8 sec. The EEG signals were amplified between 0.15 and 60 Hz (3 dB points, 16 dB per octave) with two coupled

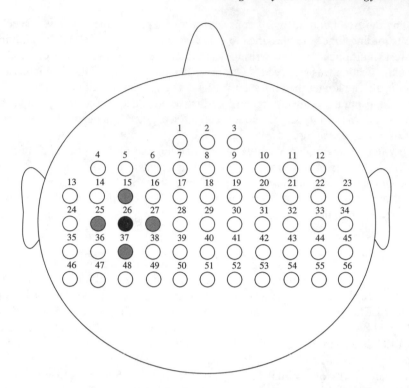

FIGURE III.4.1 Electrode positions on skull and corresponding numbering. Channel 26 corresponds approximately to position C3 in the international 10–20 system, channel 32 to C4, channel 2 to Fz, and channel 29 to Cz, respectively. (From Peters, B.O., Pfurtscheller, G., and Flyvbjerg, H., Automatic differentiation of multichannel EEG signals, *IEEE Trans. Biomed. Engin.*, 48, 111–116, 2001 © 2001 IEEE. With permission.)

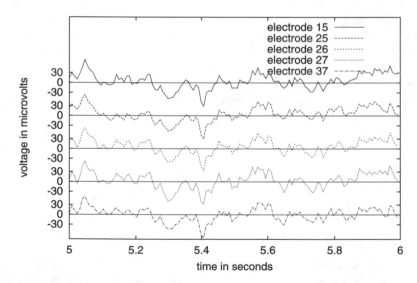

FIGURE III.4.2 Typical example of raw data for a left index finger movement in channels 15, 25, 26, 27, and 37 (cf. Figure III.4.1) for subject A4. The signals are clearly correlated.

32-channel amplifier system (BEST, Fa. Grossegger, Austria). Figure III.4.2 shows an example of raw data collected from five neighbor channels during feature extraction, i.e., during the 6th sec. Correlations between channels are clearly seen.

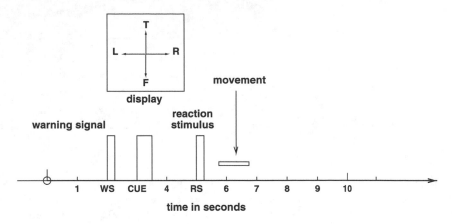

FIGURE III.4.3 Experimental protocol: 2 sec after recording was begun, a short beep (warning stimulus, WS) alerted the subject; 1 sec later, an arrow was displayed on a screen (CUE), indicating which movement had to be performed. Approximately 1 sec after a reaction stimulus (RS), the movement was performed.

TABLE III.4.1 Number of Trials Recorded for Each Subject and Each Event

Subject	L	R	F	Sum
A4	73	73	77	223
B6	55	54	64	173
B8	37	48	39	124

4.2.2 Experimental Paradigm

The experimental protocol used to obtain these EEGs is illustrated in Figure III.4.3: 2 sec after recording was begun, the subject was alerted by a short beep (*warning stimulus*, or WS). One second after the beep, an arrow pointing left, right, up, or down appeared on a screen for 300 msec, indicating that the left (left arrow) or right index finger (right), the right foot (down), or the tongue (up) soon was to be moved (CUE). Data recorded for tongue movement were mostly corrupted by the electrical signals from the movement itself, and were excluded from further analysis. For the remaining three events, we refer to L, R, and F, respectively. The subject had been instructed to move the relevant extremity in a brisk movement, but only some time *after* a second acoustic *reaction stimulus* (RS) had sounded at 5 sec.[16] The time between RS and onset of movement was 0.5–1.5 sec.

In this way, 37–77 trials were recorded for each of L, R, and F from three subjects, referred to as A4, B6, and B8 below (cf. Table III.4.1). The subjects were all healthy, right-handed students, aged 23–29, two male (A4 and B8), and one female (B6).

4.3 Signal Processing

4.3.1 Preprocessing

Classification of complex data is easier if they can be compacted with no or little loss of information by a suitable preprocessing scheme.[21] Our search for efficient data filter techniques and compression methods is described in this subsection. Ideally, for each individual subject, the algorithmic

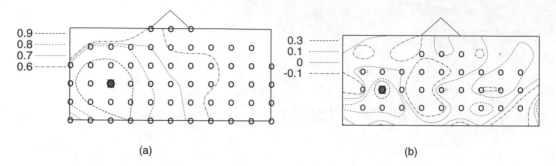

(a)   (b)

FIGURE III.4.4   Spatial correlations of EEG signals, here of channel 26 (see Figure III.4.1) with all other channels, computed from the entire duration of EEG recordings, roughly 38 min, for subject A4. (a) Correlations of raw data; (b) correlations for the Laplace filtered data. Differences above 0.02 are statistically significant.

parameters have to be adjusted, i.e., the spatial and frequential filters, and the AR model order. Since this is computationally difficult, we study the influence of the algorithmic parameters one by one, keeping the others constant, and only for one subject (A4), on the committee's classification success.

4.3.1.1 Spatial Filtering

To visualize the spatial properties of the EEG data, we computed the spatial correlations for subject A4, by averaging over the entire recording period of 2230 sec (Figure III.4.4). Because channels are correlated by 60–70% even if 15 cm apart, we searched for efficient spatial filters. With the notation $x_k(t)$ for the signal in channel k, the spatial filters were *common average reference*, or CAR $(x'_k(t) = x_k(t) - \frac{1}{56}\sum_{i=1}^{56} x_i(t))$, and *Laplace filter* $(x'_{26} = x_{26} - \frac{1}{4}\{x_{15} + x_{25} + x_{27} + x_{37}\})$. In addition, we studied a *local average technique*, or LAT (e.g., $x'_{26} = \frac{1}{5}\{x_{15} + x_{25} + x_{26} + x_{27} + x_{37}\}$).

Both CAR and Laplace filtering remove the influence of the reference electrode. Laplace filtering additionally removes any linear spatial component, and hence has some high-pass filter characteristics. On the other hand, LAT introduces a spatial low-pass filter. For both LAT and Laplace filter, only the 30 electrodes with four nearest neighbors could be analyzed, of course.

4.3.1.2 Frequency Filtering

During movement and already before, EEG activity in characteristic regions are synchronized or desynchronized in specific frequency bands.[5–7] Low- and high-frequency activity in surface EEGs is not related to movement.[22] It seems to be straightforward to filter away low and high frequencies from the signal.

We used a causal Kaiser filter[23] of length $N_K = 25$ (≈ 200 msec at our $f_s = 128$ Hz) and Kaiser parameter $\alpha_K = 2$; see Appendix. We filtered the data in the *theta* (0–6 Hz), the *alpha* (8–12 Hz), the *beta* (19–26 Hz), and the *gamma* (38–42 Hz) frequency band. The filter characteristics for these filter types are shown in Figure III.4.5.

4.3.1.3 Autoregressive Models

If brain-states classification works due to differences in the spatial frequency patterns produced by the brain during different mental tasks, the power spectrum of segments of EEG data has to be computed. It is, however, difficult to compute a reliable estimate for the power spectrum on very short time series.[24] We use here *autoregressive models* (AR) as power spectrum estimate: the time series $x(t)$ at time t is estimated by a linear combination of the p former instants, $\hat{x}(t) = \sum_{k=1}^{p} a_k^p x(t - k \cdot \Delta t)$, where $\Delta t = 1/f_s$ is the sampling time and p is called the AR model order. The AR coefficients, $(a_k^p)_{k=1\ldots p}$, are obtained by minimizing the squared error $\sum_t \{e(t)\}^2 = \sum_t \{\hat{x}(t) - x(t)\}^2$. Thus, it

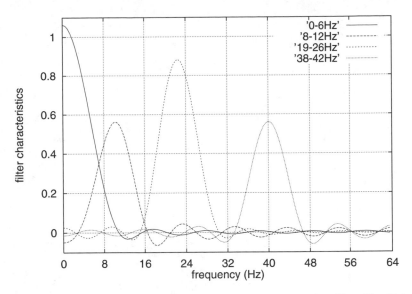

FIGURE III.4.5   Filter characteristic for different filter settings of the Kaiser filter. The Kaiser parameter was $\alpha_K = 2$, the filter length $N_K = 25$.

is not possible to study time delays or phase information with this method. By Fourier transforming the equation $x(t) = \hat{x}(t) + e(t)$, one obtains a *smooth* estimate of the power spectrum:

$$|\tilde{X}_{AR}(f)|^2 = \frac{\sigma^2}{|1 - \sum_{k=1}^{p} a_k^p \cdot e^{-i2\pi(f/f_s)\cdot k}|^2}.$$

(The term $e(t)$ describes white noise, i.e., its Fourier transform is a constant, σ.) The AR coefficients hence encode *features* of the power spectrum. A set of AR coefficients, $(a_k^p)_{k=1...p}$, which is called *feature vector*, is believed to contain all features of the EEG data in one channel, and of one trial.

The optimal AR model order (p) remains to be determined. To this end, various information theoretic methods have been developed.[24] The classification result of our classifier is already an information measure. The optimal p hence should result in a maximal classification success. We measured the classification success rate of subject A4, on Laplace-filtered segments of EEG data taken in the period from 5–6 sec. The classification was done on the coefficients of AR (p) models, with p varying from 2–50. As can be seen from Figure III.4.6, the classification success rate increases with p until $p \approx 10$, then stays at approximately the same level until $p \approx 20$, and decreases slowly for larger values of p. We consequently chose $p = 10$.

4.3.2   Artificial Neural Networks as Classifier

An artificial neural network (ANN) is a general-purpose function approximator for multi-dimensional functions of several variables.[21] The real valued input variables are the p-dimensional feature vectors (see Section 4.3.1.3). For each brain state, a *perceptron* is formed: a weighted sum of the input values is passed through a nonlinear function (we used $\tan h\, x$). The weights characterize the perceptron. In our case, the whole ANN consists of three perceptrons, one perceptron for each of the classes L, R, and F. The output of the perceptrons are supposed to be positive if a feature vector at its input belongs to the same brain state as the perceptron, and negative if the input belongs to a different brain state. An output value close to zero indicates that the perceptron does not have much of an "opinion." In this sense, the perceptron's output value indicates the "strength" of the "conviction" that a feature vector belongs (output value > 0) or does not belong (output value < 0) to "its" class.

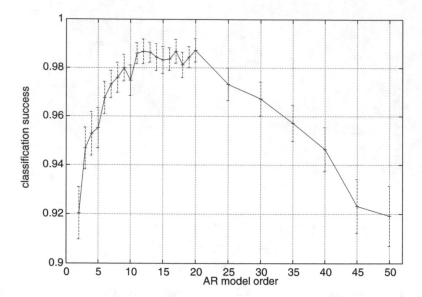

FIGURE III.4.6   Classification success vs. model order of the autoregressive models, as averaged for 100 partitions of trials into learning and test sets, for subject A4. Laplace filtered EEG data were analyzed in the period from 5–6 sec.

4.3.2.1   Training an ANN

On a subset of all feature vectors, the so-called *training set*, the weights of the ANN are adjusted as follows. The desired perceptron output is 1, if a feature vector presented as input belongs to the brain state represented by this perceptron, and −1 otherwise. The perceptron's output is computed and compared to the desired value. The sum over the squared differences between desired and actual output, taken over all three brain states and all trials in the training set, is computed as *error function*. This error is a function of all the weights that characterize the ANN and has to be minimized during training. To this end, each weight is corrected by a quantity which is proportional to the derivative(s) of the error function with respect to that weight.[21]

4.3.2.2   Early Stopping

The ability of the ANN to correctly classify feature vectors is monitored on a second, independent subset of trials, the so-called *validation set*. This monitoring is done by presenting to the ANN, one by one, the feature vectors in the validation set, computing its output, deciding which perceptron has the biggest output value, and comparing that perceptron's class label to the feature vector's class label. If they coincide, the feature vector is correctly classified. The performance of the ANN on the validation set is the number of correctly classified feature vectors divided by the size of the validation set. As the training on the training set proceeds, the classification success on the validation set will first increase, but will decrease eventually when the ANN learns the training feature vectors "by heart." This is called over-learning. In addition to the general structure, the classifier also learns random fluctuations. Since we wish to optimize the ANN's ability to give correct input-output relations on unseen data, the training of the ANN is stopped when the classification success on the *validation* set reaches a maximum.

4.3.2.3   Generalization Ability, Bootstrapping, and Architecture

The classification success on the validation set will not represent the ANN's generalization ability, because information from the trials in the validation set was used to obtain this classifier. It is hence necessary to compute the classification success rate on a third, independent *test set*. The partitioning of the available experimental trials on training, validation, and test set was done randomly. The training set and the validation set must both consist of an equal number of trials from classes L, R, and F. Otherwise, the ANN is biased toward the class from which it has seen the most feature vectors during training or validation. We used for subject A4, e.g., $50 + 50 + 50$ trials in the training set, $16 + 16 + 16$ trials in the validation set and $7 + 7 + 11$ trials in the test set, for each of the classes L, R, and F, respectively.

For each channel separately, the ANN training procedure as described above is repeated, leaving us with 30 trained ANNs, one for each of the 30 channels of Laplace-filtered EEG data (56 trained ANNs in the case of CAR filtering). In a last step, the 30 ANNs are combined in a sort of "committee of experts." The committee forming procedure is described in the following subsection. In order to reduce fluctuations due to the statistics, the random partitioning of the trials on training, validation, and test sets is repeated at least 100 times.

We have tried to use multi-layer perceptrons as classifiers in the individual EEG channels. Since the classification success rate of the committee was higher when omitting any hidden layer, we decided to use here as committee members only simple perceptrons.[25]

4.3.3   Multichannel Signal Processing: The "Committee" Method

For each individual channel, one ANN has been trained as described above. For an input feature vector from a trial in the test set, each ANN gives a three-dimensional output vector with its class label estimate. Then, a "committee" is formed. The output vectors are summed up (Figure III.4.7). An input feature vector is classified according to which component in the output vector sum is the largest, i.e., "receives the strongest vote."

The motivation for adding up the signals in this manner was our observation that committee members with strong convictions (large absolute value of a perceptron's output) tended to be right, while committee members with weak convictions (perceptron's output close to zero) more often were wrong. Thus, by weighting a vote by its strength, the emphasis is placed on votes that are correct with a high probability, while votes that are less probably correct are de-emphasized and left to cancel each other.

The committee is expected to yield a better classification accuracy than any individual channel could provide, and is a way to combine information from several channels, i.e., from different spatial regions. Choosing optimal electrode positions and the optimal number of channels for an EEG-based

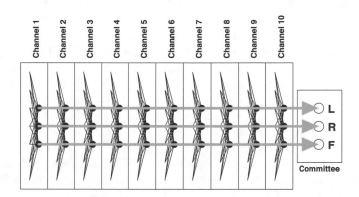

FIGURE III.4.7   A committee with 10 member ANNs is symbolized. Feature vectors from individual trials are presented to the ANNs in the ten channels. The output vectors of the ANNs in each of the member channels are summed up to yield one, the committee's, vector sum. The committee's estimate of which movement an input trial belongs to is given by the class label (L, R, or F) of that component in the vector sum which has the largest numerical value.

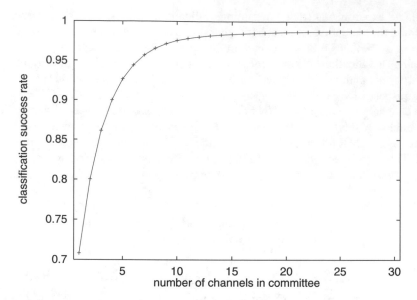

FIGURE III.4.8 Classification success averaged over trials from the validation set as function of committee size, as averaged over 5000 random partitions of the available trials in training, validation, and test set. (From Peters, B.O., Pfurtscheller, G., and Flyvbjerg, H., Automatic differentiation of multichannel EEG signals, *IEEE Trans. Biomed. Engin.*, 48, 111–116, 2001 © 2001 IEEE. With permission.)

brain-state classifier is, however, a nontrivial task. In the literature, the choice of EEG channels for this purpose is done either by hand, by competent physiologists, or by doing preliminary studies.[19] Here, we focus on how to develop a classifier that *automatically* finds channels which contain information relevant for the classification task.

According to the classification success on the trials in the validation set, the 30 channels are rank-ordered. N-member "committees" are formed from the N ANNs with the best classification success, with N ranging from 1–30. The classification success rate, averaged over the trials in the validation set, was computed as a function of the committee size N. Figure III.4.8 shows this function, averaged over 5000 partitions of the trials on the training, the validation, and the test set, with Laplace-filtered data from subject A4, analyzed on AR(10) coefficients computed in the period from 5–6 sec. No band-pass filter was applied.

The committees' classification accuracy depends on N, the number of committee members. The accuracy rises first quickly with increasing N, then very slowly for $N > 10$. The average classification success does not vary significantly for committee sizes bigger than 15; the channels above this size do not contribute any additional information. Hence, we decided to use a committee size of $N = 20$.

With the committee size and the corresponding member-ANNs chosen, our design procedure for a classifier is complete. We have a particular set of channels with associated trained ANNs, and a rule for how to combine and interpret the outputs of these ANNs. The predictive power of this committee is then tested on the trials in the test set.

The whole procedure just described was repeated 100–500 times, each time partitioning the set of trials in a random manner on training, validation, and test set, and each time initiating each ANN with new random weights before training.

A few remarks on this committee method:

- The committee forming procedure consists of five steps: (1) training the ANNs, while (2) validating the generalization ability of the ANNs to stop training at its maximum, (3) rank-ordering channels according to classification success, (4) estimating committee size, and (5) testing the committee's performance. For each of these steps, one might assume that one needs independent sets of trials. This is not the case since, apart from fluctuations, the classification success rates one would obtain for such sets in steps 3–5 are equal.[21] It is hence sufficient to use only three independent sets of trials: the training, validation, and test sets.

- There is even a possibility to use only two different sets.[26] With the help of the second derivative of the error function (Section 4.3.2.1) with respect to the weights, the generalization error of an ANN can be estimated from the training error. Once the Hessian is known, the ANNs can be pruned.

- One can easily imagine other ways to combine the output of ANNs trained on individual channels in a weighted sum, the weight of each contributor being related to its overall classification accuracy. But here we consider only weight factors 0 and 1: an ANN is either in or out of the committee.

- Which EEG channels to include in the committee was decided on the basis of the committee's classification success. Comparing all possible committees was too big a task. The ideal alternative to such a brute-force endeavor would have been to find the individual channel's output distributions and the correlation functions for outputs in different channels. On this basis, one might then decide how to combine outputs from individual channels in a manner that optimizes the signal-to-noise ratio of the combined output. This information however, cannot be extracted from our rather limited datasets. So instead, we chose the individual channel's classification success as a robust and pragmatic measure of that channel's information content, ignored correlations between channels, and included channels in the committee according to their success rates. As can be seen from Figure III.4.8, the result of this choice leaves little room for improvement, as long as committees consist of 15 or more channels chosen as described.

4.3.4   The Algorithm in Summary

In short, the algorithm that we developed reads as follows:

1. Identify the period during which relevant EEG activity takes place.
2. Fit the time series from each channel's EEG with an autoregressive model of order 10, i.e., an AR(10) model, resulting in 10 *AR coefficients* per channel and per trial.
3. For all trials and all channels, store the coefficients of the AR model determined by this fit.
4. Put an equal number of trials from each brain state to be classified into a training set (approximately 70% of all trials) and into a validation set (20%). Put the remaining trials (10%) into a test set.
5. Let k denote the number of different brain states to be classified. Then, for each channel, train an artificial neural net (ANN) with 10 inputs and k outputs on the AR-coefficients of the trials in the training set.
6. While training, monitor the ANN's generalization ability on the validation set, i.e., the ratio between the number of correctly recognized trials in the validation set and the size of the validation set itself.
7. Stop training when the generalization ability or the *classification success* on the validation set has reached a maximum.
8. Compute each ANN's performance — its classification success — on the validation set.
9. The "opinion" of one ANN is a k-vector of real numbers, the output of its output neurons. Each number expresses the strength of the ANN's conviction that the input should be classified as the neuron yielding the number. Define the "opinion" of the committee as the vectorial sum of the "opinions" of its members, and classify the input as corresponding to the vector element with the largest value.

10. Form a committee of 20 members, the 20 ANNs with the best performances on the validation set. The committee with these 20 member ANNs is the final classifier.
11. Compute its performance on the test set.

4.3.5 Computational Method

To study the impact of different filter types and their combination on the classification success rate, several hours of CPU time were needed on a SUN Sparc20 workstation for each run with one particular filter setting. In order to accelerate the response time of the classifier, we decided to port the code to a parallel computer, where the training of the ANNs can be performed simultaneously instead of serially. The response time for computing the classification success rate on the basis of, e.g., 250 partitionings of the trials on training, validation, and test set, was then, depending on the AR model order, 30 sec–10 min.

The algorithms were written in C and then translated for a Cray T3E parallel computer by the message passing interface (MPI). One "master" and 30 "slave" CPUs were used (or 56 slave nodes in the case of the CAR spatial filter), one slave node for each EEG channel. The code performed in parallel at two stages. In an initialization phase, the data input and the spatial filtering were done by the master node. The filtered time series were then passed to the slave nodes, where frequency filtering and feature extraction were done. The feature vectors were stored locally on the slave nodes. Then, the slave nodes trained their ANNs in parallel. The output vectors of the ANNs in all channels and of all trials were then collected by the master node, which constructed the committee as described above.

4.4 Results

In this section, we present the results obtained with our method. After choosing the filter settings, we present classification results for the complete time course of the experiment, then in more detail for the three movements, for spatial regions on the skull with relevant information, and for different frequency bands. At the end, we discuss the accuracy of classification success rates.

4.4.1 Filter Settings

To study the influence of spatial and band-pass filters, we analyzed EEG data of subject A4 in the time interval between 5 and 6 sec, i.e., before the onset of movements in most trials. AR(10) models were fitted to the data segments in each channel separately. Trials from all three events, L, R, and F, were classified by a committee of ANNs. As can be seen from Table III.4.2, the classification success on raw data was already 87%. Frequency filtering did not enhance the classification success.

TABLE III.4.2   Three-States Classification Results for Different Spatial and Frequential Filter Settings, Obtained for Subject A4

Subject A4 Band Pass	Spatial Filters			
	CAR	Laplace	LAT	None
0–6 Hz	97	96	78	84
8–12 Hz	97	97	77	83
19–26 Hz	97	97	83	87
38–42 Hz	92	96	77	81
none	98	98	84	87

Note: EEG data was extracted in the period of the onset of movement, between 5 and 6 sec (see Figure III.4.3). The classification was done on coefficients of AR(10) models fitted to the data in each channel (see Section 4.3.1.3). Results given are % of trials in the test set correctly classified by the committee. Differences of more than 2% are statistically significant.

It can hence be concluded that the information relevant for the classification is encoded in a very wide frequency band.

From the spatial filtering techniques, LAT yielded the smallest classification success (84%), smaller than unfiltered data. After CAR or Laplace filtering, the classification success went up significantly, to 98%. Since the results for Laplace-filtered data did not depend on which frequency band was chosen, we decided to process the EEG data with the Laplace filter. The results reported in the remainder of this chapter section were obtained with Laplace-filtered data, with no band-pass filtering, and with classification based on the coefficients of AR(10) models.

4.4.2 Time Course of Classification

The complete time course of classification is presented in Figure III.4.9. The classification success rate is at the randomness level (where it should be) before the subjects have been told which movement to perform, it is peaked at the time of the actual movement, and it is high even 2–3 sec *after* the movement.

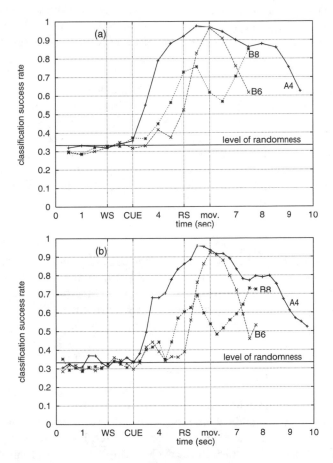

FIGURE III.4.9 Classification result during the whole experiment for subjects A4, B6, and B8. "1" on the abscissa means perfect recognition, whereas 1/3 is the level obtained when trials are recognized at random. Classification on 1-sec intervals are shown in (a), and in 500 msec intervals in (b). The results are averages of 500 partitionings of the trials on training, validation, and test set, using AR(10)-coefficients for the feature extraction. The plotting points are positioned in the middle of the corresponding time intervals. (From Peters, B.O., Pfurtscheller, G., and Flyvbjerg, H., Automatic differentiation of multichannel EEG signals, *IEEE Trans. Biomed. Engin.*, 48, 111–116, 2001 © 2001 IEEE. With permission.)

In Figure III.4.9(b), the classification is monitored in twice the time resolution of Figure III.4.9(a). The classification success rates in (b) are slightly lower than in (a) because of the shorter time windows used for classification. For all three subjects, the recognition rates differ from the randomness level immediately after the visual cue has been presented to the subject. In the time window from 3.5–4 sec, 500 msec after presenting the visual cues to the subjects, the classification reaches briefly a maximum for all subjects. But while for subject A4, the classification success rate stays around 70% until shortly before RS, for subjects B6 and B8, it drops quickly to the randomness level, before it reaches a maximum for B6 around 6.5 sec, where most of the actual movements probably take place. This second peak is less pronounced for subject B8. It is remarkable that for subject B8, there is a third, higher peak in the classification success rate at the end of the EEG recording period. The increased classification rate in subject B8 after movement-offset may be explained by the post-movement beta synchronization localized close to the corresponding primary motor area;[3,27,28] see also Section 4.4.5.

4.4.3 Classification Accuracy for Individual Movements

To investigate the classification success in more detail, we give the classification success rates of the three kinds of events separately in Figure III.4.10. The results are the number of correctly recognized trials of event X divided by the number of trials of event X in the test set. We note that these classification success rates are quite different from each other.

Subjects A4 and B8 share some common properties: The recognition of foot movements (F) and intentions of foot movements is higher than the L and R recognition. For both subjects, the recognition of right finger (R) events is higher than left finger (L) events before the movement, but after the movement this is reversed. The peak in the classification success immediately after the CUE has vanished is driven by different events for the three subjects. Whereas in subject A4, there is a jump in the classification of all three events, in subject B6 only intended finger movement classification is possible at this stage, and for subject B8 only F can be recognized well. In all subjects, the classification of both fingers decreases after the movement, before it increases again at approximately 7 sec recording time.

4.4.4 Spatial Analysis

Figure III.4.11 shows which channels were used for the committee how often at different stages of the experiment. The results were obtained by training AR(10) models to segments of EEG data, and averaging over 500 or 1000 partitionings of the trials on the three sets. The electrodes chosen for the committee were mostly grouped around the electrode positions C3 and C4 (cf. Figure III.4.1), where the motor cortex can be expected to be. There are, however, considerable inter-subject differences.

For subject A4, the spatial regions of good recognition remain the same during the time from 3.5–8 sec; the central region of the skull does not play an important role in the classification task. In subject B6, there is a peak at the frontal and central electrodes during the time period of early classification (3.5–4 sec) which vanishes during the movement. The electrodes situated frontally and parietally from the positions C3 and C4 are those included most frequently into the committee. In subject B8, the first peak of classification at time 3.5–4 sec (cf. Figure III.4.9) goes out from right regions of the skull, a phenomenon which can be explained by the higher L, contralateral, recognition at this time (cf. Figure III.4.10).

4.4.5 Frequency Analysis

To answer the question in which frequency bands most of the cortical activity is encoded, we filtered the EEG data by a causal Kaiser filter (see the Appendix) in the frequency domains 0–6 Hz (theta

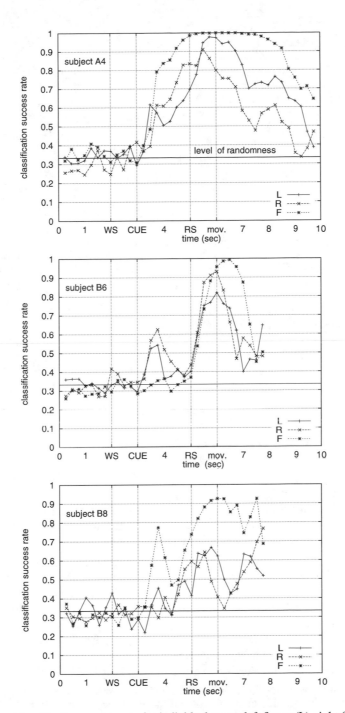

FIGURE III.4.10   Classification success rate for individual events left finger (L), right finger (R), and right foot (F), for subjects A4, B6, and B8. (From Peters, B.O., Pfurtscheller, G., and Flyvbjerg, H., Automatic differentiation of multichannel EEG signals, *IEEE Trans. Biomed. Engin.*, 48, 111–116, 2001 © 2001 IEEE. With permission.)

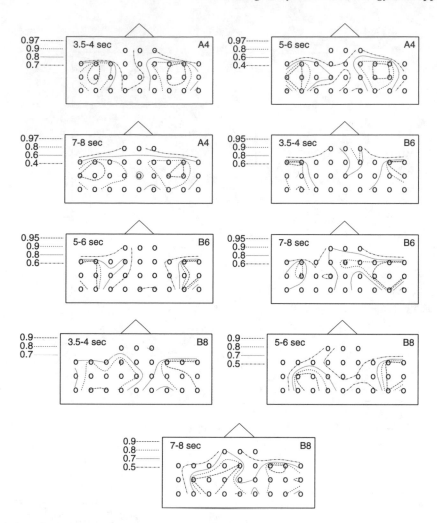

FIGURE III.4.11 Spatial distribution of the number how often channels were included into the committee. Maps were computed after 1000 partitionings of the trials into training, validation, and test set, for the time periods 3.5–4 sec, 5–6 sec, and 7–8 sec for each of the subjects A4, B6, and B8.

band), 8–12 Hz (alpha), 19–26 Hz (beta), and 38–42 Hz (gamma). Note that the frequency bands were sometimes mutually overlapping, cf. Figure III.4.5. The classification was done, again, on coefficients of AR(10) models computed during periods of 500 msec, and compared in Figure III.4.12 to the results shown in Figure III.4.9(b).

Consistently with the results presented in Table III.4.2, the classification results differ not much between the unfiltered and the band-passed data. This confirms that for all subjects the information in the EEG data is widely spread over all frequency ranges, cf. Section 4.4.1. In only a few cases, the classification success in one frequency band (gamma) is significantly higher than on the unfiltered data. The fact that this effect is not stable over time confirms our choice not to use any frequency filters. The high classification success after the movements are in all subjects due to the gamma activity.

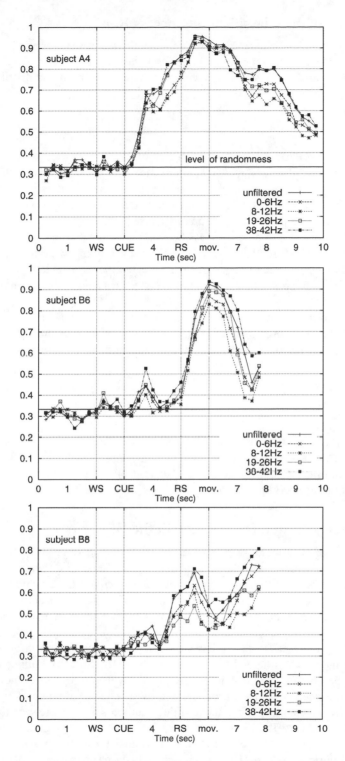

FIGURE III.4.12 Classification success rate for different filter settings of the band-pass filter, for subjects A4, B6, and B8. Individual curves for unfiltered data, and for band-passed data in the frequency ranges 0–6 Hz, 8–12 Hz, 19–26 Hz, and 38–42 Hz are shown.

4.4.6 Accuracy of the Classification Success Rate

Classification is essentially a binary decision: a trial in the test set is classified either correctly or incorrectly. The standard deviation σ around the classification success (c) on a test set of size N_{test} should hence be compared to the standard deviation estimate for a binomial distribution:

$$\sigma_e = \sqrt{\frac{c(1-c)}{N_{\text{test}}}}$$

which yields, e.g., for $c = 33.3\%$ and a test set size of 25, $\sigma_e \equiv 9.4\%$. With bootstrapping, i.e., re-partitioning of trials into the training, validation, and test sets, the standard deviation from expected value should be lower: To compute the classification success, the classification results from various permutations of trials are *averaged*. The information which is encoded in one particular dataset is thus used several times, and the individual runs are consequently not statistically independent.

Numerical simulations of a three-class classification task show that in cases when bootstrapping is applied, σ depends on both N_{test} and N_{train}, the size of the training set. For typical values of N_{test} and N_{train}, as used in the classification of EEG data here, the standard deviations with bootstrapping are reduced by a factor of approximately 2, compared to the standard deviations obtained without bootstrapping.

These results compare well with the fluctuations in the classification success of EEG data taken between 0 and 3 sec recording time (cf. Figure III.4.9), i.e., when the subjects do not know yet which movement to perform, and when in the data, there is no information relevant for the classification.

4.5   Discussion

We have presented a method for multichannel, EEG-based classification of several mental states. The information, in which spatial region(s) and in which frequency domain(s) distinguishable EEG activity occurs, is not input to our classifier. Nevertheless, our results are as good as the best results of EEG classification obtained up to now.

4.5.1   Movement-Related Information in the EEG

We analyzed what the committee based its classification on and found relevant regions of brain activity: the channels included into the committee indicate the spatial region(s) of relevant EEG activity — relevant in the sense that these regions produce *different* EEG activity during the different mental states that are considered for the classification task. The electrodes chosen for the committee were mostly those located over the motor cortex, over areas where one knows hand movement is represented.

We also found the relevant frequency bands. Surprisingly, the relevant information is encoded in all frequency bands studied here, from 0–6 Hz to 38–42 Hz. So we recommend that frequency filtering is not applied to EEG data that are to be classified with autoregressive models as done here.

One reason for the stability of classification results over frequency domains could be that with the parameters of the Kaiser filter used here, the resulting band-pass filters were sensitive in rather broad frequency ranges, cf. Figure III.4.5. Furthermore, the dominant movement-related EEG activity is a *mu*-rhythm. This mu-rhythm is not a pure sinosoidal form, but has an arch-shaped form. Consequently, its Fourier transform (FT) contains not only the basic frequency of the mu-rhythm, but also higher harmonic frequency components. Some of these higher components fall within the beta and gamma bands.[28] This could be a reason why the beta and gamma bands yield good classification results for subjects A4 and B6.

4.5.2 Comparison to Other Work

Where possible, we compared our results with other published results obtained with the same subjects. Because of the high inter-subject variability in the EEG, this provides us with the best possible test of our classification method against other methods. We also compared our EEG classification results with different experiments. The subjects that took part in this study have participated in earlier experiments[19] with a slightly different experimental paradigm; between the visual cue on the screen and the reaction stimulus, there was only 1 sec instead of 2, and the average time between reaction stimulus and onset of movement was 500 msec instead of the 1000 msec in our study, cf. Figure III.4.3. Classification was done on up to 56 EEG channels, or a subset of them, using the ERD or ERS in specific frequency bands as features, and classifying them with a so-called *Distinction Sensitive Learning Vector Quantizer*, or *DSLVQ*.[29] The *best* results for the classification of left and right index finger movements reported in Reference 19 were for:

Subject A4, all 56 electrodes, power in 10–12 Hz frequency band: 79.3 ± 2.5%
Subject B6, 11 electrodes pre-selected on the same data set by a preliminary experiment, power in 20–24 Hz frequency band: 88.6 ± 1.2%
Subject B8, 11 electrodes pre-selected by an expert, frequency band 10–12 Hz: 84.5 ± 2.2%

Pfurtscheller et al.[16] reported EEG classification results for left/right index finger movement discrimination of up to 89% as best result for a subject whose records in earlier experiments had shown significant changes in 40 Hz EEG during finger movement.

We obtained for the same subjects as in Reference 19 classification results from fitting AR(10) models to the EEG data and training our committee on the AR coefficients. The left/right index finger classification results were for subject A4 94%, for subject B6 95%, and for subject B8 91%, as measured from the time interval between 5 and 6 sec recording time.

In another experiment investigating the possibility of a brain–computer interface (BCI), Kalcher et al.[17] used the ERD to classify three types of movement (L, R, and F) in four subjects, including subject B8 of our study. After a short warning stimulus, a visual cue was presented to the subjects for 1.25 sec, during which ERD patterns in narrow frequency bands were extracted from channels C_3, C_z, and C_4. After the visual cue had vanished, the corresponding L, R, or F movement had to be performed. Approximately 1 sec later, the estimate of the DSLVQ classifier was fed back to the subject. The classification result for the three movements of that study reached 50–70% after several sessions.

With our method, we obtained for subject A4 98% correct classification for the three events, for subject B6 a maximum of 96%, and for subject B8 75% correct recognition before the movement, and even 85% in the interval from 7–8 sec (Figure III.4.9).

4.5.3 Outlook

There is some hope that a brain–computer interface (BCI) is possible technically, at least for some subjects, for judiciously selected brain states, and in cases when a brain-states classification success around 90% is sufficient. The response time of a BCI is crucial, however. We obtained the best classification success while analyzing EEG segments of length 1 sec, whereas the success rates for 500-msec windows were 2–5% lower. With this experiment, the rate with which different brain states were produced was less than 0.1 Hz, a rate too low for a BCI.

We have shown that it is possible to classify EEG according to tasks with high accuracy. However, in the experimental paradigm used to obtain our EEG data, the subjects were *told* which movement to perform. In order to further BCI development, one needs data from experiments where the *subjects* have decided which brain state from a small dictionary to produce. In such experiments, the classification results could be fed back to the subjects, until they *learn* to produce distinguishable EEG patterns. The BCI development would then be done in a combination of such a biofeedback procedure and black-box classification of EEG data as described here. Recordings from *many* EEG channels should then be used, and the time series should be Laplace filtered to de-correlate data between neighboring channels. The classification should then be done on the coefficients of autoregressive models fitted to the time series. The fact that our classifier with only simple perceptrons allowed a high classification accuracy is encouraging, since simpler classification schemes like independent component analysis might then work as well.

Acknowledgments

We thank P. Grassberger and J. Müller-Gerking for fruitful discussions. BOP thanks the Zentralinstitut für Angewandte Mathematik, Jülich, for generous allocation of Cray T3E time, approximately 100 h on 31 nodes, and for technical support. BOP also thanks the Graduate School of Biophysics of the Danish Research Academy for hospitality when finishing the article. This research was partially, supported by the "Fonds zur Förderung der wissenschaftlichen Forschung" in Austria, Project P11208-MED.

References

1. Chatrian, G.E., Petersen, M.C., and Lazarte, J.A., The blocking of the rolandic wicket rhythm and some central changes related to movement, *Electroenceph. Clin. Neurophysiol.*, 11, 447–510, 1959.
2. Kuhlman, W.N., Functional topography of the human mu rhythm, *Electroenceph. Clin. Neurophysiol.*, 44, 83–93, 1978.
3. Toro, C. et al., Event-related desynchronization and movement-related cortical potentials on the ECoG and EEG, *Electroenceph. Clin. Neurophysiol.*, 93, 380–389, 1994.
4. Peters, B.O., Pfurtscheller, G., and Flyvbjerg, H., Automatic differentiation of multichannel EEG signals, *IEEE Trans. Biomed. Engin.*, 48, 111–116, 2001.
5. Pfurtscheller, G. and Neuper, C., Event-related synchronisation of mu rhythm in the EEG over the cortical area in man, *Neurosci. Lett.*, 174, 93–96, 1994.
6. Pfurtscheller, G. and Aranibar, A., Event-related cortical desynchronization detected by power measurement of scalp EEG, *Electroenceph. Clin. Neurophysiol.*, 42, 817–826, 1977.
7. Pfurtscheller, G. and Aranibar, A., Evaluation of event-related desynchronization (ERD) preceding and following voluntary self-paced movement, *Electroenceph. Clin. Neurophysiol.*, 46, 138–146, 1979.
8. Kristeva, R., Cheyne, D., and Deecke, L., Neuromagnetic fields accompanying unilateral and bilateral voluntary movements: Topography and analysis of cortical sources, *Electroenceph. Clin. Neurophysiol.*, 81, 284–298, 1991.
9. Feige, B. et al., Neuromagnetic study of movement-related changes in rhythmic brain activity, *Brain Research*, 734, 252–260, 1996.
10. Pfurtscheller, G., Neuper, C., Flotzinger, D., and Pregenzer, M., EEG-based discrimination between imagination of right and left hand movement, *Electroenceph. Clin. Neurophysiol.*, 103, 642–651, 1997.
11. Keirn, Z.A. and Aunon, J.I., A new mode of communication between man and his surroundings, *IEEE Trans. Biomed. Engin.*, 37, 1209–1214, 1990.
12. Anderson, C.W., Devulapalli, S., and Stolz, E.A., Determining mental states from EEG signals using parallel implementations of neural networks, *Scientific Programming*, 4, 171–183, 1995.

13. Wolpaw, J.R. and McFarland, D.J., Multichannel EEG-based brain-computer communication, *Electroenceph. Clin. Neurophysiol.*, 90, 444–449, 1994.
14. Anderson, C.W., Effects of variations in neural network topology and output averaging on the discrimination of mental tasks from spontaneous electroencephalogram, *J. Intelligent Systems*, 7, 165–190, 1997.
15. Peltoranta, M. and Pfurtscheller, G., Neural network based classification of non-averaged event-related EEG responses, *Med. Biol. Engineering and Computing*, 32, 189–196, 1994.
16. Pfurtscheller, G., Flotzinger, D., and Neuper, C., Differentiation between finger, toe, and tongue movement in man based on 40Hz EEG, *Electroenceph. Clin. Neurophysiol.*, 90, 456–460, 1994.
17. Kalcher, J. et al., Graz brain-computer interface II: Towards communication between humans and computers based on online classification of three different EEG patterns, *Med. Biol. Engineering and Computing*, 34, 382–388, 1996.
18. Müller-Gerking, J., Pfurtscheller, G., and Flyvbjerg, H., Prompt classification of EEG signals based on their spatial distribution prior to movement, *Soc. Neurosci. Abstr.*, 23, 1278, 1997.
19. Pregenzer, M., Pfurtscheller, G., and Flotzinger, D., Selection of electrode positions for an EEG-based brain computer interface (BCI) *Biomedizinische Technik*, 39, 264–269, 1994.
20. Jasper, H., The ten-twenty electrode system of the international federation, *Electroenceph. Clin. Neurophysiol.*, 10, 371–375, 1958.
21. Hertz, J., Krogh, A., and Palmer, R.G., *Introduction to the Theory of Neural Computation*, Addison-Wesley, Reading, MA, 1991.
22. Nunez, P.L. and Katznelson, R., *Electrical Fields of the Brain — The Neurophysics of EEG*, Oxford University Press, New York, 1981.
23. Johnson, J.R., *Introduction to Digital Signal Processing*, Prentice-Hall International, London, 1991.
24. Childers, D.G., *Modern Spectrum Analysis*, IEEE Press, New York, 1978.
25. Peters, B.O., Pfurtscheller, G., and Flyvbjerg, H., Prompt recognition of brain states by their EEG signals, *Theory in Biosciences*, 116, 247–258, 1997.
26. Hansen, L.K., Rasmussen, C.E., Svarer, C., and Larsen, J., Adaptive regularization, in *Neural Networks for Signal Processing IV: Proc. 1994 IEEE workshop*, John Vlontzos, H.-N. Hwang, and E. Wilson, Eds., Piscataway, NJ, 1994, 78–87.
27. Neuper, C. and Pfurtscheller, G., Post-movement synchronization of beta rhythms in EEG over the cortical foot area in man, *Neurosci. Lett.*, 216, 17–20, 1996.
28. Pfurtscheller, G., Stancák, Jr., A., and Edlinger, G., On the existence of different types of central beta rhythms below 30 Hz, *Electroenceph. Clin. Neurophysiol.*, 102, 316–325, 1997.
29. Pregenzer, M., Flotzinger, D., and Pfurtscheller, G., Distinction sensitive learning vector quantizer — a noise-insensitive classification method, in *Proc. ICNN-94*, pp. 2890–2894, IEEE, Orlando, 1994.

Appendix

We present here the algorithm for computing a digital *Kaiser* band-pass filter[23] which was introduced in Section 4.3.1.2. The filtered signal $y(t)$ is given by:

$$y(t) = \sum_{n=-M}^{M} x(t-n)h(n)$$

where $N_K = 2M + 1$ is the length of the filter and $x(t)$ is the raw signal. The filter coefficients, $h(n)$, are given by:

$$h(n) = w_K(\alpha_K, n)h_d(n).$$

$h_d(n)$ is the digital Fourier transform of the ideal filter:

$$H_d(\omega) = \begin{cases} 0 & : 0 \leq |\omega| < \omega_1 \\ e^{-i\omega M} & : \omega_1 \leq |\omega| \leq \omega_2 \\ 0 & : \omega_2 < |\omega| \leq \pi \end{cases}$$

($\omega = 2\pi f/f_s$, if f_s is the sampling rate), given by:

$$h_d(n) = \begin{cases} \frac{1}{n\pi}[sin(n\omega_2) - sin(n\omega_1)] & : n \neq 0 \\ \frac{1}{\pi}(\omega_2 - \omega_1) & : n = 0 \end{cases}$$

With the Kaiser parameter (α_K), the windowing coefficients, $w_K(\alpha_K, n)$, are:

$$w_K(\alpha_K, n) = \begin{cases} \dfrac{I_o(\alpha\sqrt{1 - (n/M)^2})}{I_o(\alpha)} & : |n| < M \\ 0 & : |n| \geq M \end{cases}$$

$I_o(x)$ is the modified Bessel function:

$$I_o(x) = \sum_{k=0}^{\infty} \left[\frac{1}{k!} \left(\frac{x}{2} \right)^k \right]^2.$$

Given f_1 and f_2 as filter settings, and for given filter parameters M and α_K, one can easily compute the filter characteristic (e.g., Figure III.4.5):

$$H(f) = h(0) + \sum_{n=1}^{M} 2\, h(n)\, \cos(2\pi nf/f_s).$$

In our study, we used filter settings $\alpha_K = 2$, $M = 12$ (or $N_K = 25$), corresponding, with the sampling rate $f_s = 128$ Hz, to a time interval of length 195 msec, and lower and upper cutoff frequencies, f_1 and f_2, respectively, for analysis of the alpha (8–12 Hz), beta (19–26 Hz), gamma (38–42 Hz), and theta (0–6 Hz) band.

5

Morphological Color Image Processing: Theory and Applications

I. Andreadis
Democritus University of Thrace

Maria I. Vardavoulia
Democritus University of Thrace

G. Louverdis
Democritus University of Thrace

Ph. Tsalides
Democritus University of Thrace

5.1 Introduction

Mathematical morphology is a geometric approach to image processing that was developed as a powerful tool for shape analysis in binary and gray-scale images. The foundations of the theory of mathematical morphology were laid down in 1975 by G. Matheron, in his book entitled *Random Sets and Integral Geometry*.[1] Originally, this methodology was developed for binary and, later, gray-scale images.[2–5] Morphology has been used in a number of image processing and analysis problems, such as edge detection, noise suppression, skeletonization, and texture and shape analysis. It can be used for applications such as medical imaging, image compression, and geological image processing.[4–7]

The extension of the concepts of binary and gray-scale morphology to color images is not straightforward. When such techniques are applied independently to each primary color component of the image (component-wise morphology), there is possibility for loss or corruption of information of the image, since, in most cases, image components are highly correlated.[6–8] An alternative, perhaps more natural way to approach the problem of color morphology is to treat the color at each pixel as a *vector*. This morphological technique is called vector morphology.[7] In order to define the basic morphological operators in this approach, a concept for a supremum (or infimum) is necessary, and thus it is important to define an appropriate ordering of vectors (colors), in a selected color space.[7]

A number of different ways to order multivariate data have been proposed, since there is no generally accepted way to define ordering for this type of data in color spaces.[9-14] In marginal ordering, the color image is separated into a set of channels, in such a way that it can be fully recovered by merging the channels. Then, a univariate ordering is applied to each channel independently and the processed channels are merged to form the output image. Marginal ordering produces new color values in the output image, and thus alters the color balance of the image, which is an essential disadvantage for a variety of applications, such as morphological filtering.

In reduced ordering, a scalar parameter function is computed for each pixel of the color image and the ordering is performed according to the resulting scalar values. This procedure may lead to the existence of more than one suprema (or infima) and, thus, introduce ambiguity in the resultant data. The basic property of this type of ordering is that the 'strength' of each color value in the comparison between color values (which takes place at each computational step of a morphological operation) does not depend on the color context.

In partial ordering, the input multivariate samples are partitioned into smaller groups which are then ordered. This type of ordering may also lead to the disadvantage of multiple suprema (or infima).

In conditional ordering, the vectors are initially ordered according to the ordered values of one of their components, e.g., the first component. At the second step, vectors that have the same value for the first component are ordered according to the ordered values of another component, e.g., the second component, etc.

An extension of standard mathematical morphology, called "soft mathematical morphology,"[15-20] was introduced by Koskinen et al., in order to improve the behavior of standard morphological filters in noisy environments. In soft gray-scale morphology, the min/max operators which are used in standard morphology are substituted by weighted order statistics and the structuring element is divided into two parts: the core and the soft boundary. It has been shown that soft morphological operations retain most of the properties of standard morphological operations, but they are also less sensitive to impulse noise and to small variations in object shape.

The aim of this chapter is to present a number of typical morphological applications for color image processing. For this purpose we use an extension of mathematical morphology to color images based on a conditional ordering scheme in the HSV color space. This methodology is described in detail in Section 5.2. In Section 5.3 we present the basic concepts of soft mathematical morphology for color images. Finally, in Section 5.4, a number of morphological applications for color image processing are discussed. Experimental results demonstrate the use of mathematical morphology in the processing and analysis of color images, in applications such as noise suppression, edge detection, and skeleton extraction.

5.2 Morphological Color Image Processing

This section presents a generalization of the concepts of gray-scale morphology to color images. A vector ordering scheme in the HSV color space is described, infimum and supremum operators are defined, and the fundamental morphological operations of erosion and dilation are extracted.[14] In the following sections, this methodology will be referred to as *standard vector morphology*. The basic properties of the presented vector morphology are described and its similarities to gray-scale morphology are pointed out. The main advantages of this approach are that the defined morphological operations are vector preserving and always produce unique results. Thus, no ambiguity is introduced in the resultant data. Furthermore, standard vector morphology is compatible with gray-scale morphology, and its operations possess the same basic properties as the corresponding gray-scale operations. The presented methodology also provides improved results in many morphological applications compared to other related morphological approaches.[14]

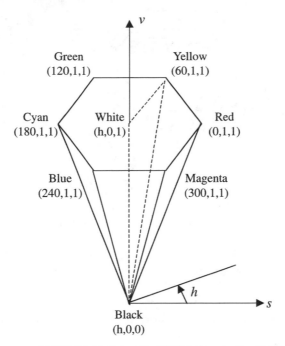

FIGURE III.5.1 The HSV color space.

5.2.1 Vector Ordering in the HSV Color Space

The classic way to represent color images is by using the Red-Green-Blue (RGB) representation. However, this type of representation has some drawbacks, since the RGB components are highly correlated and, therefore, chromatic information is not directly fit for use.[7,8] The HSV (Hue-Saturation-Value) color space has been chosen since it is intimately related to the way in which human beings perceive color. Hue represents the dominant color as perceived by an observer. Saturation refers to relative purity or the amount of white light mixed with a hue. Value is used for the amount of black that is mixed with a hue.

In Figure III.5.1 the HSV hexcone is shown. In the HSV space, a color is a vector with three components: h (Hue), s (Saturation), v (Value). The h values range from 0–360, $h \in [0, 360)$. The s and v values range from 0–1, $s \in [0, 1]$, $v \in [0, 1]$. When s decreases without changing v, it is like adding white to the initial color. When v decreases without changing s, it is like adding black to the initial color. When both s and v decrease, various tones of the initial color are composed.

The proposed vector ordering scheme is as follows:

1. Initially, vectors are ordered with respect to the third component, v. In particular, vectors are sorted from those with the smallest v to vectors with the greatest v. As can be noticed from Figure III.5.2, in this way the HSV hexcone is sliced into levels which are vertical to its height. Any lower level with a given v is less than all higher levels which correspond to greater values of v. Thus, the smallest level that has been reduced to a point is black, with $v = 0$, while the level of the base of hexcone with $v = 1$ is the greatest level.

2. Vectors having the same value of v, i.e., vectors lying on the same level, are then ordered with respect to the second component s. In particular, they are sorted from vectors with the greatest s to vectors with the smallest s. As can be noticed from Figure III.5.2, in this way each level in the HSV hexcone is sliced into homocentric hexagons. At a specific level, any external hexagon with corresponding s value is less than all hexagons included in it that correspond

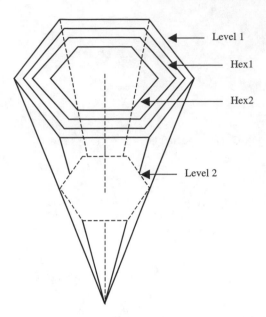

FIGURE III.5.2   Ordering of the HSV color space vectors (levels and homocentric hexagons of different order).

to smaller s values. Thus, at a specific level, the most external hexagon which corresponds to $s = 1$ is the smallest, while the hexagon that has been reduced to a point with $s = 0$ is the greatest.

3. Finally, vectors that have the same values of s and v, i.e., vectors that are included in a specific hexagon, are ordered with respect to the h component. More specifically, they are sorted from vectors with the smallest h to vectors with the greatest h.

The meaning of the ordering procedure described above is that the larger the mixture of a color with black is, the smaller the color is considered, regardless of its saturation and its hue. In the case that two colors have been mixed with the same amount of black, the smaller is the one with the less mixture of white. If two colors are equally dark, i.e., they have been mixed with the same amount of black and the same amount of white, the smaller is the one with the smaller hue value.

Based on the previous discussion, we define vector operators $<$ and $=$, for two colors $c_1(h_1, s_1, v_1)$ and $c_2(h_2, s_2, v_2)$, as follows:

$$c_1(h_1, s_1, v_1) < c_2(h_2, s_2, v_2) \Leftrightarrow \begin{cases} v_1 < v_2 \\ \quad \text{or} \\ v_1 = v_2 \text{ and } s_1 > s_2 \\ \quad \text{or} \\ v_1 = v_2 \text{ and } s_1 = s_2 \text{ and } h_1 < h_2 \end{cases} \quad \text{(III.5.1)}$$

$$c_1(h_1, s_1, v_1) \equiv c_2(h_2, s_2, v_2) \Leftrightarrow \begin{cases} v_1 = v_2 \\ \text{and} \\ s_1 = s_2 \\ \text{and} \\ h_1 = h_2 \end{cases} \quad \text{(III.5.2)}$$

5.2.2 Definitions of Infimum and Supremum Operators

Let SB_n be a subset of the HSV space, which includes n vectors $c_1(h_1, s_1, v_1)$, $c_2(h_2, s_2, v_2)$, ..., $c_n(h_n, s_n, v_n)$. Using the vector ordering procedure described in Section 5.2.1, we define the $\wedge$ infimum operator in SB_n as follows:

$$\wedge SB_n = \wedge\{c_1(h_1, s_1, v_1), c_2(h_2, s_2, v_2), \ldots, c_n(h_n, s_n, v_n)\}$$

$$= c_k(h_k, s_k, v_k): \begin{cases} v_k = \min\{v_1, v_2, \ldots, v_n\} \text{ if } \nexists\, i \neq j : v_i = v_j = \min\{v_1, v_2, \ldots, v_n\} \\ \qquad\qquad\qquad\qquad \text{or} \\ v_k = v_i = v_j = \min\{v_1, v_2, \ldots, v_n\} \text{ and } s_k = \max\{s_i, s_j\} \\ \text{If } \exists\, i \neq j : v_i = v_j = \min\{v_1, v_2, \ldots, v_n\} \text{ and } s_i \neq s_j \\ \qquad\qquad\qquad\qquad \text{or} \\ v_k = v_i = v_j = \min\{v_1, v_2, \ldots, v_n\} \text{ and } s_k = s_i = s_j \\ \text{and } h_k = \min\{h_i, h_j\} \text{ if } \exists\, i \neq j : v_i = v_j = \min\{v_1, v_2, \ldots, v_n\} \end{cases}$$

$$(\text{III.5.3})$$

with $1 \leq k \leq n, 1 \leq i, j \leq n$.

In a similar way we define the $\vee$ supremum operator in SB_n as follows:

$$\vee SB_n = \vee\{c_1(h_1, s_1, v_1), c_2(h_2, s_2, v_2), \ldots, c_n(h_n, s_n, v_n)\}$$

$$= c_k(h_k, s_k, v_k): \begin{cases} v_k = \max\{v_1, v_2, \ldots, v_n\} \text{ if } \nexists\, i \neq j : v_i = v_j = \max\{v_1, v_2, \ldots v_n\} \\ \qquad\qquad\qquad\qquad \text{or} \\ v_k = v_i = v_j = \max\{v_1, v_2, \ldots, v_n\} \text{ and } s_k = \min\{s_i, s_j\} \\ \text{if } \exists\, i \neq j : v_i = v_j = \max\{v_1, v_2, \ldots, v_n\} \text{ and } s_i \neq s_j \\ \qquad\qquad\qquad\qquad \text{or} \\ v_k = v_i = u_j = \max\{v_1, v_2, \ldots, v_n\} \text{ and } s_k = s_i = s_j \\ \text{and } h_k - \max\{h_i, h_j\} \text{ if } \exists\, i \neq j : v_i = u_j = \max\{v_1, v_2, \ldots v_n\} \end{cases}$$

$$(\text{III.5.4})$$

with $1 \leq k \leq n, 1 \leq i, j \leq n$.

It is important to notice that the application of these two operators to a particular set SB_n results in only *one* output vector which *is* included in the input set SB_n. Consequently, the presented operators are *vector preserving* since no vector (color), which is not present in the input data, is generated.[11]

From Equation III.5.3 it is obvious that the introduced infimum operator possesses the following property:

1. If SB_n is an arbitrary subset of the HSV color space which includes n vectors, $SB_n = \{c_1, c_2, \ldots, c_n\}$, then:

$$\forall\, SB_n, \exists\, c_i \in SB_n \text{ such that } \wedge(SB_n) = c_i. \qquad (\text{III.5.5})$$

Moreover, the $\wedge$ operator possesses the following two basic properties:

2. $\forall\, c_1, c_2 \in \text{HSV}$:

$$\wedge\{c_1, c_2\} = c_1 \text{ and } \wedge\{c_2, c_1\} = c_2 \Leftrightarrow c_1 = c_2. \qquad (\text{III.5.6})$$

3. $\forall\, c_1, c_2, c_3 \in \text{HSV}$:

$$\wedge\{c_1, c_2\} = c_1 \text{ and } \wedge\{c_2, c_3\} = c_2, \Rightarrow \wedge\{c_1, c_3\} = c_1. \qquad (\text{III.5.7})$$

Properties 1, 2, and 3 ensure that the $\wedge$ operator defines a total ordering relation in the HSV color space.[11] Similarly, the previous discussion applies to the $\vee$ operator which can be proven to be vector preserving and to possess properties 1, 2, and 3 as well.

5.2.3 Morphological Operators for Color Images

In the following subsections the fundamental operations of standard vector erosion and dilation for color images are defined.

5.2.3.1 Basic Definitions

Let f, g be two functions such that $f, g : R^2 \to \text{HSV}$, which is the case of a color image. We define:

$$\textit{Finite Domain of } f(z) : \text{D}[f] = \{z : f(z) \in \text{HSV}\} \qquad \text{(III.5.8)}$$

If $f(k) = c(h_{kf}, s_{kf}, v_{kf})$ and $g(k) = c(h_{kg}, s_{kg}, v_{kg}), k \in R^2$, then we define the operations of *addition* and *subtraction* as follows:

Addition:

$$f(k) + g(k) = c(h_{kf} + h_{kg}, s_{kf} + s_{kg}, v_{kf} + v_{kg}) \qquad \text{(III.5.9)}$$

Limitations:

$$\text{If } h_{kf} + h_{kg} > 360 \text{ then } h_{kf} + h_{kg} = 360$$
$$\text{If } s_{kf} + s_{kg} > 1 \text{ then } s_{kf} + s_{kg} = 1$$
$$\text{If } v_{kf} + v_{kg} > 1 \text{ then } v_{kf} + v_{kg} = 1$$

Subtraction:

$$f(k) - g(k) = c(h_{kf} - h_{kg}, s_{kf} - s_{kg}, v_{kf} - v_{kg}) \qquad \text{(III.5.10)}$$

Limitations:

$$\text{If } h_{kf} - h_{kg} < 0 \text{ then } h_{kf} - h_{kg} = 0$$
$$\text{If } s_{kf} - s_{kg} < 0 \text{ then } s_{kf} - s_{kg} = 0$$
$$\text{If } v_{kf} - v_{kg} < 0 \text{ then } v_{kf} - v_{kg} = 0$$

We can now give the following definitions:

Spatial translation of f by d :

$$f_d(z) = f(z - d) \qquad \text{(III.5.11)}$$

Offset of f by y :

$$(f + y)(z) = f(z) + y, y \in \text{HSV} \qquad \text{(III.5.12)}$$

Translation of f :

$$(f_d + y)(z) = f(z - d) + y, y \in \text{HSV} \qquad \text{(III.5.13)}$$

Reflection through the origin of spatial coordinates axes:

$$f'(z) = f(-z) \qquad \text{(III.5.14)}$$

g beneath f :

$$g \ll f \Leftrightarrow D[g] \subseteq D[f] \text{ and } g(z) \leq f(z) \ \forall z \in D[g] \qquad \text{(III.5.15)}$$

5.2.3.2 Standard Vector Erosion

Let us consider the set f to be a color image with pixel values in the HSV color space and the set g to be the *structuring element* for the vector morphological operations that will be described here.

We define *standard vector erosion* of f by g at a point x as follows:

$$(f \Theta g)(x) = \wedge \{ f(z) - g_x(z) \} \tag{III.5.16}$$

$$\text{for } z \in D[f] \cap D[g_x], x : D[g_x] \subseteq D[f] \tag{III.5.17}$$

The previous definition implies that in order to perform the standard vector erosion of an input image f by the structuring element g at a point x:

1. First, we translate spatially g by x, so that its origin is located at point x.
2. We find all *differences* between colors of the points of f with the colors of the corresponding points of the translated g, $\forall z \in D[f] \cap D[g_x]$ using the subtraction operation defined in Equation III.5.10.
3. The result of the previous step is a set of colors. We find the infimum of these colors using the infimum operator that was defined previously. This infimum color is the color of the eroded image at the point x.

In Equation III.5.17 the restriction for x implies that the vector erosion of f by g is not defined at any point x for which the domain of the translated g by x does not completely lie in the domain of f, in a similar way to gray-scale erosion. It should be noticed that vector erosion becomes the standard gray-scale erosion when the vector dimension is equal to one.

An equivalent definition of standard vector erosion follows:

$$(f \Theta g)(x) = \wedge \{ f_{-x}(z) - g(z) \} = \wedge \{ f(z + x) - g(z) \} \tag{III.5.18}$$

$$\text{for } z \in D[g]$$

In this definition, instead of translating spatially g by x, we translate spatially f by $-x$.

It is worthwhile to make some remarks about the meaning of vector erosion that point out its similarities to that of gray-scale erosion. In both cases the resultant image depends on both the shape and the values of the points of the structuring element. More specifically, the greater (brighter) the colors of the structuring element are, the more the input image is eroded. When the structuring element includes even one point with white color, $c(h, 0, 1)$, the effect of the erosion operation on the input image is maximum and all the, points in the resultant image obtain black color, $c(h, 0, 0)$.

Example 1

The following example demonstrates a case of vector erosion. The input image and the structuring element are illustrated in Figure III.5.3. In this figure f and g are two-dimensional color images $(f, g : R^2 \to \text{HSV})$ represented on the Cartesian grid. The arrows denote the origin. In particular:

$$f(1, 2) = c(0, 0.8, 0.9), x = (1, 2), D[g] = \{(0, 0), (0, 1), (-1, 0), (-1, 1)\}$$

According to Equation III.5.16:

$$
\begin{aligned}
(f \Theta g)(1, 2) = \min\{ & [f(1, 2) - g(0, 0)], [f(0, 2) - g(-1, 0)], \\
& [f(1, 3) - g(0, 1)], [f(0, 3) - g(-1, 1)] \} \\
= \min\{ & [c(0, 0.8, 0.9) - c(10, 0.4, 0.6)], [c(60, 0.5, 0.8) - c(55, 0.4, 0.3)], \\
& [c(30, 0.5, 0.7) - c(10, 0.4, 0.3)], [c(80, 0.9, 0.6) - c(75, 0.4, 0.3)] \}
\end{aligned}
$$

$$= \min\{c(0, 0.4, 0.3), c(5, 0.1, 0.5), c(20, 0.1, 0.4), c(5, 0.5, 0.3)\}$$
$$= \min\{c(5, 0.5, 0.3), c(0, 0.4, 0.3), c(20, 0.1, 0.4), c(5, 0.1, 0.5)\}$$
$$= c(5, 0.5, 0.3)$$

5.2.3.3 Standard Vector Dilation

We define *standard vector dilation* of f by g at a point x as follows:

$$(f \oplus g)(x) = \vee\{f(z) + g'_{-x}(-z)\} \tag{III.5.19}$$

$$\text{for } z \in D[f] \cap D[g'_{-x}], \quad x: D[g'_{-x}] \cap D[f] \neq \varnothing \tag{III.5.20}$$

Standard vector dilation is performed in a similar way to vector erosion, according to Equation III.5.19. The previous definition implies that in order to perform the vector dilation of an input image f by the structuring element g at a point x:

1. First, we translate spatially the reflection g' of g by $-x$, so that its origin is located at point x.
2. We find all *sums* between colors of the points of f with the colors of the corresponding points of the translated g', $\forall z \in D[f] \cap D[g'_{-x}]$ using the addition operation defined in Equation III.5.9.
3. The result of the previous step is a set of colors. We find the supremum of these colors using the supremum operator that was defined previously. This supremum color is the color of the dilated image at the point x.

It should be noticed that vector dilation becomes the standard gray-scale dilation when the vector dimension is equal to one.

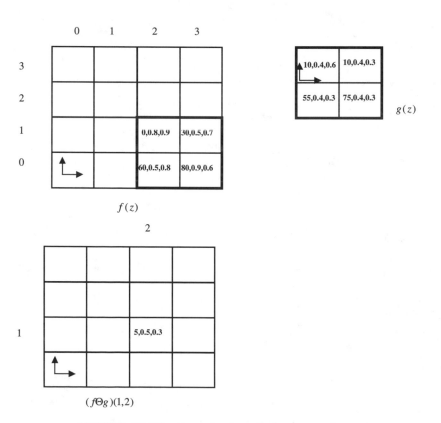

FIGURE III.5.3   Example of standard vector erosion.

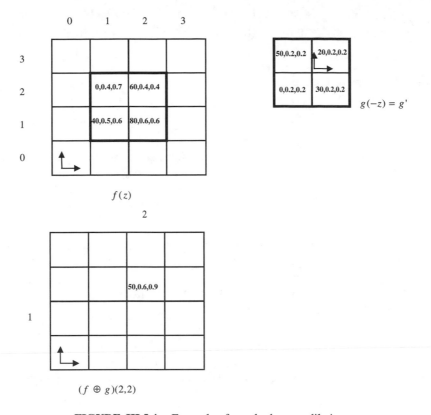

$f(z)$

$g(-z) = g'$

$(f \oplus g)(2,2)$

FIGURE III.5.4 Example of standard vector dilation.

An equivalent definition of standard vector dilation follows:

$$(f \oplus g)(x) = \vee\{f_{-x}(-z) + g(z)\} = \vee\{f(x - z) + g(z)\} \qquad \text{(III.5.21)}$$

$$\text{for } z \in D[g], x - z \in D[f]$$

In this definition, instead of translating spatially g' by $-x$, we translate spatially f' by $-x$.

Again, the meaning of vector dilation is similar to that of gray-scale dilation. As in the case of gray-scale dilation, the image that results from the application of vector dilation depends on both the shape and the values of the points of the structuring element. More specifically, the greater (brighter) the colors of the structuring element are, the more the input image is dilated.

Example 2

The following example demonstrates a case of vector dilation. The input image and the structuring element are illustrated in Figure III.5.4. In this figure f and g are two-dimensional color images $(f, g : R^2 \rightarrow \text{HSV})$ represented on the Cartesian grid. The arrows denote the origin. In particular:

$$f(2, 2) = c(60, 0.4, 0.4), x = (2, 2), D[g'] = \{(0, 0), (0, -1), (-1, 0), (-1, -1)\}$$

According to Equation III.5.19:

$$(f \oplus g)(2, 2) = \max\{[f(2, 2) + g(0, 0)], [f(1, 2) + g(-1, 0)], [f(2, 1)$$
$$+ g(0, -1)], [f(1, 1) + g(-1, 1)]\}$$
$$= \max\{[c(60, 0.4, 0.4) + c(20, 0.2, 0.2)], [c(80, 0.6, 0.6) + c(30, 0.2, 0.2)],$$
$$[c(0, 0.4, 0.7) + c(50, 0.2, 0.2)], [c(40, 0.5, 0.6) + c(0, 0.2, 0.2)]\}$$

$$= \max\{c(80, 0.6, 0.6), c(110, 0.8, 0.8), c(50, 0.6, 0.9), c(40, 0.7, 0.8)\}$$

$$= \max\{c(80, 0.6, 0.6), c(110, 0.8, 0.8), c(40, 0.7, 0.8), c(50, 0.6, 0.9)\}$$

$$= c(50, 0.6, 0.9)$$

5.2.3.4   Standard Vector Opening and Closing

Standard vector opening of f by g is defined as follows:

$$f \circ g = (f \ominus g) \oplus g \tag{III.5.22}$$

The previous definition implies that in order to perform the standard opening of an image f by the structuring element g, we have to perform a vector erosion followed by a vector dilation. *Standard vector closing* of f by g is defined as follows:

$$f \bullet g = (f \oplus g) \ominus g \tag{III.5.23}$$

According to the previous definition, in order to perform the standard closing of an image f by the structuring element g, we have to perform a vector dilation followed by a vector erosion.

5.2.4   Basic Properties of Vector Erosion and Dilation

In the following subsections we present the basic properties of the standard vector morphological operations. We approach the presented operators from a lattice theory point of view and we prove that they form an adjunction for any structuring element. For a detailed exposition on complete lattice theory, the reader may refer to References 21 and 22. Furthermore, we prove that vector erosion and dilation possess the same basic properties with gray-scale erosion and dilation, respectively.

5.2.4.1   The Adjunction Property

A set L with an ordering scheme is called a complete lattice if every subset H of L has a least upper bound (supremum) $\vee H$ and a greatest lower bound (infimum) $\wedge H$.

Let us consider $\varepsilon : L \rightarrow M$ and $\delta : M \rightarrow L$ as two operators. The pair (ε, δ) is an adjunction between L and M if:[7]

$$\delta(Y) \leq X \Leftrightarrow Y \leq \varepsilon(X), \text{ for all } X \in L, Y \in M \tag{III.5.24}$$

When a pair of operators is an *adjunction*, their composition defines an opening and a closing, and they can be used for the design of morphological filters.[7]

Let ε, δ be the erosion and the dilation operators defined in this section, respectively. We will prove that for any structuring element, the pair (ε, δ) of the presented operators is an *adjunction*.

Proof

In the case of color images, if we consider the set f to be a color image with pixel values in the HSV color space and the set g to be the *structuring element*, the Equation III.5.35 of adjunction property can be expressed as:

$$f(x) \leq (f \ominus g)(y) \Leftrightarrow (f \oplus g)(x) \leq f(y), \text{ with } x, y \in R^2, f(x) \in \text{HSV} \tag{III.5.25}$$

Let us suppose the following:

$$f(x) \leq (f \ominus g)(y)) \tag{III.5.26}$$

$$(f \ominus g)(y) = f(m) - g_y(m) \tag{III.5.27}$$

$$(f \oplus g)(x) = f(l) + g_{-x}(-l) \tag{III.5.28}$$

for a point m at the neighborhood of point y and for a point l at the neighborhood of point x.

$$(\text{III.5.26}), (\text{III.5.27}) \Rightarrow f(x) \leq f(m) - g_y(m) \qquad (\text{III.5.29})$$

$$(\text{III.5.27}) \Rightarrow f(m) - g_y(m) \leq f(y) - g_y(y) \qquad (\text{III.5.30})$$

$$(\text{III.5.29}), (\text{III.5.30}) \Rightarrow f(x) \leq f(y) - g_y(y) \qquad (\text{III.5.31})$$

For the structuring element

$$g : 0 \in g \Rightarrow g_y(y) = g_{-x}(-x) \qquad (\text{III.5.32})$$

$$(\text{III.5.31}), (\text{III.5.32}) \Rightarrow f(x) \leq f(y) - g_{-x}(-x) \Rightarrow f(x) + g_{-x}(-x) \leq f(y) \qquad (\text{III.5.33})$$

$$(\text{III.5.28}) \Rightarrow f(x) + g_{-x}(-x) \leq f(l) + g_{-x}(-l) \qquad (\text{III.5.34})$$

$$(\text{III.5.33}) - (\text{III.5.34}) \Rightarrow [f(x) + g_{-x}(-x)] - [f(x) + g_{-x}(-x)]$$

$$\leq f(y) - [f(l) + g_{-x}(-l)]$$

$$\Rightarrow (0,0,0) \leq f(y) - [f(l) + g_{-x}(-l)]$$

$$\Rightarrow \begin{cases} 0 \leq v_{yf} - (v_{lf} + v_{lg'}) \\ \qquad \text{or} \\ 0 = v_{yf} - (v_{lf} + v_{lg'}) \text{ and } 0 \geq s_{yf} - (s_{lf} + s_{lg'}) \\ \qquad \text{or} \\ 0 = v_{yf} - (v_{lf} + v_{lg'}), 0 = s_{yf} - (s_{lf} + s_{lg'}) \text{ and} \\ 0 \leq h_{yf} - (h_{lf} + h_{lg'}) \end{cases}$$

$$\Rightarrow \begin{cases} v_{yf} \geq v_{lf} + v_{lg'} \\ \qquad \text{or} \\ v_{yf} = v_{lf} + v_{lg'} \text{ and } s_{yf} \leq s_{lf} + s_{lg'} \\ \qquad \text{or} \\ v_{yf} = v_{lf} + v_{lg'}, s_{yf} = s_{lf} + s_{lg'} \text{ and} \\ h_{yf} \geq (h_{lf} + h_{lg'}) \end{cases}$$

$$\Rightarrow f(y) \geq f(l) + g_{-x}(-l) \qquad (\text{III.5.35})$$

From Equations III.5.28, III.5.35 $\Rightarrow f(y) \geq (f \oplus g)(x)$.

Therefore, the pair (ε, δ) of the proposed operators is an *adjunction* and their composition defines an opening and a closing, for any structuring element. Thus, they can be used for the design of morphological filters.

5.2.4.2 Other Properties

In the following subsections the other basic properties of vector erosion and dilation are proved.

5.2.4.2.1 *Extensivity – Anti-Extensivity*

If the origin lies inside the structuring element, standard vector erosion is *anti-extensive* and standard vector dilation is *extensive:*

$$0 \in g \Rightarrow f \ominus g \ll f \qquad (\text{III.5.36})$$

$$0 \in g \Rightarrow f \ll f \oplus g \qquad (\text{III.5.37})$$

Proof

Let us consider $D[f] \cap D[g_x] = \{z_1, z_2, \ldots, z_n\}$, with $x \in D[f \ominus g]$. Moreover, let $f(z_1) = c$ (h_{1f}, s_{1f}, v_{1f}), $f(z_2) = c(h_{2f}, s_{2f}, v_{2f}), \ldots, f(z_n) = c(h_{nf}, s_{nf}, v_{nf})$ and $g(z_1) = c(h_{1g}, s_{1g}, v_{1g})$, $g(z_2) = c(h_{2g}, s_{2g}, v_{2g}), \ldots, g(z_n) = c(h_{ng}, s_{ng}, v_{ng})$. Then, $0 \in g \Rightarrow x \in D[g_x]$.

But

$$x : D[g_x] \subseteq D[f] \Rightarrow x \in D[f] \cap D[g_x] \Rightarrow x = z_m \in D[f] \cap D[g_x],$$
$$\text{with } 1 \leq m \leq n, \Rightarrow f(x) = f(z_m) \Rightarrow f(x) = c(h_{mf}, s_{mf}, v_{mf}) \tag{III.5.38}$$

By definition:

$$(f \ominus g)(x) = c(h_{\wedge s}, s_{\wedge s}, v_{\wedge s}) \tag{III.5.39}$$

where

$$S = SBfg_x = \{f(z) - g_x(z)\}, \text{ for } z \in D[f] \cap D[g_x].$$
$$SBfg_x = \{f(z) - g_x(z)\} = \{c(h_{1f} - h_{1g}, s_{1f} - s_{1g}, v_{1f} - v_{1g}),$$
$$c(h_{2f} - h_{2g}, s_{2f} - s_{2g}, v_{2f} - v_{2g}), \ldots, c(h_{nf} - h_{ng}, s_{nf} - s_{ng}, v_{nf} - v_{ng})\}$$
$$\Rightarrow \wedge SBfg_x = c(h_{kf} - h_{kg}, s_{kf} - s_{kg}, v_{kf} - v_{kg}), \text{ with } 1 \leq k \leq n \tag{III.5.40}$$

From Equations III.5.39 and III.5.40 $\Rightarrow$

$$(f \ominus g)(x) = c(h_{kf} - h_{kg}, s_{kf} - s_{kg}, v_{kf} - v_{kg}) \tag{III.5.41}$$

$$(\text{III.5.40}) \Rightarrow \begin{cases} v_{kf} - v_{kg} < v_{mf} - v_{mg} \\ \qquad\qquad \text{or} \\ v_{kf} - v_{kg} = v_{mf} - v_{mg} \text{ and } s_{kf} - s_{kg} > s_{mf} - s_{mg} \\ \qquad\qquad \text{or} \\ v_{kf} - v_{kg} = v_{mf} - v_{mg}, s_{kf} - s_{kg} = s_{mf} - s_{mg} \text{ and} \\ h_{kf} - h_{kg} < h_{mf} - h_{mg} \end{cases} \tag{III.5.42}$$

We consider the following two cases:

1. The pixel $g(z_m)$ is black, with $v_{mg} = 0$ and $s_{mg} = 0$ and $h_{mg} \geq 0$. In this case:

$$(\text{III.5.42}) \Rightarrow \begin{cases} v_{kf} - v_{kg} < v_{mf} \\ \qquad\qquad \text{or} \\ v_{kf} - v_{kg} = v_{mf} \text{ and } s_{kf} - s_{kg} > s_{mf} \\ \qquad\qquad \text{or} \\ v_{kf} - v_{kg} = v_{mf}, s_{kf} - s_{kg} = s_{mf} \text{ and} \\ h_{kf} - h_{kg} < h_{mf} - h_{mg} < h_{mf} \end{cases} \tag{III.5.43}$$

From (III.5.38), (III.5.41), and (III.5.43) $\Rightarrow (f \ominus g)(x) < f(x)$ (III.5.44)

2. The pixel $g(z_m)$ has $v_{mg} > 0 \Rightarrow -v_{mg} < 0 \Rightarrow v_{mf} - v_{mg} < v_{mf}$ and:

$$\text{from (III.5.30)} \Rightarrow v_{kf} - v_{kg} < v_{mf} \tag{III.5.45}$$

From (III.5.38), (III.5.41), and (III.5.45) $\Rightarrow (f \ominus g)(x) < f(x)$ (III.5.46)

Thus, for the two cases, from Equations III.5.44 and III.5.46:

$$\left. \begin{array}{c} 0 \in g \Rightarrow (f \ominus g)(x) < f(x) \\ \text{and also } x : D[g_x] \subseteq D[f] \Rightarrow [f \ominus g] \subseteq D[f] \end{array} \right\} \Rightarrow f \ominus g \ll f$$

Standard vector dilation's extensivity can be proven in a similar way.

5.2.4.2.2 *Increasing – Decreasing*

Both standard vector erosion and dilation are *monotonically increasing* relative to a specific structuring element:

$$f_1 \ll f_2 \Rightarrow f_1 \ominus g \ll f_2 \ominus g \tag{III.5.47}$$

$$f_1 \ll f_2 \Rightarrow f_1 \oplus g \ll f_2 \oplus g \tag{III.5.48}$$

Proof

Let us consider $D[f_1] \cap D[g_x] = D[f_2] \cap D[g_x] = \{z_1, z_2, \ldots, z_n\}$ and $f_1(z_1) = c(h_1 f_1, s_1 f_1, v_1 f_1)$, $f_1(z_2) = c(h_2 f_1, s_2 f_1, v_2 f_1), \ldots, f_1(z_n) = c(h_n f_1, s_n f_1, v_n f_1)$. In the same way, $f_2(z_1) = c(h_1 f_2, s_1 f_2, v_1 f_2)$, $f_2(z_2) = c(h_2 f_2, s_2 f_2, v_2 f_2), \ldots, f_2(z_n) = c(h_n f_2, s_n f_2, v_n f_2)$ and $g(z_1) = c(h_{1g}, s_{1g}, v_{1g})$, $g(z_2) = c(h_{2g}, s_{2g}, v_{2g}), \ldots, g(z_n) = c(h_{ng}, s_{ng}, v_{ng})$. If $\wedge\{f_1(z) - g_x(z)\} = c(h_{k1 f_1} - h_{k1g}, s_{k1 f_1} - s_{k1g}, v_{k1 f1} - v_{k1g})$, with $1 \le k_1 \le n$, $z_{k1} \in \{z_1, \ldots, z_n\}$, and $\wedge\{f_2(z) - g_x(z)\} = c(h_{k2 f2} - h_{k2g}, s_{k2 f2} - s_{k2g}, v_{k2 f2} - v_{k2g})$, with $1 \le k_2 \le n$, $z_{k2} \in \{z_1, \ldots, z_n\}$, then, by definition:

$$(f_1 \ominus g)(x) = c(h_{k1 f1} - h_{k1g}, s_{k1 f1} - s_{k1g}, v_{k1 f1} - v_{k1g}) \tag{III.5.49}$$

$$(f_2 \ominus g)(x) = c(h_{k2 f2} - h_{k2g}, s_{k2 f2} - s_{k2g}, v_{k2 f2} - v_{k2g}) \tag{III.5.50}$$

$$(\text{III.5.49}) \Rightarrow \begin{cases} v_{k1 f1} - v_{k1g} < v_{k2 f1} - v_{k2g} \\ \qquad\qquad \text{or} \\ v_{k1 f1} - v_{k1g} = v_{k2 f1} - v_{k2g} \text{ and } s_{k1 f1} - s_{k1g} > s_{k2 f1} - s_{k1g} \\ \qquad\qquad \text{or} \\ v_{k1 f1} - v_{k1g} = v_{k2 f1} - v_{k2g},\, s_{k1 f1} - s_{k1g} = s_{k2 f1} - s_{k2g} \text{ and} \\ h_{k1 f1} - h_{k1g} < h_{k2 f1} - h_{k2g} \end{cases} \tag{III.5.51}$$

But $f_1 \ll f_2 \Rightarrow f_1(z_{k2}) < f_2(z_{k2}) \Rightarrow c(h_{k2 f1}, s_{k2\,f1}, v_{k2\,f1}) < c(h_{k2 f2}, s_{k2 f2}, v_{k2 f2})$

$$\Rightarrow \begin{cases} v_{k2 f1} < v_{k2 f2} \\ \qquad\qquad \text{or} \\ v_{k2 f1} = v_{k2 f2} \text{ and } s_{k2 f1} > s_{k2 f2} \\ \qquad\qquad \text{or} \\ v_{k2 f1} = v_{k2 f2},\, s_{k2 f1} = s_{k2 f2} \\ \qquad\qquad \text{and} \\ h_{k2 f1} < h_{k2 f2} \end{cases}$$

$$\Rightarrow \begin{cases} v_{k2 f1} - v_{k2g} < v_{k2 f2} - v_{k2g} \\ \qquad\qquad \text{or} \\ v_{k2 f1} - v_{k2g} = v_{k2 f2} - v_{k2g} \text{ and } s_{k2 f1} - s_{k2g} > s_{k2 f2} - s_{k2g} \\ \qquad\qquad \text{or} \\ v_{k2 f1} - v_{k2g} = v_{k2 f1} - v_{k2g},\, s_{k2 f1} - s_{k2g} = s_{k2 f2} - s_{k2g} \text{ and} \\ h_{k2 f1} - h_{k2g} < h_{k2 f2} - h_{k2g} \end{cases} \tag{III.5.52}$$

$$(\text{III.5.51}) \Rightarrow \begin{cases} v_{k1f1} - v_{k1g} < v_{k2f2} - v_{k2g} \\ \qquad\qquad\text{or} \\ v_{k1f1} - v_{k1g} = v_{k2f2} - v_{k2g} \text{ and } s_{k1f1} - s_{k1g} > s_{k2f2} - s_{k2g} \\ \qquad\qquad\text{or} \\ v_{k1f1} - v_{k1g} = v_{k2f2} - v_{k2g}, s_{k1f1} - s_{k1g} = s_{k2f2} = s_{k2g} \text{ and} \\ h_{k1f1} - h_{k1g} < h_{k2f2} - h_{k2g} \end{cases} \qquad (\text{III.5.53})$$

From Equations III.5.49, III.5.50, and III.5.53

$$\Rightarrow (f_1 \ominus g)(x) < (f_2 \ominus g)(x) \qquad (\text{III.5.54})$$

Let

$$x \in D[f_1 \ominus g] \qquad (\text{III.5.55})$$

By definition

$$x : D[g_x] \subseteq D[f_1] \Rightarrow y + x \in D[f_1], \forall y \in D[g] \qquad (\text{III.5.56})$$

But

$$f_1 \ll f_2 \Rightarrow D[f_1] \subseteq D[f_2] \qquad (\text{III.5.57})$$

From Equations III.5.56 and III.5.57

$$\Rightarrow y + x \in D[f_2], \forall y \in D[g]$$

$$\Rightarrow x : D[g_x] \subseteq D[f_2] \Rightarrow x \in D[f_2 \ominus g] \qquad (\text{III.5.58})$$

From Equations III.5.55 and III.5.56

$$\Rightarrow D[f_1 \ominus g] \subseteq D[f_2 \ominus g] \qquad (\text{III.5.59})$$

Finally, from Equations III.5.54 and III.5.59:

$$f_1 \ll f_2 \Rightarrow f_1 \ominus g \ll f_2 \ominus g$$

In a similar way it can be proven that standard vector dilation is monotonically increasing relative to a specific structuring element.

5.2.4.2.3   *Duality*

Standard vector erosion and dilation are *dual* operations:

$$-(f \oplus g)(x) = (-f \ominus g')(x) \qquad (\text{III.5.60})$$

Proof

$$- \vee \{f_{-x}(-z) + g(z)\} = \wedge\{-f_{-x}(-z) - g(z)\} = \wedge\{-f_{-x}(z) - g(-z)\} \qquad (\text{III.5.61})$$

If we consider:

$$\{f_{-x}(-z) + g(z)\} = SB1$$

$$\{-f_{-x}(z) - g(-z)\} = SB2 \qquad (\text{III.5.62})$$

then Equation III.5.61 implies that

$$- \vee SB1 = \wedge SB2. \tag{III.5.63}$$

By definition:

$$(f \oplus g)(x) = c(h_{\vee SB1}, s_{\vee SB1}, v_{\vee SB1}) \Rightarrow -(f \oplus g)(x)$$

$$= c(-h_{\vee SB1}, -s_{\vee SB1}, -v_{\vee SB1})$$

$$\Rightarrow -(f \oplus g)(x) = c(h_{-\vee SB1}, s_{-\vee SB1}, v_{-\vee SB1}) \Rightarrow -(f \oplus g)(x)$$

$$= c(h_{\wedge SB2}, s_{\wedge SB2}, v_{\wedge SB2})$$

$$\Rightarrow -x(f \oplus g)(x) = (-f \ominus g')(x).$$

5.2.4.2.4 *Translation Invariance*

Both standard vector erosion and dilation are *translation invariant* relative to the input image:

$$(f \ominus g)_y + k = (f_y + k) \ominus g, k \in HSV \tag{III.5.64}$$

$$(f \oplus g)_y + k = (f_y + k) \oplus g, k \in HSV \tag{III.5.65}$$

Proof

It is enough to prove that $[(f \ominus g)(x)]_y + k = [(f_y + k) \ominus g](x + y), \forall x \in D[f \ominus g]$.

$$\text{If } k = c(h, s, v) \text{ then } [(f \ominus g)(x)] + k = [(f \ominus g)(x)] + c(h, s, v)$$

$$\Rightarrow [(f \ominus g)(x)] + k = [c(h_{\wedge SBfgx}, s_{\wedge SBfgx}, v_{\wedge SBfgx})] + c(h, s, v)$$

$$\Rightarrow [(f \ominus g)(x)]_y + k = [c(h_{\wedge SBfgx}, s_{\wedge SBfgx}, v_{\wedge SBfgx})]_y + c(h, s, v)$$

$$\Rightarrow [(f \ominus g)(x)]_y + k = c(h_{\wedge [SBfgx]y}, s_{\wedge [SBfgx]y}, v_{\wedge [SBfgx]y}) + c(h, s, v)$$

$$\Rightarrow [(f \ominus g)(x)]_y + k = c(h_{\wedge SBfyg(x+y)}, s_{\wedge SBfyg(x+y)}, v_{\wedge SBfyg(x+y)}) + c(h, s, v)$$

$$\Rightarrow [(f \ominus g)(x)]_y + k = c(h_{\wedge SBfyg(x+y)} + h, s_{\wedge SBfyg(x+y)} + s, v_{\wedge SBfyg(x+y)} + v)$$

$$\Rightarrow [(f \ominus g)(x)]_y + k = c(h_{\wedge SB(fy+k)g(x+y)}, s_{\wedge SB(fy+k)g(x+y)}, v_{\wedge SB(fy+k)g(x+y)})$$

$$\Rightarrow [(f \ominus g)(x)]_y + k = [(f_y + k) \ominus g](x + y).$$

In a similar way it can be proven that standard vector dilation is translation invariant relative to the input image as well.

5.3 Soft Morphology for Color Image Processing

The initial form of mathematical morphology is usually referred to as *standard mathematical morphology* in the literature, in order to be distinguished by its later extensions. Such an extension is *soft mathematical morphology*,[15–20] which was introduced by Koskinen et al. in order to improve the behavior of standard morphological filters in noisy environments.

Standard morphological transforms take into account the geometrical shape of the objects to be analyzed. Because of this main characteristic, they have been successfully used in many image processing applications, such as biomedical and electron microscopy analysis and various computer vision applications.[15] Yet another characteristic of standard morphological transforms is that they are quite sensitive to noise in the image under process. As a result, in some applications pre-filtering to remove noise is necessary. This pre-filtering may induce distortion of the shape of objects under consideration if is not done very carefully, causing in this way the downgrading of the overall system performance.

Soft morphological transforms retain most of the desirable properties of standard morphological transforms, but they are also less sensitive to impulse noise and to small variations in object shape. In soft morphological transforms the maximum or the minimum operations, used in standard gray-scale morphology, are substituted by weighted order statistics. A weighted order statistic is a specific element of a multi-set, the members of which have been ordered. A multi-set is a collection of objects, where repetition of objects is allowed. In particular, the structuring element is divided into two parts: the *core* and the *soft boundary*. The pixels of the core participate with a weight k greater than or equal to 1, while the pixels of the soft boundary participate with weight equal to 1. Thus, the results (differences or sums) involving the pixels of the core are repeated more than once (k times), before sorting. The kth order statistic is the result of a soft morphological transform. The parameter k is called the *repetition parameter*. It must be mentioned that for $k = 1$, the soft morphological operations become the standard morphological operations.

In this section an approach to soft color image mathematical morphology is presented. This extends the standard vector morphology theory discussed in Section 5.2 in the same way that soft gray-scale morphology extends the standard gray-scale morphology theory. It intends to improve the behavior of standard vector morphological transforms in noisy environments, just as in the case of gray-scale soft morphology. It maintains the concept of splitting the structuring element in two parts: the core and the soft boundary. It also preserves the concept of the repetition parameter k, which implies that the core "weights" more than the soft boundary in the computation of the result. In this approach, the vector values included in a multi-set of HSV space vectors are ranked by means of $<$ operator defined in (1). Again the kth largest or smallest vector (order statistic) is the result of a soft vector morphological transform.

5.3.1   Soft Morphological Operators for Color Images

In the following sections $k \diamond x$ denotes the k-times repetition of item x.

Let SB_n be an arbitrary subset of the HSV space, including n vectors $c_1, c_2, \ldots, c_n$. Using the vector ordering scheme presented in subsection 5.2.1, we form the set $SB_{n(ord)}$ of the ordered values $c_{(1)}, c_{(2)}, \ldots, c_{(n)}$, i.e.:

$$SB_{n_{(ord)}} = \{c_{(1)}, c_{(2)}, \ldots, c_{(n)}\}, c_{(1)} \leq c_{(2)} \leq \cdots \leq c_{(n)}.$$

Then:

$$\min{}^{(k)} (SB_n) = c_{(k)}, 1 \leq k \leq n \tag{III.5.66}$$

and

$$\max{}^{(k)} (SB_n) = c_{(n-k+1)}, 1 \leq k \leq n \tag{III.5.67}$$

5.3.1.1   Soft Vector Erosion

Let us consider two functions $f, g : R^2 \Rightarrow$ HSV (two color images), with domains $D[f]$ and $D[g]$, respectively, where f denotes the image under process (the input image) and g denotes the structuring element. The core and the soft boundary of the structuring element $g(z)$ are denoted by $\alpha(z_\alpha)$ and $\beta(z_\beta)$, respectively, where $z_\alpha \in D[\alpha], z_\beta \in D[\beta] = D[g] \backslash D[\alpha]$ and $\backslash$ denotes the set difference. Let also $f(x) = c(h_{xf}, s_{xf}, v_{xf})$. Then *soft vector erosion* of f by g at a point x is defined as follows:

$$(f \ominus [\beta, \alpha, k])(x) = \min{}^{(k)} \{k \diamond (f(z_1) - \alpha_x(z_1))\} \cup \{f(z_2) - \beta_x(z_2)\} \tag{III.5.68}$$

$$\text{for } z_1 \in D[f] \cap D[\alpha_x], z_2 \in D[f] \cap D[\beta_x], x : D[g_x] \subseteq D[f] \tag{III.5.69}$$

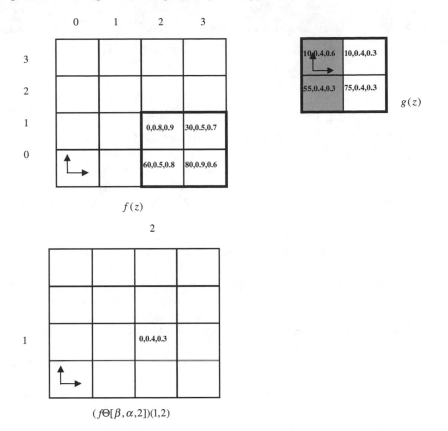

$$(f \ominus [\beta, \alpha, 2])(1,2)$$

FIGURE III.5.5   Example of soft vector erosion.

An equivalent definition for soft vector erosion is:

$$(f \ominus [\beta, \alpha, k])(x) = \min^{(k)} \{ k \diamond (f_{-x}(z_1) - \alpha(z_1)) \} \cup \{ f_{-x}(z_2) - \beta(z_2) \}$$

$$= \min^{(k)} \{ k \diamond (f(z_1 + x) - \alpha(z_1)) \} \cup \{ f(z_2 + x) - \beta(z_2) \} \qquad \text{(III.5.70)}$$

with $z_1 \in D[a]$, $z_2 \in D[\beta]$.

Example 3

Figure III.5.5 demonstrates a case of soft vector erosion. In this figure f and g are two-dimensional color images ($f, g : R^2 \rightarrow$ HSV) represented on the Cartesian grid. The shaded area is the core α. The arrows denote the origin. In particular:

$$f(1, 2) = c(0, 0.8, 0.9), \, x = (1, 2), \, D[g] = \{(0, 0), (0, 1), (-1, 0), (-1, 1)\},$$

$$D[\alpha] = \{(0, 0), (-1, 0)\}, \, D[\beta] = \{(0, 1), (-1, 1)\}.$$

According to Equation III.5.68:

$$(f \ominus [\beta, \alpha, 2])(1, 2)$$

$$= \min^{(2)} \{ 2 \diamond (f(1, 2) - g(0, 0), f(0, 2)$$

$$- g(-1, 0)) \} \cup \{ f(1, 3) - g(0, 1), f(0, 3) - g(-1, 1) \}$$

$$= \min^{(2)}\{2 \diamond (c(0, 0.8, 0.9) - c(10, 0.4, 0.6), c(60, 0.5, 0.8) - c(55, 0.4, 0.3))\}$$

$$\cup \; \{c(30, 0.5, 0.7) - c(10, 0.4, 0.3), c(80, 0.9, 0.6) - c(75, 0.4, 0.3)\}$$

$$= \min^{(2)}\{c(0, 0.4, 0.3), c(0, 0.4, 0.3), c(5, 0.1, 0.5),$$

$$c(5, 0.1, 0.5), c(20, 0.1, 0.4), c(5, 0.5, 0.3)\}$$

$$= \min^{(2)}\{c(5, 0.5, 0.3), c(0, 0.4, 0.3), c(0, 0.4, 0.3),$$

$$c(20, 0.1, 0.4), c(5, 0.1, 0.5), c(5, 0.1, 0.5)\}$$

$$= c(0, 0.4, 0.3)$$

In gray-scale morphology, soft morphological transforms preserve better shapes and small details of the objects in the image under process than the corresponding standard morphological transforms.[15–20] It is the shape and the pixel values of the selected structuring element which play very important roles in detail preservation. In general, the greater the value of the repetition parameter k, the better the detail preservation. This is a desirable feature in various applications, especially in *filtering* (noise elimination) applications.

Soft vector morphological transforms act in a similar way with their gray-scale counterparts: they are advantageous regarding shape and small detail preservation in the original color image, compared with the corresponding standard vector morphological transforms. This is shown in Figure III.5.6. In particular, Figures III.5.6(b) through (e) illustrate the effect of k in preserving small details in shapes during soft vector erosion. In particular, in Figures III.5.6(b) through (e) the eroded images for $k = 1$ through 4 are demonstrated. It can be observed that the increase of the repetition parameter k yields better detail preservation (e.g., in the lips and the teeth).

From Figures III.5.6(b)–(e) it is also obvious that the smaller the value of k, the closer the behavior of soft vector erosion is to that of the corresponding standard vector erosion, just like in the case of soft gray-scale morphology. Particularly, in the example of Figure III.5.6, it can be seen that for soft color erosion:

$$f \ominus g \leq f \ominus [\beta, \alpha, k] \leq f \ominus [\beta, \alpha, k + 1] \leq f$$

which holds for soft gray-scale erosion as well.

5.3.1.2 Soft Vector Dilation

Soft vector dilation of f by g at a point x is defined as follows:

$$(f \oplus [\beta, \alpha, k])(x) = \max^{(k)}\{k \diamond (f(z_1) + \alpha_{-x}(-z_1))\} \cup \{f(z_2) + \beta_{-x}(-z_2)\} \qquad \text{(III.5.71)}$$

$$\text{for } z_1 \in D[f] \cap D[\alpha'_{-x}], z_2 \in D[f] \cap D[\beta'_{-x}], x : D[f] \cap D[g'_{-x}] \neq \varnothing \qquad \text{(III.5.72)}$$

An equivalent definition for soft vector dilation is:

$$(f \oplus [\beta, \alpha, k])(x) = \max^{(k)}\{k \diamond (f_{-x}(-z_1) + \alpha(z_1))\} \cup \{f_{-x}(-z_2) + \beta(z_2)\}$$

$$= \max^{(k)}\{k \diamond (f(x - z_1) + \alpha(z_1))\} \cup \{f(x - z_2) + \beta(z_2)\} \qquad \text{(III.5.73)}$$

$$\text{for } z_1 \in D[\alpha], z_2 \in D[\beta], x - z_1, x - z_2 \in D[f].$$

Example 4

The following example, illustrated in Figure III.5.7, demonstrates a case of soft vector dilation:

$$f(2, 2) = c(60, 0.4, 0.4), x = (2, 2), D[g'] = \{(0, 0), (0, -1), (-1, 0), (-1, -1)\},$$

$$D[\alpha'] = \{(0, 0), (-1, 0)\}, D[\beta'] = \{(0, -1), (-1, -1)\}.$$

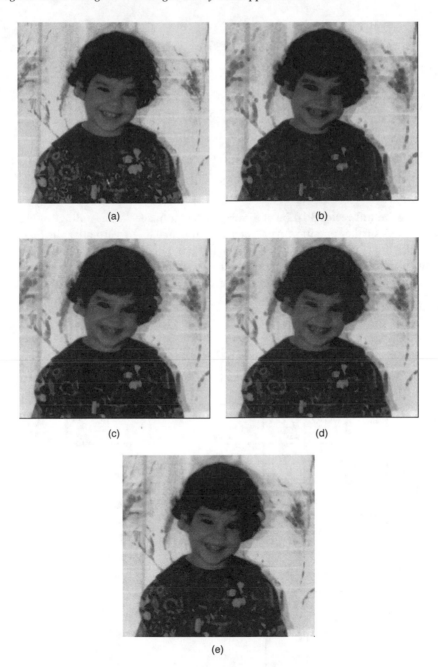

FIGURE III.5.6 (a) Original color image "Veta smiling"; (b) image after standard vector erosion by $g = [\beta, \alpha, 1]$; (c) image after soft vector erosion by $[\beta, \alpha, 2]$; (d) image after soft vector erosion by $[\beta, \alpha, 3]$; and (e) image after soft vector erosion by $[\beta, \alpha, 4]$. The structuring element g used was a flat 3×3 square of color value $c(0, 0, 0.2)$.

According to Equation III.5.71:

$$(f \oplus g)(2, 2) = \max{}^{(2)}\{2 \diamond (f(2, 2) + g(0, 0), f(1, 2) + g(-1, 0))\} \cup \{f(2, 1) + g(0, -1), f(1, 1) + g(-1, -1)\}$$

$$= \max^{(2)}\{2 \diamond (c(60, 0.4, 0.4)) + c(20, 0.2, 0.2), c(80, 0.6, 0.6)$$
$$+ c(30, 0.2, 0.2))\} \cup \{c(0, 0.4, 0.7)$$
$$+ c(50, 0.2, 0.2), c(40, 0.5, 0.6)$$
$$+ c(0, 0.2, 0.2)\}$$
$$= \max^{(2)}\{c(80, 0.6, 0.6), c(80, 0.6, 0.6), c(110, 0.8, 0.8),$$
$$c(110, 0.8, 0.8), c(50, 0.6, 0.9), c(40, 0.7, 0.8)\}$$
$$= \max^{(2)}\{c(80, 0.6, 0.6), c(80, 0.6, 0.6), c(110, 0.8, 0.8),$$
$$c(110, 0.8, 0.8), c(40, 0.7, 0.8), c(50, 0.6, 0.9)\} = c(40, 0.7, 0.8)$$

In Figure III.5.8 the soft vector dilation of the original color image "Veta smiling" of Figure III.5.6(a) for various values of the repetition parameter k is illustrated. In particular, in Figures III.5.8(a) through (d), the dilated images for $k = 1$ through 4 are demonstrated. It can be observed that, as in soft vector erosion, the increase of the repetition parameter k yields better detail preservation. Furthermore, the decrease of the repetition parameter yields results closer to that of standard vector dilation.

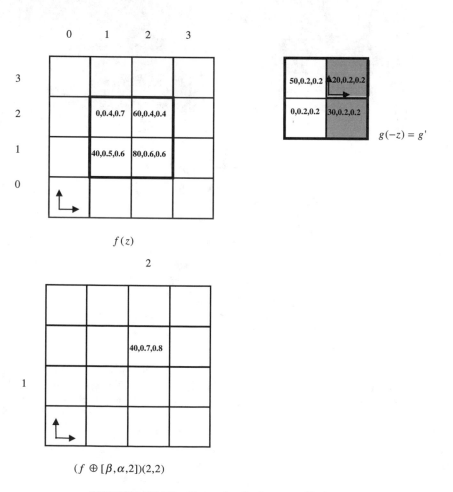

FIGURE III.5.7   Example of soft vector dilation.

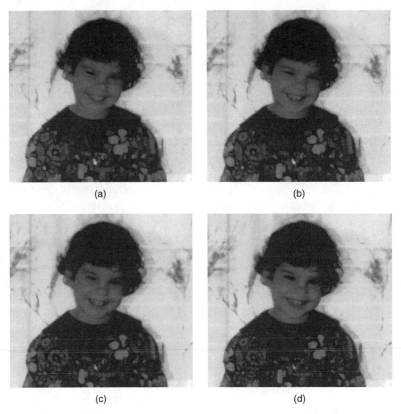

(a)   (b)

(c)   (d)

FIGURE III.5.8   The effect of soft vector dilation to the color image of Figure III.5.6(a): (a) image after standard vector dilation by $g = [\beta, \alpha, 1]$; (b) image after soft vector dilation by $\lfloor \beta, \alpha, 2\rfloor$; (c) image after soft vector dilation by $[\beta, \alpha, 3]$; and (d) image after soft vector dilation by $[\beta, \alpha, 4]$.

5.3.1.3   Soft Vector Opening and Closing

Soft vector opening of f by g at a point x is defined as follows:

$$f \circ [\beta, \alpha, k] = (f \ominus [\beta, \alpha, k]) \oplus [\beta, \alpha, k] \tag{III.5.74}$$

Soft vector closing of f by g at a point x is defined as follows:

$$f \bullet [\beta, \alpha, k] = (f \oplus [\beta, \alpha, k]) \ominus [\beta, \alpha, k] \tag{III.5.75}$$

Soft vector morphology is compatible with soft gray-scale morphology. First, as is deduced from definitions of standard vector and soft morphological operations, in the case $k = 1$ or $\alpha = g$, soft vector morphological operations are identical to the corresponding standard vector morphological operations, like in soft gray-scale morphology.

Furthermore, the constraint $k \leq \min\{\mathrm{Card}[g]/2, \mathrm{Card}[\beta]\}$ is adopted in soft vector morphology as well as in soft gray-scale morphology, in order for the nature of soft color morphological operations to be preserved.[18,19] (Card $[X]$ denotes the cardinality, i.e., the number of elements of a set X.) For instance, if $k > \mathrm{Card}[\beta]$, the nature of soft vector morphological operations is not maintained: in this case soft vector morphological operations are not affected by the soft boundary β and they are equivalent to the corresponding standard vector morphological operations of the input image f by the structuring element α.

Moreover, it is obvious that when soft vector morphological operations are applied to gray-scale images, they are reduced to the corresponding soft gray-scale morphological operations. Indeed, one must keep in mind that the operation of *subtraction (addition)* between gray-level values in the HSV color model (undefined $h, s = 0$, and v in the range $[0, 1]$) is identical to the standard arithmetic subtraction (addition) between the arithmetic equivalent values in the range $[0–255]$ of these gray-level values. In addition, the application of the proposed ordering $(\leq)$ to a set of grays in the HSV color model is identical to the application of the standard arithmetic ordering $(\leq)$ to the set of the arithmetic equivalent values in the range $[0–255]$ of these gray-level values. Therefore, the application of the $\min^{(k)} \max^{(k)}$ operator to a set of gray-level values in the HSV color model is identical to the selection of the kth smallest (largest) in the set of the arithmetic equivalent values in the range $[0–255]$ of these gray-level values. Subsequently, in the case of gray-scale f and g, the previous definitions are reformed to those of the corresponding soft gray-scale morphological operations.

Finally, the primary and secondary soft color morphological operations possess the same basic properties as their gray-scale counterparts, as is shown in the following subsection 5.3.2.

5.3.2 Basic Properties of Soft Vector Morphological Operations

5.3.2.1 Increasing Monotony

Both soft vector erosion and dilation are *monotonically increasing* relative to a specific structuring element:

$$f_1 \ll f_2 \Rightarrow f_1 \ominus [\beta, \alpha, k] \ll f_2 \ominus [\beta, \alpha, k] \tag{III.5.76}$$

$$f_1 \ll f_2 \Rightarrow f_1 \oplus [\beta, \alpha, k] \ll f_2 \oplus [\beta, \alpha, k] \tag{III.5.77}$$

Proof

$$f_1 \ll f_2 \Rightarrow \underset{x \in R^n}{f_1(x) \leq f_2(x)}$$

$$\Rightarrow \begin{cases} f_1(z_1) \leq f_2(z_1), & z_1 \in D[\alpha_x] \cap D[f_1] \\ f_1(z_2) \leq f_2(z_2), & z_2 \in D[\beta_x] \cap D[f_1] \end{cases}$$

$$\Rightarrow \begin{cases} f_1(z_1) - \alpha_x(z_1) \leq f_2(z_1) - \alpha_x(z_1), & z_1 \in D[\alpha_x] \cap D[f_1] \\ f_1(z_2) - \beta_x(z_2) \leq f_2(z_2) - \beta_x(z_2), & z_2 \in D[\beta_x] \cap D[f_1] \end{cases}$$

$$\Rightarrow \min^{(k)} \{ k \diamond (f_1(z_1) - \alpha_x(z_1)) \} \cup \{ f_1(z_2) - \beta_x(z_2) \}$$

$$\leq z_1 \in D[f_1] \cap D[\alpha_x], z_2 \in D[f_1] \cap D[\beta_x]$$

$$\leq \min^{(k)} \{ k \diamond (f_2(z_1) - \alpha_x(z_1)) \} \cup \{ f_2(z_2) - \beta_x(z_2) \}$$

$$\Rightarrow z_1 \in D[f_2] \cap D[\alpha_x], z_2 \in D[f_2] \cap D[\beta_x]$$

$$\Rightarrow \begin{cases} v_{\min^{(k)}}(MS_{f_1 gx}) < v_{\min^{(k)}}(MS_{f_2 gx}) \\ \qquad\qquad\qquad \text{or} \\ v_{\min^{(k)}}(MS_{f_1 gx}) = v_{\min^{(k)}}(MS_{f_2 gx}) \quad \text{and} \quad s_{\min^{(k)}}(MS_{f_1 gx}) > s_{\min^{(k)}}(MS_{f_2 gx}) \\ \qquad\qquad\qquad \text{or} \\ \Rightarrow \begin{cases} v_{\min^{(k)}}(MS_{f_1 gx}) = v_{\min^{(k)}}(MS_{f_2 gx}) \quad \text{and} \quad s_{\min^{(k)}}(MS_{f_1 gx}) \\ = s_{\min^{(k)}}(MS_{f_2 gx}) \quad \text{and} \quad h_{\min^{(k)}}(MS_{f_1 gx}) \leq h_{\min^{(k)}}(MS_{f_2 gx}) \end{cases} \end{cases}$$

$$\Rightarrow c(h_{\min^{(k)}}(MS_{f_1 g_x}), s_{\min^{(k)}}(MS_{f_1 g_x}), v_{\min^{(k)}}(MS_{f_1 g_x}))$$

$$\leq c(h_{\min^{(k)}}(MS_{f_2 g_x}), s_{\min^{(k)}}(MS_{f_2 g_x}), v_{\min^{(k)}}(MS_{f_2 g_x}))$$

$$\Rightarrow (f_1 \Theta[\beta, \alpha, k])(x) \leq (f_2 \Theta[\beta, \alpha, k])(x) \tag{III.5.78}$$

Furthermore, let be

$$x \in D[f_1 \Theta[\beta, \alpha, k]] \tag{III.5.79}$$

$$f_1 \ll f_2 \Rightarrow D[f_1] \subseteq D[f_2] \tag{III.5.80}$$

By definition:

$$x : D[g_x] \subseteq D[f_1] \Rightarrow y + x \in D[f_1], \ \forall y \in D[g] \tag{III.5.81}$$

$$\text{(III.5.80) and (III.5.81)} \Rightarrow y + x \in D[f_2], \ \forall y \in D[g]$$

$$\Rightarrow x : D[g_x] \subseteq D[f_2] \Rightarrow x \in D[f_2 \Theta[\beta, \alpha, k]] \tag{III.5.82}$$

$$\text{(III.5.79) and (III.5.82)} \Rightarrow D[f_1 \Theta[\beta, \alpha, k]] \subseteq D[f_2 \Theta[\beta, \alpha, k]] \tag{III.5.83}$$

From Equations III.5.78 and III.5.83: $f_1 \ll f_2 \Rightarrow f_1 \Theta[\beta, \alpha, k] \ll f_2 \Theta[\beta, \alpha, k]$.

In a similar way it can be proven that soft vector dilation is *monotonically increasing* relative to a specific structuring element as well.

Since soft vector dilation and erosion are both monotonically increasing, it is straightforward that soft vector opening and closing possess this property as well:

$$f_1 \ll f_2 \Rightarrow f_1 \circ [\beta, \alpha, k] \ll f_2 \circ [\beta, \alpha, k] \tag{III.5.84}$$

$$f_1 \ll f_2 \Rightarrow f_1 \bullet [\beta, \alpha, k] \ll f_2 \bullet [\beta, \alpha, k] \tag{III.5.85}$$

5.3.2.2 Translation Invariance

Both soft vector erosion and dilation are *translation invariant* relative to the input image:

$$(f \Theta[\beta, \alpha, k])_y + j = (f_y + j) \Theta[\beta, \alpha, k] \qquad\qquad j \in HSV \tag{III.5.86}$$

$$(f \oplus [\beta, \alpha, k])_y + j - (f_y + j) \oplus [\beta, \alpha, k] \qquad\qquad j \in HSV \tag{III.5.87}$$

Proof
Equivalently we can prove that:

$$[(f \Theta[\beta, \alpha, k])(x)]_y + j = ((f_y + j) \Theta[\beta, \alpha, k])(x + y), \forall x \in D[f \Theta[\beta, \alpha, k]].$$

If $j = c(h, s, v)$ then

$$[(f \Theta[\beta, \alpha, k])(x)]_y + j$$

$$= [c(h_{\min^{(k)}}(MS_{f g_x}), s_{\min^{(k)}}(MS_{f g_x}), v_{\min^{(k)}}(MS_{f g_x}))]_y + c(h, s, v)$$

$$= c(h_{\min^{(k)}}(MS_{(f g_x)_y}), s_{\min^{(k)}}(MS_{(f g_x)_y}), v_{\min^{(k)}}(MS_{(f g_x)_y})) + c(h, s, v)$$

$$= c(h_{\min^{(k)}}(MS_{f_y g_{(x+y)}}) + h, s_{\min^{(k)}}(MS_{f_y g_{(x+y)}}) + s, v_{\min^{(k)}}(MS_{f_y g_{(x+y)}}) + v)$$

$$= c(h_{\min^{(k)}}(MS_{(f_y+j) g_{(x+y)}}), s_{\min^{(k)}}(MS_{(f_y+j) g_{(x+y)}}), v_{\min^{(k)}}(MS_{(f_y+j) g_{(x+y)}}))$$

$$= ((f_y + j) \Theta[\beta, \alpha, k])(x + y)$$

In a similar way it can be proven that soft vector dilation is *translation invariant* relative to the input image as well.

Since soft vector dilation and erosion are both invariant under translation, it is straightforward that soft vector opening and closing are translation invariant as well:

$$(f \circ [\beta, \alpha, k])_y + j = (f_y + j) \circ [\beta, \alpha, k] \quad j \in \text{HSV} \tag{III.5.88}$$

$$(f \bullet [\beta, \alpha, k])_y + j = (f_y + j) \bullet [\beta, \alpha, k] \quad j \in \text{HSV} \tag{III.5.89}$$

5.3.2.3 Duality

Soft vector erosion and dilation are dual operations:

$$-(f \oplus [\beta, \alpha, k])(x) = (-f \ominus [\beta', \alpha', k])(x) \tag{III.5.90}$$

Proof

$$- \max{}^{(k)}\{k \diamond (f_{-x}(-z_1) + \alpha(z_1))\} \cup \{f_{-x}(-z_2) + \beta(z_2)\}$$
$$= z_1 \in D[f'_{-x}] \cap D[\alpha], z_2 \in D[f'_{-x}] \cap D[\beta]$$
$$= \min{}^{(k)}\{k \diamond (-f_{-x}(-z_1) - \alpha(z_1))\} \cup \{-f_{-x}(-z_2) - \beta(z_2)\}$$
$$= z_1 \in D[-f'_{-x}] \cap D[\alpha], z_2 \in D[-f'_{-x}] \cap D[\beta]$$
$$= \min{}^{(k)}\{k \diamond (-f_{-x}(z_1) - \alpha(-z_1))\} \cup \{-f_{-x}(z_2) - \beta(-z_2)\}$$
$$\Rightarrow z_1 \in D[-f_{-x}] \cap D[\alpha'], z_2 \in D[-f_{-x}] \cap D[\beta']$$
$$\Rightarrow -\max{}^{(k)}(MS_{f'-xg}) = \min{}^{(k)}(MS_{-f_{-x}g'}) \tag{III.5.91}$$

By definition:

$$(f \oplus [\beta, \alpha, k])(x) = c(h_{\max^{(k)}(MS_{f'-xg})}, s_{\max^{(k)}(MS_{f'-xg})}, \nu_{\max^{(k)}(MS_{f'-xg})})$$
$$\Rightarrow -(f \oplus [\beta, \alpha, k])(x) = c(-h_{\max^{(k)}(MS_{f'-xg})}, -s_{\max^{(k)}(MS_{f'-xg})}, -\nu_{\max^{(k)}(MS_{f'-xg})})$$
$$\Rightarrow -(f \oplus [\beta, \alpha, k])(x) = c(h_{-\max^{(k)}(MS_{f'-xg})}, s_{-\max^{(k)}(MS_{f'-xg})}, \nu_{-\max^{(k)}(MS_{f'-xg})})$$
$$\Rightarrow -(f \oplus [\beta, \alpha, k])(x) = c(h_{\min^{(k)}(MS_{-f_{-x}g'})}, s_{\min^{(k)}(MS_{-f_{-x}g'})}, \nu_{\min^{(k)}(MS_{-f_{-x}g'})})$$
$$\Rightarrow -(f \oplus [\beta, \alpha, k])(x) = (-f \ominus [\beta', \alpha', k])(x)$$

Soft vector opening and closing are dual operations too:

$$-(f \circ [\beta, \alpha, k]) = -f \bullet [\beta', \alpha', k] \tag{III.5.92}$$

Proof

$$-(f \circ [\beta, \alpha, k]) = -((f \ominus [\beta, \alpha, k]) \oplus [\beta, \alpha, k])$$
$$= -(f \ominus [\beta, \alpha, k]) \ominus [\beta', \alpha', k])$$
$$= (-f \oplus [\beta', \alpha', k]) \ominus [\beta', \alpha', k]) = -f \bullet [\beta', \alpha', k]$$

5.3.2.4 Extensivity – Anti-Extensivity

If the origin lies inside the core of the structuring element, soft vector erosion is *anti-extensive* and soft vector dilation is *extensive*:

$$0 \in \alpha \Rightarrow (f \ominus [\beta, \alpha, k])(x) \le f(x) \tag{III.5.93}$$

$$0 \in \alpha \Rightarrow (f \oplus [\beta, \alpha, k])(x) \ge f(x) \tag{III.5.94}$$

Proof

Let Card$[g] = n$ and Card$[\alpha] = m$. We consider two distinct cases:

1. Let

$$\text{Card}[g]/2 < \text{Card}[\beta] \Rightarrow n/2 < n - m \qquad \text{(III.5.95)}$$

Then it must be $k \leq n/2$. It is obvious that if Equation III.5.76 holds for $k = n/2$, it holds for $k < n/2$ as well. So let $k = n/2$. Also:

$$D[f] \cap D[\alpha_x] = \{x_1, x_2, \ldots, x_m\} \text{ and } D[f] \cap D[\beta_x] = \{x_{m+1}, x_{m+2}, \ldots, x_n\} \quad \text{(III.5.96)}$$

By supposition $0 \in \alpha \Rightarrow x \in D[\alpha_x] \underset{(D2)}{\Rightarrow} x = x_{e1 \leq e \leq m} \in \{x_1, x_2, \ldots, x_m\}$.

By definition:

$$(f \ominus [\beta, \alpha, k])(x) = \min^{(n/2)}(\{\underbrace{d(x_1), \ldots, d(x_1)}_{n/2 \text{ times}}, \ldots, \underbrace{d(x_m), \ldots, d(x_m)}_{n/2 \text{ times}},$$

$$d(x_{m+1}), d(x_{m+2}), \ldots, d(x_n)\}) = \min^{(n/2)}(A) \qquad \text{(III.5.97)}$$

$$\text{where } \underset{1 \leq i \leq m}{d(x_i)} = f(x_i) - \alpha(x_i) \text{ and } \underset{m+1 \leq i \leq n}{d(x_i)} = f(x_i) - \beta(x_i) \qquad \text{(III.5.98)}$$

Then:

a. In the case that $\exists \underset{1 \leq i \leq m}{x_i} : d(x_i) \leq d(x_e)$, since $d(x_i)$ is repeated $n/2$ times, it is obvious from Equations III.5.97 and III.5.98 that:

$$\min^{(n/2)}(A) \leq d(x_i) \leq d(x_e) \leq f(x_e)$$

$$\Rightarrow (f \ominus [\beta, \alpha, k])(x) \leq f(x). \qquad \text{(III.5.99)}$$

b. In the case that $d(x_e) \leq \underset{1 \leq i \leq m}{d(x_i)}$ and also $d(x_e) \leq \underset{m+1 \leq i \leq n}{d(x_i)}$, since $d(x_e)$ is repeated $n/2$ times, it is obvious from Equations III.5.97 and III.5.98 that:

$$\min^{(n/2)}(A) = d(x_e) \leq f(x_e) \Rightarrow (f \ominus [\beta, \alpha, k])(x) \leq f(x) \qquad \text{(III.5.100)}$$

c. In the case that $d(x_e) \leq \underset{1 \leq i \leq m}{d(x_i)}$ but $\exists \underset{m+1 \leq i \leq n}{d(x_i)} : d(x_i) \leq d(x_e)$, since by supposition the number of $\underset{m+1 \leq i \leq n}{d(x_i)}$ equals to $n - m > n/2$, it is obvious from Equations III.5.97 and III.5.98 that:

$$\min^{(n/2)}(A) \leq d(x_e) \leq f(x_e) \Rightarrow (f \ominus [\beta, \alpha, k])(x) \leq f(x) \qquad \text{(III.5.101)}$$

From Equations III.5.99, III.5.100, and III.5.101 it is evident that $0 \in \alpha \Rightarrow (f \ominus [\beta, \alpha, k])(x) \leq f(x)$.

2. In a similar way we can proceed in the case that Card$[\beta] < $ Card$[g]/2 \Rightarrow n - m < n/2$ Soft vector dilation's extensivity can be proven in a similar way.

On the contrary, soft vector opening is not in general anti-extensive:

$$\exists x \in R^n : (f \circ [\beta, \alpha, k])(x) > f(x) \qquad \text{(III.5.102)}$$

The following example confirms Equation III.5.102.

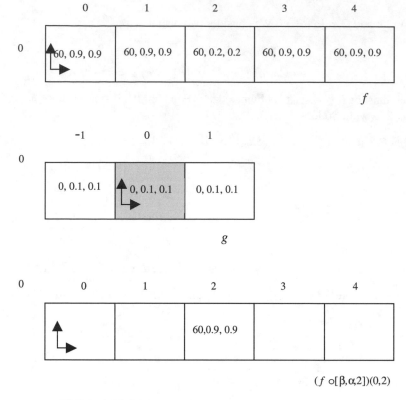

$$(f \circ [\beta, \alpha, 2])(0,2)$$

FIGURE III.5.9   Anti-extensivity of soft vector opening.

Example 5

In Figure III.5.9, for $k = 2$ and $x = (0, 2)$ notice that:

$$(f \circ [\beta, \alpha, 2])(0, 2) = c(60, 0.9, 0.9) > c(60, 0.2, 0.2) = f(0, 2).$$

In a similar way it can be proven that soft vector closing is not in general extensive:

$$\exists x \in R^n : (f \bullet [\beta, \alpha, k])(x) < f(x) \tag{III.5.103}$$

5.3.2.5 Idempotency

Like their gray-scale counterparts, soft vector opening and closing are not in general idempotent.

$$\exists f, g : R^n \to \text{HSV such that } f \circ [\beta, \alpha, k] \neq (f \circ [\beta, \alpha, k]) \circ [\beta, \alpha, k] \tag{III.5.104}$$

This is shown in the following example.

Example 6

In Figure III.5.10, for $k = 2$ notice that $f \circ [\beta, \alpha, 2] \neq (f \circ [\beta, \alpha, 2]) \circ [\beta, \alpha, 2]$.

In a similar way it can be proven that:

$$\exists f, g : R^n \to \text{HSV such that } f \bullet [\beta, \alpha, k] \neq (f \bullet [\beta, \alpha, k]) \bullet [\beta, \alpha, k] \tag{III.5.105}$$

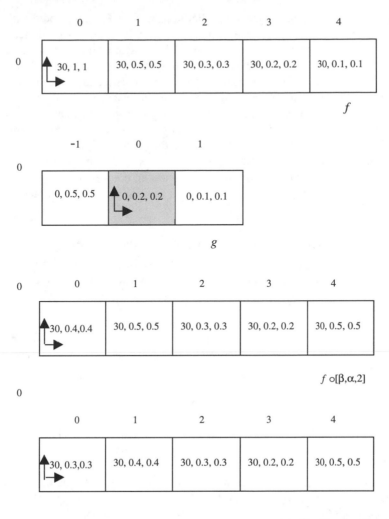

FIGURE III.5.10 Idempotency of soft vector opening and closing.

5.4 Applications

5.4.1 Noise Suppression

An application of great importance in the field of image enhancement is noise suppression. The aim of noise suppression is to eliminate noise and its effects on the original image, while distorting the image as little as possible. Standard gray-scale opening and closing have proved very useful in noise suppression applications.[4,5] They act as filters that can decrease or completely eliminate impulse noise from gray-scale images,[2,4,5,23,24] i.e., they restore the value of isolated small groups of image points that occupy extreme values compared with the values of the surrounding image points. Specifically, standard opening cuts off positive impulses while standard closing cuts off negative impulses in the image under process.

The extension of this type of morphological filtering to color images using the component-wise morphology alters the color composition and object boundaries of the image. There is a possibility that an object could be removed in one or two of the R, G, B components, but not in all of them.[6,10]

This effect is unacceptable, since objects that should be removed remain, and the filtering produces new colors not present in the input image. The morphological operators introduced in this chapter are vector preserving and can be used for this type of filtering in all three frames H, S, and V. From the definitions of these operators it is clear that the output vector (color) at each point in the image is one of the input vectors (colors), and therefore no new colors are introduced.

Impulse noise exists in many practical applications and can be generated by various sources, including many man-made phenomena such as unprotected switches, industrial machines, and car ignition systems. This type of noise is the classical salt-and-pepper noise for gray-scale images. However, a significant problem in the study of the effect of noise in color image processing is the lack of a generally accepted multivariate impulse noise model. A number of such models have been introduced recently in order to assist in the development of color image filters. The color impulse noise model that has been used in our experiments has been proposed by Plataniotis et al.[25] This impulse noise model is as follows:

$$
c_n = \begin{cases}
c_o = (h, s, v) & \text{with probability } (1 - p) \\
(d, s, v) & \text{with probability } p_1 p \\
(s, d, v) & \text{with probability } p_2 p \\
(h, s, d) & \text{with probability } p_3 p \\
(d, d, d) & \text{with probability } p_s p
\end{cases}
\tag{III.5.106}
$$

where c_n is the noisy signal, c_o is the original vector (the noise-free color vector), and d is the impulse value. Additionally, p is the degree of the impulse noise contamination in the image, and:

$$
p_s = 1 - p_1 - p_2 - p_3
\tag{III.5.107}
$$

with $p_1 + p_2 + p_3 \leq 1$. The impulse value d can be either positive or negative. Since our operators are introduced in the HSV color space, the impulse noise model is considered in the HSV space as well.

We have already mentioned that standard gray-scale opening and closing have proved very useful in filtering applications since they act as filters that can decrease or completely eliminate impulse noise from gray-scale images. However, soft gray-scale opening and closing perform better than their standard gray-scale counterparts under noisy conditions. Particularly, soft gray-scale opening can decrease or completely eliminate both positive and negative impulses. The same holds for soft gray-scale closing as well. In addition, the greater the value of the repetition parameter k, the greater the impulse noise removal capability is.[5,16,18] Soft vector opening and closing possess, as well, the same advantage: they remove impulse noise from color images better than their standard vector counterparts.

Figures III.5.11, III.5.12, and III.5.13 illustrate the fact that soft vector morphological transforms achieve improved performance under noisy conditions in comparison with the corresponding standard vector morphological transforms.

Specifically, in Figure III.5.11(b) the color image "Veta and her deer" of Figure III.5.11(a) contaminated by 6% color impulse noise is depicted. Here $p_1 = p_2 = p_3 = 0$ and $p_S = 1$, i.e., the impulses have been introduced in all three vector components h, s, and v as well. In Figure III.5.11(c) it is shown that standard vector closing suppresses only negative impulses while enlarging positive impulses. On the contrary, as can be observed from Figures III.5.11(d) and (e), soft vector closing decreases or removes both positive and negative impulses, just like its gray-scale counterpart. In Figure III.5.11(d) and (e) the soft closed images for $k = 2, k = 4$ are demonstrated, respectively. From these figures it can also be observed that the increase of the repetition parameter k increases the noise removal and the detail preservation capability.

FIGURE III.5.11   (a) Original color image "Veta and her deer"; (b) image corrupted by 6% positive and negative HSV impulse noise with $p_1 = p_2 = p_3 = 0$ and $p_s = 1$; (c) resulting image after standard vector closing by $g = [\beta, \alpha, 1]$; (d) resulting image after soft vector closing by $[\beta, \alpha, 2]$; and (e) resulting image after soft vector closing by $[\beta, \alpha, 4]$.

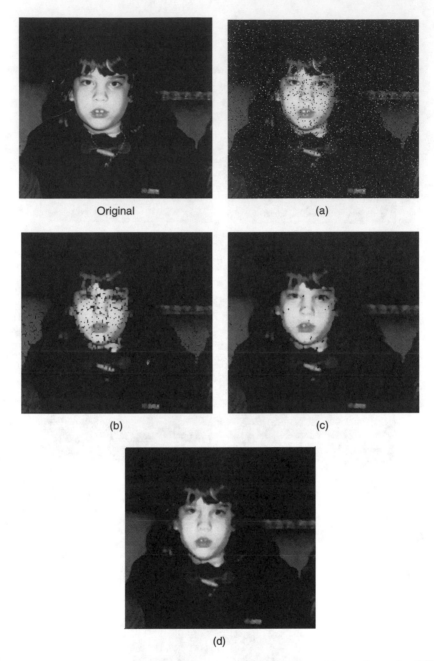

FIGURE III.5.12 (a) Original color image "Veta" corrupted by 6% positive and negative HSV impulse noise with $p_1 = p_2 = p_3 = 0$ and $p_s = 1$; (b) resulting image after standard vector opening by $g = [\beta, \alpha, 1]$; (c) resulting image after soft vector opening by $[\beta, \alpha, 2]$; and (d) resulting image after soft vector opening by $[\beta, \alpha, 4]$.

Figure III.5.12 demonstrates that soft vector opening removes impulse noise from color images better than the corresponding standard vector opening. As can be seen in Figure III.5.12(c), standard vector opening removes only positive impulses from the contaminated image, while it causes the enlargement of the negative impulses, in contrast to soft vector opening, that reduces or completely

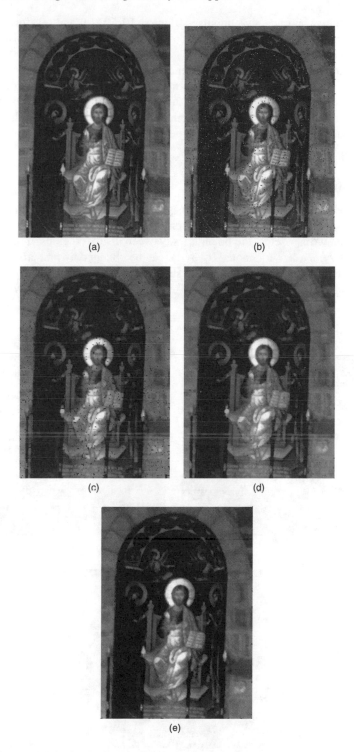

(a)

(b)

(c)

(d)

(e)

FIGURE III.5.13   (a) Original color image "Byzantine"; (b) image degraded by 3% positive and negative HSV impulse noise with $p_1 = p_2 = p_3 = 0.2$ and $p_s = 0.4$; (c) resulting image after standard vector opening by $g = [\beta, \alpha, 1]$; (d) resulting image after soft vector opening by $[\beta, \alpha, 2]$; and (e) resulting image after soft vector opening by $[\beta, \alpha, 4]$.

eliminates both positive and negative impulses [Figures III.5.12(d) and (e), for $k = 2$ and $k = 4$, respectively]. Again the increase of the repetition parameter k increases the noise removal and the detail preservation capability as well.

Figure III.5.13 demonstrates the effect of the repetition parameter k in color impulse noise removal, in the case that $p_s < 1$, i.e., when positive and negative impulses may be introduced only in just one of the vector components. In Figures III.5.13(c) through (e) the soft opened images for $k = 1$, $k = 2, k = 4$ are demonstrated, respectively. From Figure III.5.13 it is evident that even in this case soft vectors morphological transforms yield better results in color impulse noise suppression than their standard vector counterparts.

5.4.2   Boundary Extraction

A typical technique that is used in the area of image analysis is *boundary extraction*. The detection of the edges in an image constitutes a typical first stage in various image processing applications. With this technique, we can detect the boundaries between distinct regions of an image. For binary images, the boundary of a set A, denoted by $\beta(A)$, can be obtained by first eroding A by a suitable structuring element B and then performing the set difference between A and its erosion. We may use this morphological expression for boundary extraction in color images. If we consider f to be

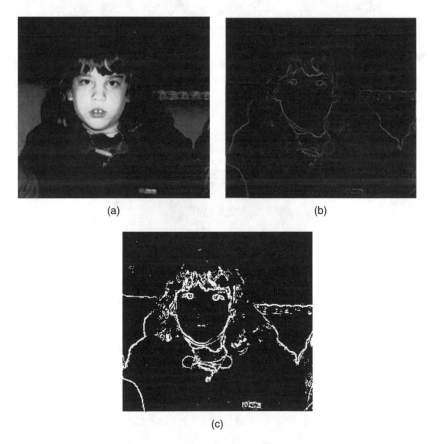

FIGURE III.5.14   (a) Original color image of Figure III.5.12, (b) edges of the image, and (c) binary equivalent of the gray-scale image of (b) after thresholding (threshold value $\nu = 0.15$).

a color image, g to be the structuring element, and $\beta(f, g)$ to be the resultant image of the color boundary extraction algorithm, we can write the following expression:

$$\beta(f, g) = f - (f \ominus g) \qquad \text{(III.5.108)}$$

If the previous expression is applied to a color image f by means of vector erosion, vector dilation, and the operation of vector subtraction defined in previous Section 5.2, then some of the edges of the image may not be detected. This problem occurs since the previous morphological boundary extractor cannot detect borderlines between two areas in the original image which have the same value of the v component, ignoring in this way edges that must be detected due to the different values of the h or s components.

In order to improve the performance of the edge detector defined in Equation III.5.108, to extract edges that must be detected due to the different values of the h or s components as well, a new vector edge detection algorithm was introduced.[26] Figure III.5.14 demonstrates the application of this algorithm in a color image. In Figure III.5.14(a) the original image "Veta" is shown. In Figure III.5.14(b) the result of the boundary extraction algorithm is demonstrated, and in (c) the image after thresholding is depicted.

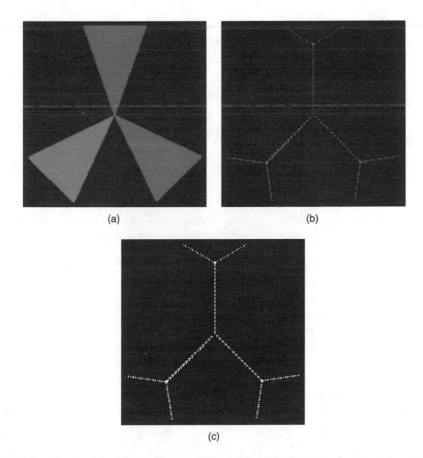

(a)　　　　　　(b)

(c)

FIGURE III.5.15 Application of the skeletonization algorithm: (a) original image; (b) skeleton of the image; and (c) thresholded result of the algorithm.

5.4.3   Color Image Skeletonization

Another interesting problem in image processing is finding a thinned replica of an image to use either in a recognition algorithm or for data compression. A commonly employed thinning procedure is *skeletonization*, which is based on the concept of maximal discs. A maximal disc is a maximum sized disc that just fits inside the object. The union of the centers of all these discs constitutes the skeleton of the object.[4,5,23]

The concept of skeleton was introduced by Blum. Binary morphological transforms have been used to extract skeletons of binary images. Lantuejoul showed that the skeleton of an object can be morphologically expressed, in terms of erosions and openings.[27] If we denote with $S(A)$ the skeleton of A, it can be expressed as follows:

$$S(A) = \bigcup_{k=0}^{K} S_k(A) \tag{III.5.109}$$

with:

$$S_k(A) = \{(A\ominus kB) - [(A\ominus kB)\circ B]\} \tag{III.5.110}$$

where B is a structuring element and $(A\ominus kB)$ indicates k successive erosions of A:

$$(A\ominus kB) = ((\ldots(A\ominus B)\ominus B)\ominus \ldots)\ominus B \tag{III.5.111}$$

k times and K is the last iteration before A erodes to an empty set. That is:

$$K = \max\{k : (A\ominus kB) \neq \varnothing\} \tag{III.5.112}$$

This skeletonization algorithm has been modified in order to accommodate color images, by using the presented standard operators.[28] The color skeletonization algorithm has been tested using a variety of images. In Figure III.5.15 an example of the application of the algorithm is shown. The input color image is shown in Figure III.5.15(a) and the results of the algorithm in (b). The thresholded skeleton image is depicted in (c).

References

1. Matheron, G., *Random Sets and Integral Geometry*, John Wiley & Sons, New York, 1975.
2. Haralick, R.M., Sternberg, S.R., and Zhuang, X., Image analysis using mathematical morphology, *IEEE Trans. Pattern Anal. Machine Intell.*, 9, 532, 1987.
3. Heijmans, H.J.A.M., *Morphological Image Operators*, Academic Press, Boston, 1994.
4. Serra, J., *Image Analysis and Mathematical Morphology*, Academic Press, London, 1982.
5. Soille, P., *Morphological Image Analysis—Principles and Applications*, Springer, Berlin, 1999.
6. Comer, M.L. and Delp, E.J., Morphological operations, in *The Color Image Processing Handbook*, Sangwine and Horne, Eds., Chapman & Hall, 1998.
7. Goutsias, J., Heijmans H.J.A.M., and Sivakumar, K., Morphological operators for image sequences, *Computer Vision and Image Understanding*, 62, 326, 1995.
8. Astola, J., Haavisto, P., and Neuvo, Y., Vector median filters, in *Proc. IEEE*, 78, 678, 1990.
9. Barnett, V., The ordering of multivariate data, *J. R. Statist. Soc. A*, 139, 318, 1976.
10. Goutsias, J. and Heijmans, H.J.A.M., Fundamenta Morphologicae Mathematicae, *Fundamenta Informaticae*, 1, 1999.
11. Talbot, H., Evans, C., and Jones, R., Complete ordering and multivariate mathematical morphology, in *Mathematical Morphology and its Applications to Image and Signal Processing*, Heijmans and Roerdink, Eds., Kluwer Academic Publishers, Amsterdam, 27, 1998.

12. Tang, K., Astola, J., and Neuvo, Y., Multivariate ordering and filtering with application to color image processing, in *Proc. 6th European Signal Proc. Conf. EUSIPCO-92*, Elsevier Science Publishers B.V., Belgium, 1992, 1481.

13. Vardavoulia, M.I., Andreadis, I., and Tsalides, Ph., Vector ordering and morphological operations for colour image processing: fundamentals and applications, to appear, *Pattern Analysis and Applications*.

14. Louverdis, G. et al., A new approach to morphological color image processing, to appear, *Pattern Recognition*.

15. Koskinen, L., Astola, J., and Neuvo, Y., Soft morphological filters, in *Proc. SPIE Symp. Image Algebra and Morphological Image Proc. II*, San Diego, 1991, 262.

16. Koskinen, L. and Astola, J., Soft morphological filters: a robust morphological filtering method, *J. Electronic Imaging*, 3, 60, 1994.

17. Kuosmanen, P., Koskinen, L., and Astola, J., Analysis and extensions of soft morphological filters, in *Proc. SPIE/IS & T Symp. Electronic Imaging: Science and Technology*, San Jose, 1993, 41.

18. Kuosmanen, P. and Astola, J., Soft morphological filtering, *J. Mathematical Imaging and Vision*, 5, 231, 1995.

19. Pu, C.C. and Shih, F.Y., Threshold decomposition of gray-scale soft morphology into binary soft morphology, *Graph. Models Image Process.*, 57, 522, 1995.

20. Shih, F.Y. and Pu, C.C., Analysis of the properties of soft morphological filtering using threshold decomposition, *IEEE Trans. Signal Processing*, 43, 538, 1995.

21. Birkhoff, G., *Lattice Theory*, 3rd ed., Vol. 25 of American Mathematical Society Colloquium Publications, American Mathematical Society, Providence, 1967.

22. Meyer-Nieberg, P., *Banach Lattices*, Springer-Verlag, Berlin, 1991.

23. Giardina, C.R. and Dougherty, E.R., *Morphological Methods in Image and Signal Processing*, Prentice-Hall, New York, 1988.

24. Sternberg, S.R., Gray-scale morphology, *Comput. Vision Graph. Image Proc.*, 35, 333, 1986.

25. Plataniotis, K.N., Androutsos, D., and Venetsanopoulos, A.N., Adaptive fuzzy systems for multichannel signal processing, in *Proc. IEEE*, 87, 1601, 1999.

26. Vardavoulia, M.I., Andreadis, I., and Tsalides, Ph., Morphological colour image segmentation, in *Proc. 12th Scandinavian Con.Image Analysis (SCIA '01)*, Bergen, 2001, 236.

27. Lantuejoul, C., Skeletonization in quantitative metallography, in *Issues of Digital Image Processing*, Sijthoff and Noordhoff, 1980.

28. Andreadis, I. et al., Color image skeletonisation, in *Proc. of EUSIPCO 2000*, Tampere, 4, 2000, 2389.

6

The Future of Artificial Neural Networks and Speech Recognition

Qianhui Liang
University of Florida

John G. Harris
University of Florida

As PC clock speeds accelerate past 2 gigahertz, few everyday applications actually require these kinds of processor speeds. Of course, some users involved with intense game playing or video editing will always demand more speed, but typical users surfing the Web, editing documents, and sending email are already content with today's PC speeds. The next killer application that will require mainstream users to push for even faster speeds will likely be speech recognition. Speech is the ultimate man-machine interface and very comfortable for humans but extremely processor-intensive for computers. Speech recognition, speech analysis and synthesis, and speech understanding have seen rapid growth during the past decade and will be maturing in the next few decades. This chapter will overview the state of the art in speech recognition and, in particular, the use of artificial neural networks (ANN) with speech recognition.

Speech recognition is a fundamental problem for any speech processing system. Work regarding this discipline is very active and has made continuous advances over the years. As early as the 1960s, artificial neural networks were designed for speech recognition due to their powerful learning capability, nonlinear mapping functionality, and natural parallelism. However, since automatic speech recognition systems are expected to achieve adequate performance for a wide range of speakers, ANNs by themselves failed to achieve adequate performance. Because of the variability of speech and the fundamental nonstationary nature of speech, ANNs by themselves are inadequate for general speech recognition without the use of sound statistical models. Since the 1980s, the statistical methods of Hidden Markov Modeling (HMM) have been drawing great attention and have become the dominant approach, especially for large vocabulary, continuous speech recognition. Current commercial HMM-based speech recognition systems achieve adequate performance on small vocabularies or on large vocabularies with a single known speaker in low-noise environments.

One way to reduce the current performance barriers is to form hybrid systems that combine HMMs and ANNs. These two models complement one another and the combination is reported to improve the recognition performance. In order to summarize this development in the speech recognition field and the use of ANN in speech recognition, this chapter provides a succinct review that covers the progress made in the speech recognition domain involving ANNs. After a short overview of speech recognition in Section 6.1, artificial neural networks are discussed in detail in Section 6.2. The section includes ANN basic concepts, and ANN variants used in speech recognition, and places where ANNs have been used in the solution of the overall speech recognition problem. Section 6.3 discusses the basic aspects of HMMs and describes their weaknesses, ultimately motivating the hybrid ANN/HMM approaches discussed in Section 6.4. Section 6.5 discusses the key hurdle remaining for speech recognition — that of robust recognition in the presence of noise or other speakers.

6.1 Introduction to Speech Recognition

Speech recognition generally includes two parts: recognizing the spoken word and comprehending its meaning.[40] However, since the latter part is usually subsumed by the topic of spoken language understanding, we hereafter refer to the former as speech recognition. A general description of speech recognition is the mapping from a recorded segment of a speech signal to a sequence of words. A typical speech recognition system is comprised of four components: signal preprocessing, feature extraction, recognizer, and the language model. Figure III.6.1 shows a system block diagram. The first two components mainly involve acoustic analysis, while the fourth component concerns linguistic constraints of different levels, such as syntax, semantics, and pragmatics. Preprocessing, such as filtering noise, windowing, and pre-emphasis prepares the signal for further processing. Proper acoustic features are extracted by the feature extraction component, such that the relevant information can be easily captured and the complexity of the recognizer can be reduced. As is true for all pattern recognition problems, a good feature set should be invariant to aspects that are unrelated to the recognition, such as pitch or environmental noise. The two types of features most often used are Mel-frequency cepstrum coefficients (MFCC) and linear perceptual prediction coefficients (LPC). The recognizer decides the class of each word in the sequence with the help of the language model. The backward arrow from the language model to the recognizer (Figure III.6.1) denotes the feedback during training.

Five important questions must be addressed when discussing particular speech recognition systems:

1. *Is the system speaker-dependent or speaker-independent?* A speaker-dependent recognizer is restricted to the single speaker who is used to train the system. Only this speech is used to train the system and typically only this speech can be reliably recognized. In contrast, a

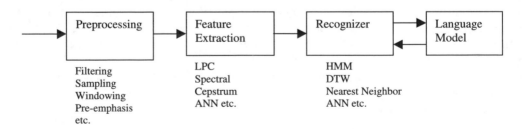

FIGURE III.6.1   Diagram of a typical speech recognition system.

speaker-independent recognizer is much more flexible as far as speakers are concerned. Here, recordings from multiple users in a population are used as training data. The system is expected to respond to utterances from many people, regardless of whether their speech was used in the training process.

2. *Is the system designed for isolated-word recognition or continuous speech recognition?* The second concern is whether the system is used to recognize isolated words or continuous speech. One extreme is isolated-word recognition (IWR), which is the easiest case. The speaker inputs the words one by one with a sufficient gap of silence between each word (usually a minimum of 200 msc,[12] so that the gaps between words are significantly longer than the gaps within a word. The other extreme of the complexity is continuous speech recognition (CSR), in which the speaker utters the speech in an unconstrained manner. In CSR, the recognizer would be able to identify the boundaries between two acoustic signals as well as deal with various possible co-articulations of acoustic units in a row. The latter problem obviously brings statistical characteristics into the recognition. In between is connected-word recognition, which is based on matching a continuously pronounced sequence with a template by concatenating individual words.

3. *What is the size of the vocabulary?* The performance of the recognition certainly bears a relation to the size of the vocabulary. The smaller the vocabulary is, the better performance the recognition system potentially can achieve. Surveying the current CSR systems, the vocabulary usually stays under 5000 words for reasonable performance.[38]

4. *What is the generality of the system; in other words, is it task-specific?* Currently, all research prototype speaker-independent CSR systems are applied to specific domains,[9] e.g., travel planning, urban exploration, and office management.

5. *How robust is the system?* Finally, we have to know in what environment the system works. Must it be in a noise-free environment? If not, what level of background noise can be tolerated? A related issue is the room acoustics, where echoes may easily confuse a system, depending on the location of the microphone relative to the speaker. Section 6.5 will discuss a few of these issues.

Research in automatic speech recognition (ASR) can be traced back to as early as the 1950s,[11,17,29,39] and many different approaches have been explored. One of the early approaches was the so-called "artificial intelligence approach," which incorporates both acoustic-phonetic information and pattern-recognition techniques.[40] Based on the understanding of the recognition task as an automatic learning and adaptive process, there has been a trend to apply ANN models from the artificial intelligence area to speech recognition since the 1980s. However, since ANNs are a relatively new technology, a multi-layer perceptron (MLP) recognition system using error back-propagation is limited to small vocabulary, isolated-word recognition.[6] The suitable form of a pure ANN model for continuous large-vocabulary speech recognition systems is not yet possible. After a major switch of ASR research from template-based to statistics-based approaches in the 1980s, the success of the statistical modeling methods, mainly HMMs, gave rise to a significant breakthrough in large-vocabulary, continuous-speech-recognition systems.[29] A fundamental belief about this approach is that speech recognition is a probability problem by nature, and therefore it cannot be solved by a deterministic view of the problem. A statistical model both of acoustic and of language is necessary. Researchers have been trying to combine ANNs with traditional HMMs to attack the recognition task. Good hybrid models have been established and improvement on performance has been observed.

6.2 Artificial Neural Networks

This section provides the background on ANN models for speech recognition.

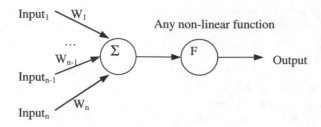

FIGURE III.6.2   Diagram of a typical neuron.

6.2.1   Concepts

An artificial neural network is a structure composed of a number of interconnected units (artificial neurons). Each unit has an input/output characteristic and implements a local computation or function. The output of any unit is determined by its I/O characteristic, its interconnection to other units, and (possibly) external inputs. Figure III.6.2 shows a typical artificial neuron. The network usually develops an overall functionality through one or more forms of training.[48]

From the above definition, we can observe several key aspects of ANNs. The overall computational model is composed of an interconnection of different units. So the functionality of the ANN depends on how these units are connected, i.e., the topology of the network. As far as the topology is concerned, a network could be recurrent or nonrecurrent, and static or dynamic. The behavior of the ANN also depends on the function that its units implement. The most widely used function is a sigmoid squashing function. It is semi-linear and can be expressed in a closed form. A third aspect of the ANN is the input and output. Inputs to the ANN should be chosen very carefully. When designing a neural network for a particular task, in our case, pattern recognition, we have to consider what are the key features that would truly derive the desired output. Also we think about how to make the output invariant to the possible diversity of extraneous factors. These are some of the considerable decisions a designer has to make. For a pattern recognition task, the output is usually much easier to choose. Training is another important aspect of ANNs. An ANN is usually formed through training. If we think of an ANN as a mapping from the input to the output, then training is to teach the network in the way that sample input/output mappings prescribe or to let the network determine how to classify the sample input.

6.2.1.1   Network Topologies

ANNs can be categorized as either recurrent or nonrecurrent. Figure III.6.3 depicts an example of a nonrecurrent network, while Figure III.6.4 illustrates an example of a recurrent network. Nonrecurrent, also known as feedforward, networks contain no closed interconnection paths, i.e., for any unit there is no cyclic path. In contrast, a recurrent network allows arbitrary connection between two units. And this allows more complex temporal dynamics.[48] An ANN can be layered (similar-structured), or competitive (interconnect-structured).

Many ANN architectures are composed of layers of units. The units of layer (i) receive outputs of units of layer $(i - 1)$ as their input, and produce output for units in layer $(i + 1)$. If i equals 0, the units of that layer will receive external input. Different layers are mutually exclusive. Layer 0 is also known as the input layer and the final layer is known as the output layer. Any layers in between the input and output layers are called hidden layers. See Figure III.6.3 for an example of a layered network. A multi-layer perceptron (MLP), or feedforward ANN, is a nonrecurrent layered network, in which each unit k is governed by an activation rule. For a more mathematical illustration, see Reference 12. A special case of an MLP is a network with only one layer, called a perceptron.

The recurrent networks are those with closed loops. Some of these networks are suitable for auto-associative addressable memory and optimization and constraint satisfaction applications.[12] There

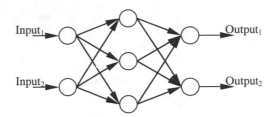

FIGURE III.6.3 An example of a nonrecurrent network.

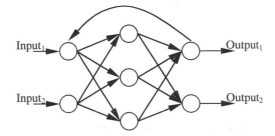

FIGURE III.6.4 An example of a recurrent network.

do not have to be any explicit input or output units in a recurrent network. The mapping is from one state to another, which is composed of the outputs of all units in the network. The recurrent network has memory and can remember unit outputs and network states. A Hopfield network is a popular type of recurrent network. Several variables defined in Hopfield's paradigm are as follows: o_i, the output state of the ith unit. Then O is the vector representing the state of the entire network; α_i, the activation threshold of the ith unit; w_{ij}, the interaction weight, or the strength of the connection between the ith unit and the jth unit.

Two important variations of the MLP network structures are radial basis function (RBF) networks and time delay neural networks (TDDN). In each case, it is observed that some neuron is responsible for some local features or is tuned to some region of the input space. An RBF network is a two-layered network. The single hidden layer units are used to carry localized information and the output layer in most cases contain linear units. The activation function of the hidden units uses an inner product operation, thus it is radially symmetric. RBFs are often used in pattern classification, and have been used in speech processing.[12,41]

Considering the high variability of speech signals, network input shift invariance is desired. This means we want to pick out the wanted input information despite a time-shift of its location. The TDNN is such an extension to MLP, which has memory of the past inputs and is spatial-temporal-relationship sensitive. The TDNN could be regarded as a collection of MLPs; each is a time-shifted version of another. As a result, the TDNN is invariant as far as time-shifts are concerned.

6.2.1.2 Learning and Training

In supervised learning, a set of "typical" input/output mappings form the training set (H). H provides significant information on the association between the input and the output data. In comparison to unsupervised learning, the elements of H are inputs or network states instead of a mapping. A mutually exclusive set of additional mapping tasks form the testing set. This set is used to test the performance of the ANN after training has completed. Therefore, the whole idea is H should be typical enough to help the network establish a positive relationship between the training set and the test set.

For some neural networks, it is very hard to form nonlinearly separable training set mappings.[48] One suitable training algorithm for the MLP is gradient descent training (GDT) using sigmoidal activation functions. Given the net activation of the unit, defined as Equation III.6.1, where $\underline{w}^{\mathrm{T}}$ is the transpose of the weight vector and $\underline{i}$ is the input vector, and activation function f, the total error over training set H is then given by Equation III.6.2, where e^p is the error for the pth pattern and n is the number of sample mappings in H. After several steps of derivation, we get the gradient of the pattern error with respect to weight, $\frac{d(e^p)}{d(w_k)}$, in Equation III.6.3, which is used to adjust $\underline{w}$, where o^p is the output by the pth input and t^p is the target of the pth input.

$$net = \underline{w}^T \underline{i} \qquad \text{(III.6.1)}$$

$$E = \sum_{p=1}^{n} (e^p)^2 \qquad \text{(III.6.2)}$$

$$\frac{d(e^p)}{d(w_k)} = (o^p - t^p)\left[\frac{df(net^p)}{dnet^p}\right] i_k^p \qquad \text{(III.6.3)}$$

An example of unsupervised training for MLP is the Kohonen self-organizing network. For a recurrent structure, a typical supervised learning Hopfield network is the content-addressable memory (CAM).

The training in TDNN is through the modified GDT. Training is to compute the weight corrections corresponding to the delayed input signals without the need of pre-processing or alignment. There are several additional aspects of TDNN learning: the training set must include a variety of shifted inputs corresponding to each desired output; hidden units look for the significant information in the current input as well as in its time-shifted versions; and the corresponding weights of the shifted versions tend to be the same for similar recognition tasks.

6.2.2   ANN Applied to Speech Recognition

Speech recognition involves two major modeling components. One is acoustic modeling and the other is temporal modeling. ANNs are able to perform acoustic modeling quite naturally. They can learn arbitrarily complicated functions, are easier to generalize, more discriminant, and more noise-tolerant. However, it is not yet known how to design ANNs for temporal modeling effectively. That may be the only and biggest barrier to ANNs accomplishing large-vocabulary continuous speech recognition by themselves. The pros and cons of ANNs for speech recognition will be discussed in a later section.

There are various ways that an ANN can be used for speech recognition. Referring back to Figure III.6.1, the ANN is a method for feature extraction, and recognition, as well as a possible language model. The ANN can also be used with dynamic time warping (DTW) to handle some amount of time variability. We will give a detailed discussion on how ANNs have been applied in various stages of speech recognition and its performance. Lippmann[35] gives a good review on ANN speech recognition systems before 1989. The bottom line, however, is that ANNs perform poorly by themselves as a speech recognizer primarily because they lack detailed modeling of the statistical variability of speech.

6.2.2.1   Why ANNs Can be Good for Speech Recognition

Bourlard has given several reasons for using ANNs in his book.[6] Here are the major points.

1. They combine multiple constraints and sources of evidence using a simple criterion without any explicit statistical assumptions.
2. They can learn over time.
3. They can generate any kind of nonlinear functions of the input.
4. They have a very flexible architecture, which can easily accommodate contextual inputs and feedback.

6.2.2.2   How ANNs are Used in Speech Recognition

6.2.2.2.1   *Fixed Length Speech Recognition*

6.2.2.2.1.1   *Recognition of Vowels and Phonemes*   By the beginning of the 1990s, simple phoneme classification by ANNs was very well researched. Both static ANNs, such as MLPs, and dynamic ANNs, such as TDNNs, were investigated. In 1988, experiments by Huang and Lippmann demonstrated that MLPs could form complex decision surfaces for phoneme recognition. In the same year, Burr[7] applied static ANNs, into which speech signals are input all at once, to rhyming English letters "B, C, D ... " recognition with a good result. One variant of the static approach is scaling the input to a predefined fixed length, either by linear scaling or nonlinear scaling. Dynamic programming would be needed for nonlinear scaling.[33]

On the other hand, the TDNN — well known for its economy on weights, hierarchy of delays, and shift invariance — was demonstrated to have excellent results for three vowels and three consonants recognition by Waibel.[58] Later, they scaled the TDNN up to recognize 18 Japanese consonants with a modular approach. Each TDNN was trained to recognize a small subset of phonemes; then these TDNNs were connected into a larger network through a light "glue" training process. The integrated TDNN achieved a better performance than a relatively advanced HMM-based recognizer for the phoneme recognition task. At the same time, recurrent networks (which are theoretically more powerful than feedforward networks and capable of representing temporal sequence) were also investigated for phoneme classification. Unlike the TDNN, recurrent networks represent temporal sequences with recurrent connections. Watrous[59] applied the RN to some basic discrimination tasks and obtained good results.

Lippmann[36] reviewed several speech recognition experiments on various ANNs and some traditional classifiers around that time. Comparisons show that ANNs perform comparatively or even better under their testing conditions. Five types of classifiers were considered: traditional parametric probabilistic methods, global-discriminant NN classifiers, local-discriminant NN classifiers, nearest-neighbor classifiers, and rule-forming classifiers. The traditional parametric methods are poor when the assumed class distribution is not in accord with the true distribution or when there is not sufficient training data. The output from NN classifiers is supposed to provide accurate approximation to Bayesian probabilities,[36] which would be used in higher-level analysis.

The first set of experiments included in his review is to evaluate eight NN classifiers and six conventional classifiers for speaker-dependent digit and vowel recognition tasks. Classification performance did not make a distinguishable difference among the classifiers tested. A second series of experiments is for a more complex speaker-independent phoneme recognition task. An RBF was demonstrated to need far less parameters to achieve a comparable error rate as the Gaussian mixture classifier, which falls into the conventional classifier category. This result is not surprising because the RBF minimizes the overall error rate without approximating the distribution functions separately. In 1992, Bengio[4] applied an RBF network to phoneme recognition of the TIMIT database and confirmed that RBF learns 3 times faster than MLP, which was first reported by Renals and Rohwer.[41] Bengio also proposed an improvement by combining RBF units and MLP units into one recurrent network.

6.2.2.2.1.2   *Recognition of Words*   Among more complicated experiments are word recognition tasks. Since there are temporal variations within word recognition tasks, we have good reason to believe that dynamic networks would be preferred to static networks. TDNNs were applied to word recognition.[34] The TDNN achieved an error rate of 9.5%. In 1987, Tank and Hopfield[54] proposed a "Time Concentration" network representing words by a weighted sum of delayed evidence, with proportional dispersion, until the end of the word. Unnikrishnan[56] also reported good results for this network. Franzini et al.[14] compared the RN with the MLP. Their results showed that no significant performance improvement was gained through RN. This suggests that the recognition network needs some memory, no matter in what form.

6.2.2.2.2   *Continuous Speech Recognition*

The temporal variability in continuous speech is high. Comparison experiments for two front-end components in continuous speech recognition systems were made by Renals et al.[42] in 1990. Their experiments used the RBF and the HMM as phoneme modeling methods. Both speaker-dependent and speaker-independent recognition is tested upon two databases, respectively. As mentioned before, the RBF is a feedforward network featuring hidden nodes with Gaussian transfer functions and output nodes with linear function. If the transformed outputs are assumed to approximate the posterior probabilities $P(\text{class}|\text{input})$, then the probabilities are transformed to a segment-wide hypothesis by the Viterbi algorithm. This lattice of phoneme probabilities, together with phoneme transitional costs computed by counting from the test data, comprise an ANN front end. The results showed that for speaker-dependent recognition on a task-dependent database, the neural network model front end as a Gaussian classifier performs the best without grammar analysis. For speaker-independent recognition on the TIMIT database, RBF performs better than HMM on the segment level, although both gave poor performance.

The Multi-State Time Delay Neural Network (MS-TDNN) of Haffner and Waibel[23] extended Waibel's work in 1989 for phoneme recognition to continuous speech recognition. The second hidden layer is no longer integrated over time to give a final output. Instead, these are regarded as a sequence of state scores. Time alignment is used to find the best state sequence. They also described a better training algorithm than the standard back-propagation algorithm, in which a training error only occurs when the output of an incorrect word is larger than the output of the correct word by a margin m. And the word is learned in the context of its proceeding word and following word.

6.2.2.2.3   *ANNs in Feature Computation*

In machine speech recognition, one of the fundamental problems is extracting features that are highly sensitive to the phonetic variability and less sensitive to other extraneous variability. In other words, phonetically equivalent input should be closely clustered in the feature space, while phonetically distinct input should be in different clusters. Some work was done to map from the original acoustic feature space to a phonetic feature space so that more distinction could be made as far as phonetics are concerned.[62] Labeled training data was used to generate features that enhance speech recognition. In their early study, they used a linear transformation component of the acoustic-to-phonetic transformation. In particular, to enhance the cepstral coefficients, linear discriminant analysis (LDA) was applied. Under their test conditions, the speaker-independent isolated-word-recognition error rate dropped from 30 to 16% using HMMs. Their later approach involved replacing LDA with more capable nonlinear ANNs. Several ANNs were investigated for the processing. Both TDNN and recurrent NN were trained as phonetic "classifiers." Since these networks have memory, some dynamic features were remembered in the internal representation. The performance of systems using the output of memory networks is superior to those using static features only. An alternative in their paper is to use a memoryless feedforward network while letting the HMM have a combination of static and dynamic features. Zahorian's results showed only a slight improvement in performance and a great economy on feature dimensions.

ANNs were also used as phonetic feature extractors. In Waibel's experiment,[57] certain units in the first and second hidden units in TDNNs detected phonetic features during training. As mentioned in his paper, some units were able to detect a falling second formant at a frequency of about 1600 Hz; another unit was able to detect a steady second formant at around 1800 Hz. Later, Bimbot, in his experiments with French phoneme recognition, used a TDNN to discriminate between certain phonetic features. In this case, binary and ternary distinction was made by the TDNN. It was also reported that TDNNs were effective in detecting modes of articulation, such as nasal and nonnasal, fricative and nonfricative, voice and nonvoiced. However, features related to the position of the articulation could not be detected.

6.3 HMMs for Speech Recognition

A speech recognition problem may look similar to a speech analysis problem, which is to identify a speech signal. Template matching of acoustic-phonetics in speech analysis could also be applied to speech recognition for isolated or connected speech. However, things are different in continuous speech recognition due to the tremendous variability in pronunciation. For the same speech, different speakers or even the same speaker at different times, would produce the speech signal quite differently. The speed, accent, and co-articulation affect the signal output. Therefore, the performance requirement is specified in a statistical sense, such as the mean square error. Quite naturally, a statistical model would be needed to take care of the desired performance in the statistical sense and to find a rigorous answer to the speech recognition problem. An HMM model could be decomposed into an acoustic model and a language model.[28]

6.3.1 Mathematical Foundation

The Hidden Markov Model (HMM) is based on a stochastic process model called the Markov chain. The Markov chain models a class of random processes that involve a minimum amount of memory without being completely memoryless. Let $X_1, X_2, \ldots, X_n$ be a sequence of random variables taking their values in the same finite alphabet. Bayes's formula tells us:

$$P(X_1, X_2, \ldots, X_n) = \prod_{i=1}^{n} P(X_i | X_1, X_2, \ldots, X_{i-1}) \tag{III.6.4}$$

The random variables are said to form a Markov chain, if:

$$P(X_i | X_1, X_2, \ldots, X_{i-1}) = P(X_i | X_{i-1}) \tag{III.6.5}$$

for all i. As a result, the following is true for the Markov chain:

$$P(X_1, X_2, \ldots, X_n) = \prod_{i=1}^{n} P(X_i | X_{i-1}) \tag{III.6.6}$$

If we think of the values of X_i as a state, then a Markov chain is a finite state process with transitions between states specified by the function $P(x'|x)$, where $X_i = x'$ and $X_{i-1} = x$.

To have more flexibility in the random process, multiple transitions are allowed for pairs of states. A probability is assigned to each transition respectively. And a different output symbol would be generated whenever a particular transition is taken. More particularly, when an unknown speech waveform is converted by the pre-processor and feature extractor into a sequence of feature vectors, $Y = y_1, y_2, \ldots, y_r$. The HMM model of the speech $M = m_1, m_2, \ldots, m_n$ consists of a sequence of HMM models for words. The recognition system is to determine the most probable sequence $\hat{M}$ given the input vector string. Bayes's rule allows us to compute $\hat{M}$ as the following:

$$\hat{M} = \arg\max_m P(M|Y) = \arg\max_m \frac{P(M)P(Y|M)}{P(Y)} \tag{III.6.7}$$

This equation illustrates the most fundamental idea of HMMs. To find the most likely word sequence, we would have to maximize the product of $P(M)$ and $P(Y|M)$ when Y can be regarded as a constant. $P(M)$ is called the prior probability that is independent of the observed vectors and is decided by the language model. $P(M|Y)$ is the probability of observing the sequence, given a certain sequence of words. This probability is decided by the acoustic model that is discussed in the following subsection.

6.3.2   Acoustic Model

The value of $P(Y|M)$ must be calculated for all possible pairings of M with Y; the calculation must be done in real-time. Thus, to compute $P(Y|M)$, we need a statistical acoustic model of the speaker's interactions with the acoustic processor. The model would include the way the speaker pronounces the word sequence, the room noise, reverberation, the microphone placement, and characteristics and acoustic processing. Each word would be decomposed to a sequence of sound units, such as phones or phonemes, by a pronunciation dictionary. Each sound unit has a corresponding HMM model. With a hierarchical scheme, only HMM models for limited basic sound units are necessary to form the composite HMM models for a long speech segment by simply concatenating all the HMM models of the lower level. For example, word HMM models are a concatenation of phones or phonemes. HMM models and sentence models are a concatenation of word models.

By the above composite model, $P(Y|M)$ then is able to be calculated. Given a probability distribution of transitions between states $p(s'|s)$ and an output probability distribution $q(y|s, s')$ associated with transitions from state s to state s', the probability of observing an HMM output vector string $y_1, y_2, \ldots, y_k$ is then given by:

$$P(y_1, y_1, \ldots, y_k) = \sum_{s_1,\ldots,s_k} \prod_{i=1} p(s_i|s_{i-1}) q(y_i|s_{i-1}, s_i) \qquad (III.6.8)$$

where s represents the states in the finite state process and the sum extends over all possible state paths of length k in the composite model. This is sometimes approximated as:

$$P(y_1, y_1, \ldots, y_k) = \max \prod_{i=1} p(s_i|s_{i-1}) q(y_i|y_{i-1}, y_i) \qquad (III.6.9)$$

and called the "Viterbi" approximation, which is said to perform recognition without loss of much performance. Acoustic models can be of continuous density, discrete density, or semi-continuous density. Many state-of-the-art systems prefer continuous density modeling to the other two modeling approaches because the number of parameters used for the HMM observation distribution could easily be adapted to the amount of training data available.[18] To obtain more accurate recognition results, different models could be established and trained for different genders. These models are trained upon the nonoverlapping subset of the training data.

6.3.3   Language Model

In Equation III.6.7, we notice that we have to compute *a priori* probability $P(M)$ for every word string M. $P(M)$ is the probability that a speaker wishes to speak a particular segment, which is defined by the language model. $P(M)$ can be factorized into the following form by the Bayes formula:

$$P(M) = \prod_{i=1}^{n} P(m_i|m_1, \ldots, m_{i-1}) \qquad (III.6.10)$$

Since the number of possible word sequences from a medium-sized vocabulary would be huge, only an estimation of the above equation is required. One of the possible estimates is to classify the history of words uttered into corresponding classes, $\Phi(m_1, m_2, \ldots, m_{i-1})$. Equation III.6.10 actually becomes:

$$P(M) = \prod_{i=1}^{n} P(m_i|\Phi(m_1, \ldots, m_{i-1})) \qquad (III.6.11)$$

A language model copes with how to classify the history, i.e., to get corresponding $\Phi(w_1, w_2, \ldots, w_{i-1})$, and how to calculate $P(m_i|\Phi(m_1, \ldots, m_{i-1}))$.

6.3.4 Weaknesses of HMMs

Although HMMs are the state-of-the-art speech recognition approach, they suffer from several weaknesses. For example, HMMs make some assumptions which are not true for speech signals, such as feature vectors extracted from one phonetic frame and the adjacent ones of the same segment are uncorrelated; all probabilities depend solely on the current state. Other limitations include: likelihood maximization training algorithm, which inappropriately assumes the correctness of the model and implies bad discrimination; and suboptimal modeling accuracy for discrete, continuous, or semi-continuous HMM probability density models. Tebelskis[55] has given a more detailed explanation in his dissertation.

In order to overcome these weaknesses, HMMs usually rely heavily on context-dependent phone models. The context-dependent model requires so many parameters that they must be extensively shared. This, in turn, asks for a very complicated mechanism.[27]

Luckily, ANNs have features that can complement the weaknesses of the HMMs. They do not make the assumption of decorrelation, and they are more discriminant and require less feature dimension. Sets of controlled experiments have shown that the use of ANNs together with HMMs can significantly improve the recognition performance. Also, this new combination paradigm is based on two frameworks, both of which have solid theoretical support. So we could expect it to bring new breakthroughs to the development of speech recognition systems.

6.4 ANNs Combined with HMMs

Although ANNs have shown promise in speech recognition, their success has only been in the simpler problems of phoneme recognition or isolated-word recognition. As the temporal variability of continuous speech recognition continues to increase, the invariance is localized and the global classification is an integration of localized decisions. Dynamic ANNs were demonstrated to have some temporal integration ability as long as the operation can be described statically.[55] However, it is not appropriate to represent the widely variable speech with static structured networks. Therefore, in their current form, ANNs are still not applicable by themselves to solve the continuous speech recognition problem effectively.

On the other hand, in HMMs, time-sequence information is well represented by the state transition with learned probabilities. This provides a good reason to combine ANNs and HMMs for continuous large-vocabulary speech recognition. Besides, ANNs are discriminant-based pattern classifiers, which try to minimize the likelihood for the wrong class as well as to maximize the likelihood of the correct class. HMMs basically try to maximize the likelihood for the correct class alone. Therefore, a combination of HMMs and ANNs will enhance the classifier by including the discriminant-based measurement. In the next section, we provide a description of HMMs as a background before we describe the completely new hybrid platform of a hybrid NN-HMM.

6.4.1 Implementation of Components of HMMs

Early research on the integration of ANNs and HMMs involved implementation of components of HMM models by ANNs. Early in 1987, Lippmann designed a neural network to implement the Viterbi algorithm, called the Viterbi net. A sequence of feature vectors was input to the Viterbi net one at a time. The output was the cumulative score along the search path. This structure was able to recognize isolated words using a comparison among several parallel networks. But this system was still not able to perform continuous speech recognition. Among other experiments is Bridle's Alphanet. Alphanet uses a neural net to compute the forward probability of particle sequence y_j^t, which ends with state j.

6.4.2 Generating Probabilities Used in the Underlying HMM Models

Many laboratories that published reports on continuous speech recognition used ANNs to generate emission probabilities that could be used by the HMM models to obtain $P(Y|W)$. This method for combination aimed at taking advantage of the best of both worlds and increasing the overall recognition accuracy. ANNs are well suited for this job because they accept continuous input and do not suffer from quantization error. They also don't make any assumption about the parametric distribution function.[55]

Different strategies can be used to design the neural networks. Tebelskis provided a nice summary of different designs in his dissertation. The simplest approach is to map the input frame directly to the emission symbol output, (so-called frame training) where training is accomplished frame by frame. This approach was extensively investigated by Philips, ICSI, and SRI. Experiments on it gave excellent results on the Resource Management database.[6,37,43] Simple MLPs were used in these experiments. Franzini et al.[15] also studied frame training in which they plugged the neural network into an existing HMM to represent emission probabilities. This network is for continuous probabilities rather than binary classes. Good word recognition results were obtained in their experiments.

Another training level is segment training. The neural network receives input from an entire segment of speech instead of a single frame or a fixed window of frames. This allows correlation information between frames of the segment to be properly captured. However, to take advantage of this, the speech must first be segmented. The TDNN was initially used to integrate the partial scores from the whole duration of the phoneme by Waibel,[58] but its input assumes the length of the window to be 15 frames. Austin[2] proposed a SNN (Segment Neural Network), which is a full-featured frame training neural network for large vocabulary continuous speech recognition. The network is trained for recognizing phonemes from variable-duration segments. The system achieved an 11.6% error rate on the RM database and later it was improved to a rate of 9% by including training for incorrect segments.

The next step of extension is word level training. The feature vectors of the entire word are provided to the network. The network is trained to optimize word recognition performance. Sakoe et al.[46] implemented such a neural network called the Dynamic Programming Neural Network (DPNN). In the DPNN, hidden layers represent states and the output layer represents words. An alignment path is obtained through DTW, and the output unit integrates the activation of all the hidden units along the alignment path. It achieved 99.3% word accuracy in their experiments. Haffner[22] also integrated DTW into TDNN and proposed the Multi-State TDNN (MS-TDNN), which has one more hidden layer and which has a hierarchy of time delays. It was applied to a database of spoken letters and achieved an average accuracy of 93.6%. Later, this model was improved by Hild and Waibel[25] and provided even better results.

The last extension mentioned in the dissertation is global optimization. Its purpose is to relax the rigidities in a system so the performance is less handicapped by false assumptions. Efforts regarding global optimization were also made by Bengio.[4]

6.4.3 Context-Dependent Models

Today, context-dependent phone models are probably the most widely used speech-unit models in any usable and competitive speech recognition systems. The aim of introducing phonetic context dependency is to model the phones in context. The context that a phone sits in greatly influences its acoustical representation in continuous speech. One way to handle the variance of a phone is to have a model to represent a phone in context. It is generally known that context-dependent models perform better than context-independent models. Phonetic Context both in HMMs and in ANNs must be considered for optimal performance.

6.4.3.1 Context-Dependent HMMs

The first context-dependent phoneme HMM systems were proposed in 1984.[49] Triphone models were used successfully. A triphone model includes the immediate left and right phonetic contexts in HMMs. Training for the parameters in these models are upon the data in that specific context. There are several problems with the training, such as insufficient training data relative to the number of parameters and so-called "unseen data." (See Reference 31 for detailed explanation.)

Kershaw summarized the tripone models into three categories. The first is the backed-off HMM models. Left-context biphones, right-context biphones, and monophones are trained. If a particular HMM does not exist during the decoding process, triphones are backed-off to right-context biphones, which are backed-off to left-context biphones, which are backed-off to monophones. The second is the Deleted Interpolation model. In this model, triphone, biphone, and monophone HMMs are trained for every context available in the training data. The overall observation probability is the weighed sum of observed probabilities. The third is the Generalized Model. The main idea is that a common model, which requires fewer parameters, is possible for several similar contexts of different HMM models. In this way, context-dependent models can be clustered into generalized context models by determining the similarities among context-dependent HMMs.

6.4.3.2 Context-Dependent ANN/HMM Hybrid Systems

Work regarding context-dependent hybrid systems was jointly carried out by SRI, ICSI, and Stanford.[51,52] The purpose of context-dependent models, in general, is to account for co-articulation effects. Therefore, models must be trained in different contexts. In hybrid systems, where ANNs are used to estimate the posterior probabilities, ANNs are used to generate the probability density of the feature vector, given the phone class and its context class. Accordingly, the HMM probabilities are given by the following equation:

$$p(y(t)|q_j, c_k) = \frac{\Pr(q_j|y(t), c_k)\,p(y(t)|c_k)}{\Pr(q|c_k)} \qquad \text{(III.6.12)}$$

where $y(t)$ is the feature vector, q_j is the phone class, and c_k is the context. $p(y(t)|c_k)$, in turn, is given as:

$$p(y(t)|c_k) = \frac{\Pr(c_k|y(t))\,p(y(t))}{\Pr(c_k)} \qquad \text{(III.6.13)}$$

Based on the above two equations, it is easy to see how the estimation of $p(y(t)|q_j, c_k)$ proceeds. For $\Pr(q_j|y(t), c_k)$, there will be the same number of MLPs as that of context classes, and each will be trained by a set of non-overlapping data only under the corresponding context. For $\Pr(c_k|y(t))$, a separate context-independent MLP will be trained to estimate the posterior probability of a context class given the feature vector. $\Pr(c_k)$, $\Pr(q|c_k)$ can be estimated from the context language model.

The context-independent MLP is trained first, and the result of the training is used to initialize the context-dependent MLPs. The training of context-dependent MLPs is similar to the context-independent case. As mentioned in Kershaw's dissertation, to simplify the process, input-to-hidden layer weights are kept fixed during the context-dependent training stage.

Experiments were performed on speaker-independent continuous-read speech from the DARPA Resource Management database. In Table III.6.1, the recognition error and number of parameters used are listed. An average reduction of error rate of about 23.5% was reported.[52]

In comparison to the factorization in the previous method, another alternative would be to express $\Pr(q_j, c_k|y(t))$ as a product of the monophone posterior probability and a new conditional:

$$\Pr(q_j, c_k|y(t)) = \Pr(q_j|y(t))\,\Pr(c_k|y(t), q_j) \qquad \text{(III.6.14)}$$

TABLE III.6.1   Performance on Speaker-Independent Continuous Speech Recognition from DARPA RMDB

Task	CI MLP	CD MLP	% red^n WER
Feb91	5.8	4.7	19.0
Sep92	10.9	7.6	21.1
Sep92.sd	9.5	6.6	30.5
# Parms	300K	2150K	

Hence, the triphone model would be expressed as:

$$p(y(t)|c_k^L, q_j, c_k^R) = \frac{\Pr(c_k^L, q_j, c_l^R|y(t))}{\Pr(c_k^L, q_j, c_k^R)} \qquad (\text{III.6.15})$$

where c_k^L, c_l^R are the left and right context class of state q_j, respectively.

6.4.3.3   Context-Dependent SNN/HMM Hybrid System

This hybrid system differs from the previous one in that a segment neural network (SNN) was introduced to alleviate the restriction of HMMs by the conditional-independence assumptions. Research in such systems was described in several papers.[63–65] Two types of networks are trained. One is a context-independent SNN, a neural network that classifies the whole phonetic segment as a unit, instead of a concatenation of independent frames. The SNN is supposed to extract additional correlation information used to increase the accuracy of the context model. A typical SNN takes as input a fixed length of feature vectors for the phonetic segment, which is obtained after warping of variable-length vectors by DCT. The output is the posterior probability of a phonetic class given the acoustic signal.

The other type of network is a left-context network. To classify 53 phonemes in a particular left context, 53 networks are trained. After receiving output from both types of networks, a re-scoring is done. The overall score would be a weighted average of both the text-independent SNN output and left-context networks output. Experiments on the DARPA 1000-word Resource Management Task of February 1989 showed a reduction of 17.5% in the error rate compared to that of the context-independent network.

6.4.3.4   Context-Dependent Recurrent Network/HMM Hybrid System

Some of the approaches of nonrecurrent networks in the hybrid systems described in the previous section could be extended to recurrent networks. The factorization of conditional context class probabilities can be used as a solution to the problem of implementing phonetic context dependency. An efficient tree-based method for producing a context-dependent system was proposed using continuous density statistics estimated from the acoustic data.[31] An implementation of such a context-dependent hybrid system was also presented. The performance of a larger context-dependent RNN outperformed a traditional RNN with monophone outputs. Besides, in order to shorten the training process, special hardware had been constructed.[32] However, with this hardware, the training time is still considerable and the performance also deteriorates. So, an existing well-trained monophone model (the RNN) was used instead, and training was done on a standard workstation rather than by special hardware by Kershaw.

The RNN context-dependent factorization for the observation likelihood is given by the following equation (Bayes's rule is applied twice):

$$p(y(t)|q_i(t), c_{j|i}(t)) = \frac{\Pr(q_i(t)|y(t)) \Pr(c_{j|i}(t)|y(t), q_i(t))}{\Pr(q_i(t), c_{j|i}(t))} p(u(t)) \qquad (\text{III.6.16})$$

where $c_{j|i}$ represent a phone class q_i in context class j.

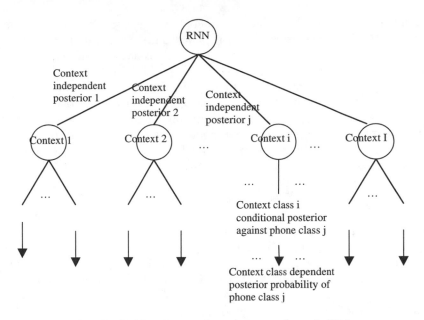

FIGURE III.6.5   Context-dependent posterior probabilities.

The context class selection is through a decision-tree approach. The training is accomplished with the modular method shown in Figure III.6.5, rather than with an output layer containing context-dependent targets.

The context class conditional posteriors are estimated by a set of single-layer perceptrons. The modular approach results in rapid training with good context-class classification performance. It was reported that a set of context networks (for each RNN) could be built in reasonable time on a regular machine, given a decision tree. The context-dependent phone posteriors could be computed efficiently compared to the baseline RNN system.

6.4.4   Advances in Large-Vocabulary Continuous Speech Recognition

Large-vocabulary continuous speech recognition (LVCSR) is the ultimate purpose of speech recognition research. Such systems usually have a vocabulary of more than 1000 words. The technology today is at the point where the usefulness of LVCSR becomes the major concern. In a whole range of application areas, ranging from less demanding applications such as domain-specific dictation and interactive database accessing, to more demanding tasks such as multimedia information retrieval, people are looking forward to the appearance of usable systems. After 1996, various speaker-dependent off-the-shelf products have become available, allowing for medium vocabularies with very clean speech. On the other hand, much more effort is still needed to make robust and dependable systems ubiquitous. Considerable progress continues to be made in hybrid systems for large-vocabulary continuous speech processing. Advances are seen in the improved training techniques, improvement of the robustness, algorithms for audio segmentation, acoustic model adaptation, etc. Two of the obstacles to improvement of LVCSR hybrid connectionist systems are: tremendous training time of acoustic models and poor scalability of connectionist acoustic models to larger systems. Hereafter, we try to review major research efforts of training and scalability made for LVCSR hybrid systems.

6.4.4.1   Hierarchical Architecture of Neural Networks

Several connectionist systems generate the emission probability estimation for context-dependent classes as a product of network outputs. In Cohen's approach,[8] separated sets of context-dependent output layers are used. A set of networks classifies eight different broad classes of left and right contexts. Alternatively, context-dependent probability could be modeled based on the factorization of the probability $P(phone\_context|data)$ into $P(phone|data)*P(context|phone, data)$.

Kershaw[30] adopted a modular approach to incorporate context-dependent phone classes in a connectionist-HMM hybrid speech recognition system. Their basic framework is an extension of Bourlard,[5] by applying a recurrent network as the acoustic model for the HMM framework. Single-layer MLP networks are used to classify different context classes given the phone class and the feature vectors. They are trained on the state vectors corresponding to each monophone to classify the context classes. The parameters of context-dependent modules are estimated by embedded training. An RNN in a context-independent model generates frame-by-frame monophone posterior probabilities. Results for the context-independent system showed an average reduction of 16% in error rate on the test set. The network still uses the same fixed-sized observation vectors to estimate the posterior probabilities for all phones. The decoding speed is also more than doubled.

Fritsch[16] also proposed a so-called "ACID/HMM" for the large-vocabulary continuous speech recognition framework. He argued that "a hierarchical organization of the acoustic model is crucial for competitive performance". The Agglomerate Clustering scheme based on the Information Divergence Hierarchy of Neural Networks (HNN) forms soft decision trees for hierarchical classification. It is reported that this approach avoids scalability problems due to nonuniform prior distributions and could easily be integrated.

Antoniou[1] decomposed neural networks into small pieces. Each piece would be trained to estimate a single phone. The sizes of the feature vectors are not the same across the modularized network. Instead, the size is selected to account for the phone that the network tries to classify. Their motivation of decomposing modeling tasks into subtasks is if one MLP is coupled with one state per HMM, the dimensionality of the observation vector has to be increased, which means the training is done in a high multidimensional feature space. As a result, the data sparsity problems worsen. Their approach works as follows. For each of 39 phones, a detector MLP with single output is associated with one phone. Further, these 39 phones were divided into seven broad classes, so that more concise indication of left or right context is provided. Experiments on the TIMIT database showed the phone and word recognition rates improved and are better than the best context-dependent systems in the literature.

6.4.4.2 One Large Network with More Training Data

In comparison with the above section, Ellis et al.[13] investigated the relationship between the training, error rate, and size of training set, and size of the neural network. Standard architecture and training methods work effectively with simple tasks. On the other hand, for large tasks, when an amount of training data is required, competing with the performance of HMMs becomes challenging.

Ellis has proposed to keep the network architecture the same while increasing both the network size and the amount of training data. Doubling the training data and system size, the performance improvement diminished in terms of error rate reduction for the large systems. In order to meet the intensive computation demand, a vector microprocessor has been developed. The following table shows the word error percentage for the hybrid system with neural networks with different numbers of hidden units. Their surprising observation is that improvements can be realized by either increasing the size of the network or the size of training data. Improvements continue to be significant before some limit is reached. Table III.6.2 shows the diminishing of returns for this strategy.

6.4.4.3 Multi-Source

Different features of speech signals are potentially able to provide complementary information useful to improve the recognition performance of some ambiguously articulated sounds. This would be especially useful in the continuous speech case, for which co-articulation frequently occurs. The nice property about artificial neural networks is that they are able to incorporate heterogeneous information from different sources without assumption on their statistical distribution. Some research work has

TABLE III.6.2   Experiment on a Single Large Network by Ellis

# Hidden Units	Training Set Size, hours			
	9.25	18.5	37	74
500	42.8%	41.0%	40.2%	39.2%
1000	41.8%	38.8%	36.5%	36.9%
2000	40.4%	37.2%	35.6%	36.9%
4000	40.3%	37.4%	33.9%	33.7%

been performed to find out the potential that a neural network could make the recognition more accurate. Gemello[19,20] proposed a Multi-Source NN (MSNN) for making parallel use of different features in speech recognition and provided a good review of the results. These features are extracted separately from the signal. We will look at some key parts of his approach.

The three categories of features are basic features, additive features, and alternative features. Basic features are the standard features in speech recognition — MFCC was the choice made in Gemello's system. Gemello also defined a set of additive features, which are not independent but could be used in addition to the basic features. Gravity centers, regarded as one of the additive features in Gemello's system, indicate frequencies of great energy concentrations and are related to vocal tract resonance; their computation is much simpler and robust.[20] Another additive feature is frequency derivatives (taking the derivative of the power spectrum with respect to frequency). First-order derivatives were computed from normalized power spectrum vectors. Then the second-order derivatives were computed based on the first-order derivatives. Both sets of results were grouped into a smaller number of subclasses, which compose the frequency derivative features. Alternative features are those that are sufficiently informative and could be used instead of the basic features. Two alternative features mentioned are RASTA-PLP features and Ear Model features. RASTA-PLP was introduced by Hermansky[24] and used to emulate the insensitivity of human hearing to slowly varying stimuli. The Ear Model transforms acoustic signals into perceptual representations of human hearing systems.

The input layer of the network is made up of frames. Within each frame, there are three or more blocks of features. The first block is for basic features, followed by one or more blocks for additive features, if any, and the last one is for alternative features. There are two hidden layers. The first hidden layer could be divided along time dimensions and source dimensions. The second hidden layer performs the integration of the local output from the previous units.

The telephonic test corpora used for testing are made up of continuous telephone speech with a vocabulary of 9400 words. The results showed that additive features improve the performance by 11.6% for gravity centers and 6.3% for frequency derivatives. Alternative features contribute to the accuracy of recognition by systems with basic features. Adding Ear Model features exhibits a 14.4% improvement in performance.

6.5   ANN in Robust Speech Recognition Systems

One of the big challenges of speech recognition is the robustness of the systems. Robustness in speech recognition can be defined as minimal, graceful degradation in performance due to changes in input conditions caused by different microphones, room acoustics, background or channel noise, different speakers, or other small systematic changes in the acoustic signal.[9] Though good research progress has been made in this area, the present systems are still far from robust. The performance reported was mostly obtained under restrictive conditions, and would significantly degrade with a small change of the room condition, the microphone type, the way people talk (e.g., pauses between two words vs. no pauses between in continuous speech, close talking vs. remote talking). Therefore, speech recognition robustness needs to be addressed.

It is crucial to understand the source of variability in speech recognition. In Cole's paper,[9] some major sources of variability in speech signals were described and explained. Some of the many sources are the speaker and the nature of the task, the physical environment, and the communication channel between user and machine. There are several ways to improve the robustness of the system. In the early 1990s, different NN architectures were proposed to add robustness to the variance with speech uttered by a speaker. More recent research is focusing on noise-corrupted speech. According to Sankar,[47] noise reduction techniques fall into three categories. The first uses mathematical models, which have *a priori* knowledge about the signal and noise.[53,60] The second uses "intuitive" predetermined features of noise and speech, such as spectral or cepstrum. The third uses neural network models, e.g., MLPs,[50,61] or the RBF neural network[47] (as a nonlinear model) which is able to map from the noise to the noise-free domain.

One of the interesting aspects of noise reduction in speech recognition regards car noise. The key application that is stimulating research of various noise reduction techniques[47] is trying to automatically understand conversations within a noisy car, particularly given the great penetration of cell phones. An analysis and classification of car noise helps to improve the SNR. Sankar[47] has presented a robust speech recognition technique for voice-activated dialing. Noise reduction techniques by Linear Spectral Subtraction (LSS) and Nonlinear Spectral Subtraction (NSS)[21,60] are applied. Then in the homomorphic front end, cepstral features are Mel-frequency scale warped, lifted, and FFTed.

For the RBF neural network, two topologies were studied. The first structure creates as many Radial Basis neurons as there are input training vectors. The second is more economical with neurons and creates the Radial Basis layer one neuron at a time. The result showed that the second structure provided better recognition accuracy and NSS improved the performance by 25% over no subtraction, combined with a rectangular lifter. Only 1% improvement was reported using NSS over LSS.

In a hybrid system, both ANN and HMM models are made robust in order to force the overall system to be robust. In 2000, Cong[10] proposed a SMQ/HMM_SVQ/HMM)/MLP robust speaker-independent isolated-word speech recognition system, which combines dual split matrix quantization (SMQ) and split vector quantization (SVQ) modeling stochastic sequences and the nonlinear neural networks. The two quantization methodologies for this hybrid system were proposed to divide the speech parameters into matrices operating in both the time and frequency domains. The noise spectrum is therefore localized in the frequency domain. This system contains robust HMM models. The inputs to HMMs are obtained from a given word at different SNR levels, each of which is quantized by the dual SMQ/SVQ quantization pair. The MLP classification is also robust. For any input word, the assigned hybrid probability measures (vector/matrix quantized and HMM classified) to an input word allow a "pattern" of the hybrid probability measure to be generated. Simulation tests indicate that this method outperforms the traditional VQ/HMM and MQ/HMM by 98 and 95.8% at 20-dB and 5-dB SNR levels, respectively.

6.6   Conclusion

In this chapter, we argued the promise of ANNs in speech recognition by illustrating basic concepts of ANNs, reviewing some recent applications and advances of ANNs in speech recognition, and analyzing the different roles ANNs play in these application systems. We justify that ANN/HMM hybrid systems show superiority according to recent reported results in the literature. Strong evidence can also be given in theory, such as capability of nonlinear mapping, being discriminant-based, capability of integration of features from heterogeneous input without any assumptions about the statistical distribution, parallel computation functionality, etc. We also summarized the challenges that current systems have to face, including the robustness, task-dependent, real-time, and sponta-neous. In the coming era, we should observe continuous efforts on application of ANN to speech recognition and better techniques that make ANN suitable for temporal models, particularly with their combination with HMMs. It is clear that robust speech recognition devices of the future will use tremendous amounts of computation, and thus will require continued advances in computer speed.

References

1. Antoniou, C., Modular neural networks exploit large acoustic context through broad-class posteriors for continuous speech recognition, *Acoustics, Speech, and Signal Processing, Proc. IEEE Int. Conf.*, 1, 505–508, 2001.
2. Austin, S., Zavaliagkos, G., Makhoul, J., and Schwartz, R., Speech recognition using segmental neural nets, *Acoustics, Speech, and Signal Processing, IEEE Int. Conf.*, 1, 625–628, 1992.
3. Barnard, E., Optimization for Training Neural Networks, *IEEE Trans. Neural Networks*, 3(2), March 1992.
4. Bengio, Y., De Mori, R., Flammia, G., and Kompe, R., Global optimization of a neural network—Hidden Markov Model, *IEEE Trans. Neural Networks*, 3(2), 252–9, March 1992.
5. Bourlard H. and Morgan, N., Continuous speech recognition by connectionist statistical methods, *IEEE Trans. Neural Networks*, 4(6), 893–909, 1993.
6. Bourlard, H. and Morgan, N., *Connectionist Speech Recognition: A Hybrid Approach*, Kluwer Academic Publishers, Boston/Dordrecht/London, 1994.
7. Burr, D., Experiments on Neural Net Recognition of Spoken and Written Text, in *IEEE Trans. Acoustics, Speech and Signal Processing*, 3, 1162–1168.
8. Cohen, M., Franco, H., Morgan, N., Rumelhart, D., and Abrash, V., Context-Dependent Multiple Distribution Phonetic Modeling with MLPs in 'NIPS 5', 1992.
9. Cole, R. et al., The Challenge of Spoken Language Systems: Research Directions for the Nineties, *IEEE Trans. SAP*, 3(1), January 1995.
10. Cong, L., Asghar, S., and Cong, B., Robust speech recognition using neural networks and hidden Markov models, *Proc. Int. Conf. Information Technology: Coding and Computing*, 2000, 350–354.
11. Davis, K.H., Biddulph, R., and Balashek, S., Automatic recognition of spoken digits, *J. Acoust. Soc. Am.*, 24, 637–642, 1952, Cited in Reference 29.
12. Deller, J.R., Proakis, J.G., and Hansen, J.H.L., *Discrete-Time Processing of Speech Signals*, Macmillan Publishing Company, 1993.
13. Ellis D. and Morgan, N., Size Matters, An empirical study of neural network training for large vocabulary continuous speech recognition, *Acoustics, Speech, and Signal Processing*, 1999. Proc. 1999 IEEE Int. Conf. vol. 2 , 1999, 1013–1016.
14. Franzini, M., Witbrock, M., and Lee, K.F., Speaker-independent recognition of connected utterances using recurrent and non-recurrent neural networks, in *Proc. Int. Jt. Conf. on Neural Networks*, 1989.
15. Franzini, M., Lee, K.F., and Waibel, A., A connectionist Viterbi training: a new hybrid method for continuous speech recognition, *Acoustics, Speech, and Signal Processing, 1990 Int. Conf.*, 1990, 425–428, vol. 1.
16. Fritsch, J., ACID/HNN: A framework for hierarchical connectionist acoustic modeling, automatic speech recognition and understanding, *Proc. 1997 IEEE Workshop*, 1997, 164–171.

17. Fry, D.B. and Dennes, P., Theoretical aspects of mechanical speech recognition, The design and operation of the mechanical speech recognizer at University College London, *J. British Inst. Radio Engr.*, 19(4), 211–229, 1959.

18. Gauvain, J.L. and Lamel, L., Large-vocabulary continuous speech recognition: advances and applications, *Proc. IEEE*, 88(8), 1181–1200, 2000.

19. Gemello, Albesano, D., and Mana, F., Multi-source neural networks for speech recognition, *Neural Networks, Int. Jt. Conf.*, 5, 2946–2949, 1999.

20. Gemello, Albesano, D., and Mana, F., CSELT hybrid HMM/neural networks technology for continuous speech recognition, Neural Networks, 2000. IJCNN 2000, *Proc. IEEE-INNS-ENNS Int. Jt. Conf.*, 5, 103–108, 2000.

21. Gupta, S.K., Soong, F., Hiami-Cohen, R., High-accuracy connected digit recognition for mobile applications, *Proc. ICASSP*, vol. 1, 1996.

22. Haffner, P., Franzini, M., and Waibel, A., Integrating time alignment and connectionist networks for high performance continuous speech recognition, in *Proc. IEEE Int. Conf. ASSP*, 1991.

23. Haffner, P. and Waibel, A., Multi-state time delay neural networks for continuous speech recognition, in *Adv. Neural Information Proc. Syst.*, 4, 135–142, 1992.

24. Hermansky, H. and Morgan, N., RASTA Processing of speech, *IEEE Trans. Speech Audio Process*, 4(2), October 1994.

25. Hild, H. and Waibel, A., Connected letter recognition with a multi-state time delay neural network, in *Adv. Neural Information Proc. Syst.*, Hanson, S., Cowan, J., and Giles, C.L., Eds., Morgan Kaufmann Publishers, 1993.

26. Huang, W.M. and Lippmann, R., Neural net and traditional classifiers, *Neural Information Processing Systems*, D. Anderson. (ed.) New York: American Institute of Physics, 1988, pp. 387–396.

27. Hwang, M.Y., Huang, X.D., and Alleva, F., Predicting unseen triphones with senomes, in *Proc. IEEE Int. Conf. Acoustics, Speech and Signal Processing*, 1993, Anerson, D. (Ed.), 387–396, American Institute of Physics, New York.

28. Jelinek, F., *Statistical Methods for Speech Recognition*, The MIT Press, Cambridge, MA.

29. Juang, B.H. et al., The past, present, and future of speech processing, *IEEE Signal Processing Magazine*, May 1998.

30. Kershaw, D., Hochberg, M., and Robinson, T., Incorporating Context-Dependent Classes in a Hybrid Recurrent Network-HMM Speech Recognition System. F-INFENG TR217, Cambridge University Engineering Department, May 1995.

31. Kershaw, D.J., Phonetic Context-Dependency in a Hybrid ANN/HMM Speech Recognition System, Ph.D. dissertation at St. John's College, University of Cambridge.

32. Kohn, P. and Bilmes, J., Ring Array Processor (RAP): Software Users Manual Version 1.0. Technical Report TR-90-049, Int. Comp. Sci. Inst., Oct. 1990.

33. Krause, A. and Hackbarth, H., Scaly artificial neural networks for speaker-independent recognition of isolated words, in *Proc. IEEE Int. Conf. Acoustics, Speech and Signal Processing*, 1, 21–24, Glasgow, 1989.

34. Lang, K., Waibel, A., and Hinton, G., A time-delay neural network architecture for isolated word recognition, *Neural Networks*, 3(1), 23–43, 1990.

35. Lippmann, R.P., Review of neural networks for speech recognition, *Neural Computation*, 1, 1–38, 1989.

36. Lippmann, R.P., A critical overview of neural network pattern classifiers. Neural networks for signal processing, *Proc. 1991 IEEE Workshop*, 1991, 266–275.

37. Morgan, N. and Bourlard, H., Continuous speech recognition using multilayer perceptrons with Hidden Markov Models, *Proc. IEEE Int. Conf. Acoustics, Speech and Signal Processing*, 1990.

38. Morgan, N. and Bourlard, H., Continuous speech recognition, *IEEE Signal Processing Magazine*, May 1995.

39. Olson, H.F. and Belar, H., Phonetic typewriter, *J. Acoust. Soc. Am.*, 28(6), 1072–1081, 1956. Cited in Reference 29.

40. Rabiner, L.R., *Fundamentals of Speech Recognition*, Prentice-Hall, 1993.

41. Renals, S. and Rohwer, R., Learning phoneme recognition using neural networks, in *Proc. IEEE Int. Conf. Acoustics, Speech and Signal Processing*, Glasgow, 1989, 413–416.

42. Renals, S. McKelvie, D., and McInnes, F., A comparative study of continuous speech recognition using neural networks and Hidden Markov Models, in *Proc. IEEE Int. Conf. Acoustics, Speech and Signal Processing*, 1991.

43. Renals, S., Morgan, N., Cohen, M., and Franco, H., Connectionist probability estimation in the DECIPHER speech recognition system, in *Proc. IEEE Int. Conf. Acoustics, Speech and Signal Processing*, 1992.

44. Ruan, H. and Sankar, R., Applying neural network to robust keyword spotting in speech recognition application, Neural Networks, 1995. *Proc., IEEE Int. Conf.*, 5, 2882–2886, 1995.

45. Ruehl, H.W., Dobler, S., Weith J., and Meyer, P., Speech recognition in the noisy car environment, *Speech Communication*, 10, 1991.

46. Sakoe, H., Isotani, R., Yoshida, K., Iso, K.I., and Watanabe, T., Speaker-independent word recognition using dynamic programming neural networks, *Acoustics, Speech, and Signal Processing, Int. Conf.*, 1, 29–32, 1989.

47. Sankar, R. and Sethi, N.S., Robust speech recognition techniques using a radial basis function neural network for mobile application, Southeastcon '97. Engineering New Century, *Proc. IEEE*, 1997, 87–91.

48. Schalkoff, R.J., *Artificial Neural Networks*, McGraw-Hill, 1997.

49. Schwartz, R., Improved Hidden Markov Modeling of phonemes for continuous speech recognition, in *Int. Conf. ASSP*, II, 35.6.1–4, March 1984.

50. Sirigos, J., Fakotakis, N., and Kokkinakis, G., Improving environmental robustness of speech recognition using neural networks, *Digital Signal Proc. Proceed., 1997*. DSP 97, 1997 13th Int. Conf., 2, 575–578, 1997.

51. Cohen, M., Franco, H., Morgan, N., Rumelhart, D., and Abrash, V., Context-dependent multiple distribution phonetic modeling with MLPs, in *Adv. Neural Information Proc. Syst.*, Vol. 5, The MIT Press, Cambridge, MA, April 1990, 77–80.

52. Franco, H., Cohen, M., Morgan, N., Rumelhart, D., and Abrash, V., Context-dependent connectionist probability estimation in hybrid Hidden Markov Model-neural net speech recognition system, *Computer Speech and Language*, 8, 211–222, 1994.

53. Tamura, S. and Waibel, A., Noise reduction using connectionist models, *Proc. IEEE ICASSP*, 553–556, 1988.

54. Tank, D. and Hopfield, J., Neural computation by concentrating information in time, *Proc. Nat. Acad. Sci. USA*, 84, 1896–1900, April 1987.

55. Tebelskis, J., Speech Recognition using Neural Nets, Ph.D. dissertation, School of Computer Science, Carnegie Mellon University, Pittsburgh, PA, 1995.

56. Unnikrishnan, K., Hopfield, J., and Tank, D., Learning Time-Delayed Connections in a Speech Recognition Circuit, *Neural Networks for Computing Conference*, Snowbird, Utah, 1988.

57. Waibel, A., Hanazawa, T., Hinton, G., Shikano, K., and Lang, K., ATR Interpreting Telephony Res. Labs., Osaka, Japan, *Int. Conf. ASSP*, 1, 107–110, 1988.

58. Waibel, A., Sawai, H., and Shikano, K., Modularity and scaling in large phonemic, *IEEE Trans.*, 37(12), 1888–1898, 1989.

59. Watrous, H.R., Speech Recognition using Connectionist Networks, Ph.D. thesis, University of Pennsylvania.

60. Yang, R. and Haavisto, P., An improved noise compensation algorithm for speech recognition in noise, *Proc. IEEE ICASSP*, 1, 49–52, 1996.

61. Yuk, D., Che, C., and Flanagan, J., Robust speech recognition using maximum likelihood neural networks and continuous density Hidden Markov Models, automatic speech recognition and understanding, *Proc., 1997 IEEE Workshop*, 1997, 474–481.

62. Zahorian, S.A., Qian D., and Jagharghi, A.J., Acoustic-phonetic transformations for improved speaker-independent isolated word recognition, *ICASSP-91., 1991 Int. Conf.*, 1, 561–564, 1991.

63. Zavaliagkos, G., Zhoa, Y., Schwartz, R., and Makhoul J., A hybrid segmental neural net/Hidden Markov Model system for continuous speech recognition, *IEEE Trans. SAP*, 2(1, Part II), 151–9, January 1994.

64. Zhoa, Y., Schwartz, R., Sroka, J., and Makhoul, J., Hierarchical mixtures of experts methodology applied to continuous speech recognition, in *Adv. Neural Information Proc. Syst.*, The MIT Press, 7, 859–65, 1995.

65. Zhoa, Y., Schwartz, R., Sroka, J., and Makhoul, J., Hierarchical mixtures of experts methodology applied to continuous speech recognition, in *Int. Conf. Acoustics, Speech and Signal Processing*, 5, 3443–3446, May 1995.

7

Advanced Neural-Based Systems for 3D Scene Understanding

Gian Luca Foresti
University of Udine

7.1   Introduction

An Intelligent System (IS) can be defined as a system with the following characteristics:[1–17] it uses intelligence to interact with the world; it learns during its existence; it continually acts, and by acting it reaches its objectives more often than pure chance indicates; it consumes energy for its internal processes and in order to act.

Intelligent systems perceive, reason and plan, act, and learn from previous experience. The design and analysis of intelligent systems builds on work in artificial intelligence,[1] cognitive science,[2,4] computational neuroscience,[2,4] machine learning,[6,9,10] simulation and control,[5,8,11] animation, computer vision,[9,10] real-time computing,[3,10] distributed systems theory,[14,15] ecological systems,[16,17] economics,[12,13] mechatronics and bionics,[11] and large-scale software systems.[7]

An IS has to have an objective; it has to be able to check if its last action was favorable, if it resulted in getting nearer to its objective or not. To reach its objective it has to select its response. A simple way to select a response is to select one that was favorable in a similar previous situation. It must be able to *learn*.[19–27] Since the same response sometimes is favorable and sometimes fails, it

0-8493-1121-7/03/$0.00+$1.50

has to be able to recall in which situation the response was favorable, and in which it was not. Therefore, it stores situations, responses, and results.

This chapter focuses on the learning aspects of a specific class of intelligent systems: neural-based systems for 3D scene understanding.

7.1.1 3D Scene Understanding

The desire to endow computers with robust visual capabilities to interpret 3D scenes is a common objective of the research community in Computer Vision.[28–40] However, 3D scene understanding is a very difficult task, and few complete systems have been developed until now.[28,29,32,40] The main difficulties of 3D scene interpretation consist of (1) high computational requirements of current recognition methods, (2) limited domains where the set of objects to be recognized should be predefined, and (3) inadequate testing and performance characterization of research systems. It is heartening to note that all of these issues are beginning to receive attention from vision researchers. Most existing systems have been developed and tested on synthetic images and/or on indoor scenes for experimental use by researchers.[28,31,33–35] However, if real and/or outdoor scenes are considered, the interpretation task becomes more and more complex.[28,29,32,37–40] The variations in scene illumination, shadows, weather, noise, etc. affect the quality of the acquired images, and therefore the features extracted from them.[41] The use of other sensing modalities,[36] e.g., range cameras, tactile sensors, color cameras, etc., either to replace or to augment the information returned from the intensity sensor, are currently being investigated.

Several researchers suggest that the problem of 3D scene interpretation can be partially solved by considering it as composed by several sub-problems, e.g., recovery of the structure of the scene, 3D object recognition, 3D object modeling and representation, etc., which can be solved separately.[28,31] 3D object recognition is the central sub-problem of 3D scene interpretation. The basic idea in 3D object recognition is to derive a representation of the objects present in the scene, and match this representation with those of a set of stored models. An object is generally represented by a set of descriptive primitives (DPs).[42] There are several kinds of DPs used in the literature, but the most common are: points,[43] lines,[44] surfaces,[45] and volume elements.[46] Matching techniques are based on the building of object graphs whose nodes represent DPs and arcs represent relations between primitives.[42]

The interpretation of a 3D scene presented in this chapter is based on the decomposition of the scene into different objects belonging to a database of known object models, e.g., roads, vehicles (car, bus, lorry, motorcycle), pedestrians, buildings, vegetation, etc. The scene is observed by a visual camera (i.e., optical, infrared, range, acoustic, etc.) (Figure III.7.1) and the obtained image is classified by means of an artificial neural network (ANN).[47] Each class corresponds to one of the predefined object models stored into the database. Shape descriptors are extracted from image patterns, and used as descriptive primitives for representing objects to be recognized and for learning ANNs.

7.1.1.1 Neural Networks

In recent years, many researchers have eschewed traditional parametric classification techniques, e.g., minimum distance classification, etc., in favor of newer, nonparametric techniques, particularly decision trees[48–50] and neural networks.[47] These techniques learn to classify instances by mapping points in feature space onto categories without explicitly characterizing the data in terms of parametrized distributions. As a result, they avoid making assumptions about the data that, when violated, can cause parametric techniques to fail. A decision tree (DT) is a recursive structure composed of two kind of nodes: (1) internal nodes having child nodes, each of which is also a DT; and (2) terminal nodes (*leaf nodes*) having no child nodes. The root node divides the feature space into subsets (*splitting*), assigning each subset to each child node. The splitting process is continued until each leaf node corresponds to one class. The main problem of DTs is that they require an exhaustive search through

FIGURE III.7.1 (a) Optical black and white image of a traffic scene, (b) optical color image of a pedestrian in an outdoor scene, (c) infrared image of a car on a road, and (d) acoustic image of a metallic object on the seabottom.

a list of arbitrarily generated splits to find the best one, and they have difficulty when input data are not linearly separable.[47]

Neural networks (NNs) with hidden layers, such as a multi-layer perceptron (MLP), handle this problem, but the exact number of hidden neurons and the connectivity between layers must be specified before the training phase. Moreover, there is no practical guarantee that correct weights will be found for a given number of neurons and a particular training set.[47] In many cases, the number of hidden neurons is chosen by trial. Both NNs and DTs subdivide, during the learning phase, the decision space into regions delimited by hyperplanes. However, DTs produce a set of hyperplanes orthogonal to the space axis,[49] while NNs produce more general hyperplanes.[47] A new neural network model, called a neural tree (NT),[51–58] provides a solution for the main limitations of both DTs and NNs.

7.2 Shape Descriptors

Shape descriptors are feature vectors[59] extracted from image patterns. They can be used by classification and coding systems to synthesize the morphological properties of objects, to identify them, or to transmit their appearances. Shape descriptors either preserve or do not preserve the information related to the pattern examined; e.g., in a coding system, the main task of a shape descriptor is to provide sufficient and compact information to recover the original visual information after transmission. On the other hand, if the goal is to assign an observed pattern to a predefined class, it is more important to obtain the descriptor's information about the shape aspects that allow one to differentiate an object from other objects.

A wide class of shape descriptors have been proposed in the literature. A survey by Pavlidis[60] proposed a distinction between internal and external shape descriptors: the latter considers object boundaries as primary features, whereas the former takes into account the whole area of a shape. Another possible classification consists of the capability of reconstructing an original shape from a vectorial descriptor.

7.2.1 Shape Descriptors Based on Mathematical Morphology

Mathematical morphology was introduced by Matheron[61] and Serra[62] and provides a conceptual basis for facing image processing with nonlinear tools. In the context of binary images, a morphological operation can be described as the interaction of a binary object (represented as a set of points) with another set of simpler shapes, i.e., the structuring elements (SE). The philosophy of shape descriptors based on MM is to transform images progressively by applying morphological operators, and by analyzing and keeping track of the loss of information which occurs by successive steps. Usually, the only degree of freedom that is left to the designer is the possibility of selecting the SE and the sequence of morphological operators to be applied at each step.

The two main shape descriptors based on mathematical morphology are the morphological skeleton transform[64] and the pattern spectrum.[65,66]

Morphological skeleton (MS) is an information-preserving shape descriptor. The principle of operation for obtaining the MS of a binary pattern is to keep track of the information loss through successive transformations (e.g., erosion) until it converges to an empty set. The set of images processed in this way forms a sequence. Each image of the sequence is analyzed by evaluating to what extent it can represent a transformation (e.g., dilation) of the corresponding shrunk image. The points in the image space where, at a given step, a difference is observed between an image and a morphological transformation of it (e.g., erosion followed by dilation, i.e., opening), are stored and constitute the morphological skeleton. Therefore, an MS corresponds to a set of image points associated with a value representing the step at which such points have been detected. Starting from this set, it is possible to perform inverse transformations, which allow one to recover the exact original pattern shape. An MS does not allow one to distinguish the content of an image, e.g., if shape and noise information are mixed. Therefore, filtering steps are required to filter out noise, before extracting the MS.

The algorithm for extracting the skeleton considers an image X as input and obtains as output an image representation $R(X)$ corresponding to the morphological skeleton. The algorithm can be written as follows:

a. Initialization: $n = 0$, $S_{-1}(X) = X$.
b. Iterations of the following steps:

  1. $S_n(X) = (S_{n-1}(X) \theta B)$;
  2. **if** $S_n(X) = \varnothing$ **then** $R_{Nx}(X) = S_{Nx-1}(X)$; Stop;
  3. $\gamma_n(X) = S_n(X) \oplus B$;
  4. $R_n(X) = S_{n-1}(X) - \gamma_n(X)$;
  5. $n = n + 1$; if $n > N_x$ then Stop, else go to 1;

where $\gamma(X)$ is the classical opening and $-$ is the set difference operator.[67] The skeleton is provided as the combination of representations at intermediate steps, i.e., $SK_x(B) = \cup_n R_n(X)$. The algorithm (i.e., step 4) is based on the assumption that the opening of a set is completely contained in X.

The pattern spectrum (called also *pecstrum*)[65,66] is an information-nonpreserving descriptor. It provides information about the way the area of an object is distributed from its details to the coarse object structure. Such information can be obtained by evaluating the results of successive morphological openings of the object by a structuring element whose size increases.

Let us define $\psi^1(X)$ and $\psi^2(X)$ as two morphology operators working on a binary set X. A measure $m(\ )$ can be associated with such operators; it describes the modifications to X performed by the operators. The two operators have the characteristic that $\psi^1(X) \supseteq \psi^2(X)$. It is also required that $X \supseteq \psi^2(X)$, but this does not necessarily hold for $\psi^1(X)$.[65] Shape description can be defined as a mapping of X into a real value obtained as a function characteristic of $\psi^1(X)$ and $\psi^2(X)$, i.e.,

$$X \rightarrow m[f(\psi^1(X), \psi^2(X)].\tag{III.7.1}$$

If $m(X)$ is the measure of the area of X and if f is chosen as the ordered difference between the sets, i.e., $f(x, y) = x - y$, then $m[f(\psi^1(X), \psi^2(X))] > 0$ by definition.

The morphological skeleton and the pecstrum can be strictly related. The definition of a pecstrum leads to a method for computing it that basically coincides with the one previously described for an MS. The main difference is that the information about image points corresponding to the difference obtained at a certain step between an image of a sequence and the dilation of the shrunk image is not kept, but only the total number of changed points is regarded as a shape feature. The loss of spatial information is the reason why a pecstrum is said to be a "nonpreserving" descriptor. The advantage of pecstrum over the morphological skeleton is the use of less memory. A pecstrum shares with a morphological skeleton the drawback of hypothesizing perfect (i.e., noiseless) shapes as input data.

7.2.2 Shape Descriptors Based on Invariant Moments

Geometric moments are the most common used object descriptors.[67–75] Let $m_{p,q}$ be the geometric moment of order $(p+q)$:

$$m_{p,q} = \sum_{i=1}^{M} \sum_{j=1}^{N} x_i^p y_j^q I(x_i, y_j), \ (p, q = 0, 1, 2, \ldots)\tag{III.7.2a}$$

where $I(x, y)$ represents the input image. However, this definition of moment of order $(p+q)$ is not invariant to changes in scale, rotation, and translation. To this end, to generate translation invariants, the $(p+q)$-th central moment should be considered:

$$v_{p,q} = \sum_{i=1}^{M} \sum_{j=1}^{N} (x_i - x_0)^p (y_j - y_0)^q I(x_i, y_j), \ (p, q = 0, 1, 2, \ldots)\tag{III.7.2b}$$

where (x_0, y_0) represents the center of mass of the input image whose coordinates are defined as:

$$x_0 = \frac{m_{1,0}}{m_{0,0}} \quad \text{and} \quad y_0 = \frac{m_{0,1}}{m_{0,0}}.$$

To introduce scale invariance, the central moment should be normalized with the use of the term introduced by Hu:[68]

$$\mu_{p,q} = \frac{v_{p,q}}{(m_{0,0})^\beta}\tag{III.7.2c}$$

where $\beta = 1 + \frac{p+q}{2}$. By applying the theory of algebraic invariants to the scale normalized invariants (Equation III.7.2c), it is possible to obtain the second- and third-order moment invariants which are invariant under rotation as well as translation and scale change. Let $\phi_1, \ldots, \phi_7$ be these moment invariants, i.e.;

$$\Phi_1 = \mu_{2,0} + \mu_{0,2} \tag{III.7.3a}$$

$$\Phi_2 = (\mu_{2,0} + \mu_{0,2})^2 + 4\mu_{1,1}^2 \tag{III.7.3b}$$

$$\Phi_3 = (\mu_{3,0} + 3\mu_{1,2})^2 + (3\mu_{2,1} - \mu_{0,3})^2 \tag{III.7.3c}$$

$$\Phi_4 = (\mu_{3,0} + \mu_{1,2})^2 + (\mu_{2,1} + \mu_{0,3})^2 \tag{III.7.3d}$$

$$\Phi_5 = (\mu_{3,0} - 3\mu_{1,2})(\mu_{3,0} + \mu_{1,2})[(\mu_{3,0} + \mu_{1,2})^2 - 3(\mu_{2,1} + \mu_{0,3})^2]$$
$$+(3\mu_{2,1} - \mu_{0,3})(\mu_{2,1} + \mu_{0,3})[3(\mu_{3,0} + \mu_{1,2})^2 - (\mu_{2,1} + \mu_{0,3})^2] \tag{III.7.3e}$$

$$\Phi_6 = (\mu_{2,0} - \mu_{0,2})(\mu_{3,0} + \mu_{1,2})^2 - (\mu_{2,1} + \mu_{0,3})^2$$
$$+4\mu_{1,1}[(\mu_{3,0} + \mu_{1,2})(\mu_{2,1} + \mu_{0,3})] \tag{III.7.3f}$$

$$\Phi_7 = 3(\mu_{2,1} + \mu_{0,3})(\mu_{3,0} - \mu_{1,2})[(\mu_{3,0} + \mu_{1,2})^2] - [3(\mu_{2,1} + \mu_{0,3})^2]$$
$$-(\mu_{3,0} - 3\mu_{1,2})(\mu_{2,1} + \mu_{0,3})[3(\mu_{3,0} + \mu_{1,2})^2 - (\mu_{2,1} + \mu_{0,3})^2] \tag{III.7.3g}$$

In order to introduce contrast invariance and to reduce the large range in these moments, the Maitra moment invariants have been considered:[69]

$$\beta_1 = \frac{\sqrt{\phi_2}}{\phi_1} \tag{III.7.4a}$$

$$\beta_2 = \frac{\phi_3 \cdot m_{0,0}}{\phi_1 \cdot \phi_2} \tag{III.7.4b}$$

$$\beta_3 = \frac{\phi_4}{\phi_3} \tag{III.7.4c}$$

$$\beta_4 = \frac{\sqrt{\phi_5}}{\phi_4} \tag{III.7.4d}$$

$$\beta_5 = \frac{\phi_6}{\phi_1 \cdot \phi_4} \tag{III.7.4e}$$

$$\beta_6 = \frac{\phi_7}{\phi_5} \tag{III.7.4f}$$

Other sets of moments have been introduced in the literature. These include Legendre,[72] rotational and complex moments,[71] and Zernike moments.[70,72–75] In particular, the Zernike moments have been shown to be superior to others.

7.3 Advanced Neural Networks

In the past few years, a variety of new models have been proposed to attempt to improve the performance of neural networks in solving a given problem. The aspects particularly requiring improvements are learning time and generalization ability.

It has been shown both theoretically and experimentally[76–79] that small networks exhibit better generalization performance than large networks. On the other hand, large networks require less learning time and should exhibit a certain degree of fault tolerance under damage conditions.[79]

An interesting hybrid neural model, called neural tree (NT), has been recently introduced by Utgoff.[80] Its creation was motivated by the combination of advantages of classical neural networks (NNs) and decision trees (DTs).

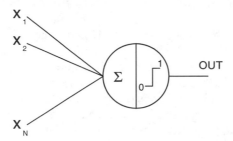

FIGURE III.7.2   General model of the perceptron with N input and one output.

DTs are based on information theory, while NNs rely on specific training algorithms. Both NNs and DTs are devices that learn a concept representation from instances of the concept itself by induction. In particular, they subdivide, during the learning phase, the decision space into regions delimited by hyperplanes: DTs produce hyperplanes orthogonal to the space axis, while NNs produce more general hyperplanes.

NNs have solved a wide class of classification problems and are able to work in a parallel way. NNs with hidden layers, such as a multi-layer perceptron (MLP),[77] are able to work when input data are not linearly separable, but the exact number of hidden neurons and the connectivity between layers must be specified before the training phase. Moreover, it is not guaranteed that correct weights will be found for a given number of neurons and a particular training set. In many cases, the number of hidden neurons is chosen by trial-and-error minimization.

DTs represent another popular approach to pattern recognition.[48,80] A DT is a recursive structure composed of two kind of nodes: (1) internal nodes having child nodes, each of which is also a DT, and (2) terminal nodes (*leaf nodes*) having no child nodes. The root of the tree divides the feature space into subsets (*splitting*), assigning each subset to each child node. The splitting process is continued until each leaf node corresponds to one class. The main problem of DTs is that they require an exhaustive search through a list of arbitrarily generated splits to find the best split.[80] However, this process is computationally inefficient and the obtained solution could be not close to the optimal solution.

NTs provides a solution for the main limitations of both DTs and NNs.

The first NT, called perceptron tree, was presented in 1988 by Utgoff.[51] The architecture of this NT is composed of a DT whose internal nodes are represented by attribute tests and leaf nodes by perceptrons. An attribute test is a single node of a DT which tests a given attribute against a threshold value. The learning algorithm starts by training a perceptron at the root node. Figure III.7.2 shows the classical perceptron model, where the output can assume two possible values:

$$\text{Out} = \begin{cases} 1 & \text{if } \sum_{i=1}^{N} \mathbf{x}_i \mathbf{w}_i > 0 \\ 0 & \text{otherwise} \end{cases} \tag{III.7.5}$$

where $\mathbf{x} = [x_1, \ldots, x_N]$ is the input vector (i.e., feature vector) and $\mathbf{w} = [w_1, \ldots, w_N]$ is the weight vector.

If the input patterns are not linearly separable, then they are split into two groups. The perceptron node is replaced by a decision node with an attribute test and two children nodes, one to group the input patterns above the threshold and the other to group the remaining patterns. This process is iterated until all training patterns are classified. The main disadvantage is that when a perceptron node is replaced by a decision node, all the training work done by the perceptron is lost. Then, this approach is limited to binary class problems. The main advantage is that the structure of the tree is determined dynamically and that such a tree is guaranteed to be smaller than a DT.

Sirat and Nadal present a new binary structure of NT for two-class problems, and then extend this approach to multi-class problems.[52]

The training process is similar to that presented by Utgoff:[51] the root node of the tree, a perceptron node without hidden layers, is trained with all the training patterns. Then it is used to divide the training patterns into K groups, where K is the number of output classes. An exhaustive search is done to find the best division. Each group is used as input for a new perceptron (*child node*) at the next level of the tree. This process is iterated recursively until all the subtrees have been terminated by leaf nodes (e.g., the patterns of a subgroup have the same classification). The classification process requires presenting unseen patterns to the root node of the NT. The root node determines the activation values for each pattern and the pattern is passed to the child node with higher activation value. The process stops when a leaf node is reached: this node reports the final classification of the pattern. An important drawback of this approach consists in the *a priori* determination of the maximum size of the tree: if the tree exceeded that size, it cannot be generated. A large tree could be polarized by the training data and the result could be not very general.[81]

To this end, Sakar and Mammone[53,82] introduced a new approach. The tree is allowed to grow without size limits, and a pruning phase is introduced at the end of the training phase to create a more generalized tree. Li et al.[54] developed an adaptive neural network, which allows the tree to grow and shrink during use. New nodes are created when needed and existing nodes are deleted when rarely used. A new general strategy for improving the learning capability of an NT and reducing its complexity is presented in Reference 83. Each node of the tree is learned by taking into account that it is followed by two son nodes. In this way, a tree better optimized in terms of average depth and generalization performances is obtained. Despite these promising early beginnings, neural tree research has focused on simple models. Few works have considered complex pattern recognition problems.[55–58,84–87]

In this chapter, we introduce a new NT model, which improves upon the standard neural tree algorithms[51–54] in terms of classification performances. A *multiple-hypothesis* procedure is applied at each node of the tree in order to find the best path to be followed during the classification phase, i.e., the path which maximizes the probability of correct decision.

7.3.1 Neural Tree Model

The proposed new NT model is a decision tree whose nodes are generalized perceptrons without hidden layers and with activation function characterized by a sigmoidal behavior.[57,58] Each perceptron takes $(N + 1)$ input (the first N input values, $p_1, \ldots, p_N$, represent selected samples of the pattern, and the last input p_{N+1} represents the bias value with weight θ_i) and generates K outputs, called activation values (Figure III.7.3). The weight matrix W will be as follows:

$$W = \begin{bmatrix} w_{11} & w_{12} & \ldots & w_{1,N} & \theta_1 \\ w_{21} & \ldots & \ldots & \ldots & \ldots \\ \ldots & \ldots & \ldots & \ldots & \ldots \\ w_{K1} & \ldots & \ldots & w_{K,N} & \theta_N \end{bmatrix} \tag{III.7.6}$$

The activation function of the ith neuron generated by the qth input pattern will be:

$$o_i^q = \frac{1}{e^{-\sum_{j=1}^{N} w_{ij} P_j^q - \theta_i}} \tag{III.7.7}$$

Let us observe that the weights θ_i of each neuron are elements of the weight matrix W, and they will be modified during the training phase, as each other element of the same matrix.[77]

The NT structure is characterized by several levels, $l \in [1, L]$ (L is the depth of the tree) (Figure III.7.4). The root node is on level $l = 1$. A node n_h on level l may have links at maximum

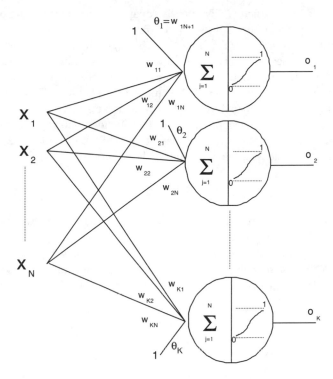

FIGURE III.7.3 Generalized perceptron model with $(N + 1)$ input and K output.

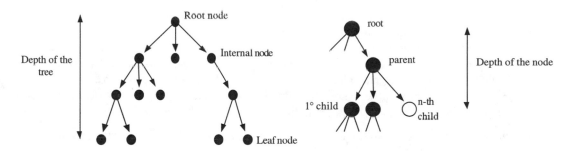

FIGURE III.7.4 General scheme of the NT architecture.

to $k \leq K$ nodes on level $(l + 1)$ which are the children of n_h. No structural constraints are specified for the nodes within the same level: they are not ordered as a one-dimensional array. The goal of the proposed NT architecture will be reached by performing a convenient training process (*learning phase*) using well-defined patterns, then performing the recognition of each pattern of a testing element (*classification phase*).

7.3.1.1 Learning Algorithm

To train the NT we have developed the following algorithm. Let $\Omega = \{\omega_1, \ldots, \omega_M\}$ be the set of the problem classes and let $TS_{root} = \{(p^q, \omega_{iq}) \mid p_q \in \Re^{N+1}, \omega_{i_q} \in \Omega, q = 1, \ldots, Q\}$ be the training set, where Q is the number of considered patterns.

1. Create a single perceptron (*root node creation rule*). Set $TS = TS_{root}$.
2. Start training the current perceptron by applying the *perceptron training rule*.[52,53]

3. If the *TS* is correctly classified (*TS* is linearly separable), the current perceptron ends with M children leaf nodes, one for each class. Go to (6).
4. If the *TS* is split into M groups ($M > 1$), $TS_1, \ldots, TS_M$, a new level of M children perceptron nodes is created, and each TS_i is assigned to the corresponding child node for the further training process. Note that a pattern is assigned to the group depending on the highest output value and that one or more groups could be empty. Go to (2) with $TS = TS_i$.
5. If the *TS* cannot be divided in any group (the perceptron repeats the same classification of the parent node), a *splitting rule* is applied.[53] The *TS* is divided into at least two groups, TS_r and TS_l, which are assigned to corresponding perceptron child nodes. Go to (2) with $TS = TS_r$.
6. If all current nodes are leaves, the algorithm ends. Otherwise go to (2) to train the remaining perceptrons.

Let us explain in more detail the procedures applied in the proposed learning algorithm.

Root node creation rule. It is composed by a perceptron without hidden layers and with matrix of weights W initialized in a random way.

Perceptron training rule. The perceptron is trained with the patterns of the training set until the average error $\bar{\delta}$ computed on the all output neurons and on the all patterns keeps in the range $[\hat{\delta} - toler, \hat{\delta} + toler]$ for more than a given number *wait* of epochs. The average error is defined as:

$$\bar{\delta} = \frac{1}{QM} \sum_{q=1}^{Q} \sum_{i=1}^{M} \delta_i^q \qquad \text{(III.7.8)}$$

where $\delta_i^q = t_i^q - o_i^q, \forall i = 1, \ldots, M$ and $t_i^q = \begin{cases} 1 & \text{if } i = i_q \\ 0 & \text{otherwise} \end{cases}$ is the target vector.

The threshold *toler* should be selected close to 0 (e.g., 10^{-6}), while the threshold *wait* should be chosen according to the complexity of the problem (e.g., $wait \in 10,50$).[57]

Splitting rule. The splitting rule consists of adding a new node, called *splitting node*, to the NT when the perceptron, after *wait* epochs, repeats the same classification of the parent node. Such a node should satisfy some constraints: (1) the TS should be divided into at least two subsets, (2) the cardinality of these subsets should be similar, (3) the splitting should be performed on the features with maximum variance (i.e., the more discriminant features).

Once the NT has been successfully trained, it can be used to classify unseen patterns (*test set*).

7.3.1.2 Classification Algorithm

A multiple-hypothesis approach is used to improve the classification performances of the standard NT:[3,8,9]

1. Present a pattern to the root of the tree.
2. Move toward the tree in a top-down way following the path suggested by the classification (maximum activation value) given by each considered node.
3. When a leaf node is reached, label the pattern with the classification of the current node.

The standard classification method sometimes ends up with wrong conclusions, e.g., the pattern is close to one of the decision hyperplanes. In fact, if this situation occurs, the activation vector $A_n = (a_{n1}, \ldots, a_{nK})$ in one or more internal nodes of the NT reached by the pattern during the classification is characterized by one or more values closed to the highest value. Let us define this node as a *doubt* node d_n. To this end, multiple paths should be taken into account to find the best classification (Figure III.7.5).

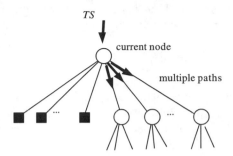

FIGURE III.7.5   Multiple paths followed when a doubt node is reached.

A node is considered doubt if the absolute difference between the maximum activation value and at least one of the other activation values is lower than a fixed threshold *th*. This is equivalent to select multiple paths to be followed by appropriately backtracking into the tree. A gain function G is assigned to each node and a technique to search for the path which maximizes this gain is proposed.

In the proposed model, we make the assumption that at a given internal node n_h of the NT are assigned K probabilities, $P_{h1}, \ldots, P_{hK}$ (one for each child node or problem class), which represent the probabilities that the associated path reaches a correct classification. These probabilities are conditional probabilities. P_{hk} is the probability that the path passing for the k children node of the n_h node reaches the correct classification, given that we have followed the path that leads to the n_h node. This probability is computed proportionally to the values of the activation vector A_n obtained at the n_h node:

$$P_{hk} = \frac{a_k}{\sum\limits_{i=1}^{K} a_{ki}}.$$

The gain function is computed as follows. Let us consider that we have followed the jth path and we have reached a final node of the NT. Let us also assume that m_j is the number of nodes of the path (we do not consider the last node which is a leaf node and not a decision node). We have two exclusive events: (1) path j that we have followed is correct, and (2) path j is wrong. If the probabilities of these two events are P_{Cj} and P_{Wj}, respectively, then $(P_{Cj} + P_{Wj}) = 1$. The cost in the case of a correct path is $L_{Cj} = m_j$ and the cost in the case of a wrong path is $L_{Wj} = (m_j + B_j)$, where B_j is the cost associated with a backtracking into the NT in order to find the correct path. In particular, B_j is the number of NT levels that must be considered to back up to the first doubt node of the considered path j. The correct conclusion is reached because the path from the root node to the first doubt node is certainly correct. It is possible to express the gain for the path j of the NT as follows:

$$G_j = L_{Cj} \cdot P_{Cj} + L_{Wj} \cdot P_{Wj} = P_{Cj}m_j + (1 - P_{Cj}) \cdot (m_j + B_j)$$
$$= m_j + B_j \cdot (1 - P_{Cj}) \tag{III.7.9}$$

The gain of the NT can be computed by weighting the gain given in Equation III.7.9 with the posterior probability P_{Pj} that the path j is followed, and then building up the following sum:

$$G = \sum_{j=1}^{J} P_{Pj} \cdot G_j = \sum_{j=1}^{J} P_{Pj} \cdot [m_j - B_j \cdot (1 - P_{Cj})] \tag{III.7.10}$$

where J is the number of possible paths. For a given NT, all components of Equation III.7.9 except for B_j are known. In order to maximize the gain G, the backtracking cost for each reachable leaf node must be minimized. According to Equation 25 in Reference 88, to achieve the minimum value of the backtracking cost B_j, we must solve the following optimization problem:

$$\min \left[\sum_{i=1}^{q} h_i \cdot (1 - Pb_{i+1}) \right]$$ (III.7.11)

where $Pb_{q+1} = 0$, $h(n_i)$ is the height of the node n_i, i.e., the number of doubt nodes between that node and a leaf node, Pb_i is the probability that the first wrong decision took place either at node n_i or at a doubt node which follows n_i in the path j, and q is the number of doubt nodes at which we back up. It is demonstrated in[88] that there is one (or possibly more) backtracking procedures with the minimum cost. It is demonstrated also that it is not necessary to calculate the costs of all possible backtracking procedures: the backtracking process which results in the minimum cost has some properties the application of which reduces the number of cost evaluations significantly.

Let $LN = (ln_1, \ldots, ln_N)$ be the list of leaf nodes (each one associated with a given path) that minimize for a given input pattern the gain function G. Since each leaf node l_i is associated with a given class c_i, e.g., $LN = (ln_1, \ldots, ln_N) \Rightarrow C = (c_1, \ldots, c_N)$, the final pattern classification is assigned to the class with the highest number of occurrences in the list C.

7.4   Range Image Understanding

7.4.1   Range Image Segmentation — State of the Art

A range image is a function $f(x, y)$ where x is the row in the image, y is the column, and $f(x, y)$ is some value which corresponds to the depth z. The range images we work with are 128×128 pixels in size, and each $f(x, y)$ is represented by an 8-bit range value: small range values represent more distant depth, and large range values correspond to pixels closed to the sensor (Figure. III.7.6). A variety of sensors are available today for extracting range data; they include radar, laser scanners, and numerical simulators. Range images, referred to as depth maps in the sequel, are particularly interesting when a 3D shape must be recognized. This is useful in robot vision as well as terrain recognition and 3D geological layer reconstruction and modeling.

Depth maps can also be calculated using several numerical methods which combine one or more plane projections in order to reconstruct the third dimension of a scene. These methods include shape

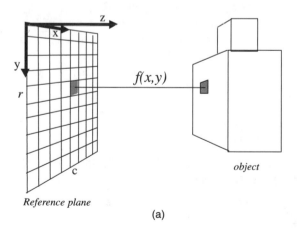
(a)

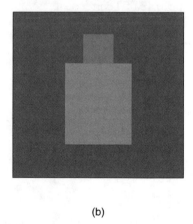
(b)

FIGURE III.7.6   (a) Geometric representation of a range image of a regular object as a matrix where the elements represent the distance between points of the surfaces defining the shape of the object, and (b) image representation of the corresponding range image.

from shading, shape from texture, and stereographic projections. All of them receive as input one or more 2D projections and produce as output a depth map of a scene. Even if cases exist in which a good solution is provided by these methods, generally the computing procedure is long and the result is uncertain.

Segmentation of depth maps has been attempted by several authors in recent years. Significant work in the field has been done by Besl and Jain.[89] They use the mean curvature H and the Gaussian curvature K to obtain an initial segmentation of small primitive regions (*seed regions*). These regions are then grown and merged using a variable-order surface-fitting (allowed by a smoothness or approximation error constraint). Pieroni and Tripathy[8] attempted to apply the differential geometry to surfaces which have been triangularized. Triangles are chosen as basic elements because their vertices enable easy calculation of derivatives and curvatures.

The segmentation of an image into regions of similar morphological properties is essential for higher-level computer vision processing. These regions can be used to generate a representation of images by defining attributes and relations between regions. Such a representation can in turn be used in recognition, planning, and manipulation tasks. Object recognition methods involve comparison with a stored model.[91,92] Detailed representation is needed to allow for the discrimination among similar objects. Coarse representation is usually sufficient for other tasks, i.e., obstacle avoidance or object manipulation.[93]

A number of researchers have based their work on simple surface primitives (e.g., planar, spherical, etc.). This is due to the difficulties associated with manipulation and parameter stability of general quadratic functions in the presence of noise. Surface representation is the most widely used representation for range data and can be readily obtained by additional processing of the output of our segmentation algorithm. For example, surface representation is the most suitable representation for hold-site determination, since it provides us with the accessible surfaces in the object and their description.

Several solutions for range image segmentation have been proposed in the literature. These techniques fall into three classes: (1) region growing or clustering, (2) edge detection, and (3) hybrid segmentation techniques. Region growing techniques attempt to group pixels into connected regions based on similarity of the surface properties. Clustering methods partition the image into several clusters of connected pixels based on the similarity of surface properties. Both approaches require *a priori* knowledge about the number of surfaces present in the image. Edge detection techniques locate pixels that lie on the boundary between two regions. There are three basic primitive types of edges in range images: step, roof, and smoothed edges. Most of the research work has been concentrated on the first two types of edges. Step edges correspond to discontinuities in depth values, while roof edges correspond to points where surface normals are discontinuous. Detection of roof and smooth edges is very difficult because they do not correspond to large depth variation and tend to hide in noise.

Most region-based segmentation methods require that region pixels be well approximated by planar or quadratic surfaces.[89–97] Some authors fitted planar surfaces extracted by region growing to segment range image data.[91,93] Other approaches are geared toward isolated cylinders in range data as a step to segmentation.[94] Herbert and Ponce segmented planes, cylinders, and cones from range data.[95] Oshima and Shirai[96] used planes, spheres, cylinders, and cones. Sethi et al. used ellipsoids to segment range images into meaningful regions.[97] Recently, Trucco and Fisher presented a range image segmentation method which partitions the range data into homogeneous surface patches using estimates of the sign of the mean and Gaussian curvatures.[98] Generally, the methods based on the computation of the H and K parameters are computationally expensive and very sensitive to noise.

A variety of methods are available for detecting edges in range images. Step edges are usually detected using techniques similar to those of intensity images.[99] Most of these approaches are based on first and second gradient computation which is sensitive to noise. Pre-filtering the data may help in some cases, but this introduces uncertainty in the location of the edge.

Detection of roof and smooth edges is a more difficult problem. Inokuchi et al.[100] present an edge detection algorithm for range data using a ring operator. The ring operator extracts a 1D periodic function of depth values that surround a given pixel. The function is transformed to the frequency domain using an FTT algorithm. Jump and roof edges are then detected by examining the frequency components of the ring surrounding the pixel. A multi-scale approach to extract information for planar contours was proposed in Reference 101. The curve is repeatedly smoothed with Gaussian masks, and features are tracked in different scales. Brady et al.[102] extract and analyze a dense set of 3D lines of curvatures to detect and describe surface intersections. This method is computationally complex, and takes a long time to even describe simple objects. Recently, Lee and Pavlidis[103] applied residual analysis to detect step edges in range data using a regularized smoothing filter. They use the difference between the input image and the smoothed image to detect edge points.

The hybrid methods refer to the combination of region-based and edge-based methods in order to overcome the problems of over-segmentation and under-segmentation. Yokoya and Levine[104] use the combination of edge detection and curvature clustering to segment a range image. The range image of an object is divided into surface primitives which are homogeneous in their intrinsic differential geometric properties and do not contain discontinuity in their depth or surface orientation. By combining the H and K parameters and examining their signs, an initial region-based segmentation (e.g., a curvature sign map) is obtained. Two initial edge-based segmentations are also computed from the partial derivatives and depth values. The three initial image maps are then combined to produce the final range image segmentation.

Jain and Nadabar[105] proposed a hybrid segmentation method which combines the initial region boundary detection method of global clustering and a Markov Random Field model-based boundary detection. Recently, Ghosal and Mehrotra presented an integrated neural-net-based approach to the segmentation of range images into distinct surfaces.[106] A two-stage model is proposed which extracts local surface features at each image point and groups pixels via local interactions among different features. Daugman's projection neural nets (DPNNs) and Kohonen's self-organizing neural nets (KSNNs) are used for the feature extraction and region growing, respectively.

7.4.2　Surface Feature Extraction

The main objective of this section is to explain how surface features can be extracted and transformed into a pattern useful for training the NT at the second stage. This is a rather important step, as an inconsistent presentation of patterns to the NT will inhibit the convergence of the algorithm toward a useful solution.[58,77] Given a range image ($f(x, y)$), we define a numerical sensor which can be applied to extract the surface features and build up the pattern $p(x, y)$ associated with the pixel with coordinates (x, y).

The following constraints are considered:

1. The pattern $p(x, y)$ is built up according to the z value of the (x, y) pixel and those of its neighborhood points.
2. The pattern must contain information to distinguish between the six classes of surfaces.
3. The surfaces to be classified can have arbitrary orientations, curvatures, and scales.
4. The range sensor introduces noise in the input images. To this end, the feature extraction criterion should be robust to noise, and as much as possible invariant to rotations (with respect to axes parallel to the z axis), translations, and to variations in curvature and direction of the normal and scale.

Surfaces are sampled by means of a numerical sensor which organizes the extracted features into patterns[58] useful for training and testing neural networks. The sensor would select surface features from a $l \times l$ image window l (l can assume only odd values) centered in the pixel (x, y) (Figure III.7.7a). Let M be an $l \times l$ matrix containing the extracted elements. First, the coefficients

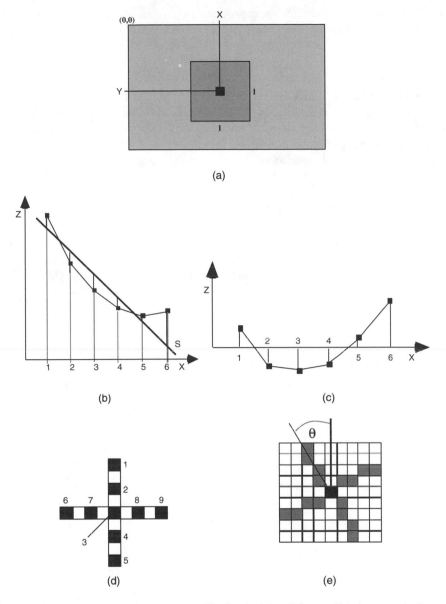

FIGURE III.7.7   (a) $l \times l$ image window centered in the pixel (x, y) from which the numerical sensor selects the elements of the feature pattern $p(x, y)$; (b) regression plane with equation $z = ax + by + c$; (c) computation of the distance between the original surface point (x, y) and its projection on the regression plane; (d) numerical sensor with 8 arms; and (e) rotation of the sensor's arms in order to obtain invariance to surface rotation with respect to the z axis.

a, b, and c of the regression plane (Figure III.7.7b) with equation $z = ax + by + c$ are computed by minimizing the least square error (LSE), i.e.:

$$\frac{1}{l^2} \sum_{x=1}^{l} \sum_{y=1}^{l} [ax + by + c - M(x, y)]^2 \qquad \text{(III.7.12)}$$

Then a new matrix $\hat{M}$, where each element represents the distance between the original point and its projection on the regression plane (Figure III.7.7c), is obtained as follows:

$$\hat{M}(x, y) = M(x, y) - ax - by - c \quad \forall x = 1, \ldots, l \ \ y = 1, \ldots, l \tag{III.7.13}$$

Finally, the values of the $\hat{M}$ matrix are normalized in the range [0,1], and the pattern $p(x, y)$ is built up by aligning the rows of the $\hat{M}$ matrix. The obtained patterns result invariant to translations, curvature variations, and small variations of the normal direction in the (x, y) point.[58] In order to obtain a pattern invariant to rotations around axes parallel to the z axis, the following procedure has been developed. The following elements of the $\hat{M}$ matrix are used in order to build up a training or test pattern, $p(x, y)$:

$$p[i, j] = \hat{M}[i + (v \cdot step), j + (q \cdot step)] \tag{III.7.14}$$

where $v = -2, 1, 0, 1, 2$, $q = -2, 1, 0, 1, 2$, $step = 1, 2, 3, \ldots$, and i, j are the indices of the central point in case of training specimens, and any pixel in case of a test range data map. Then, the eight vectors below will be calculated according with the scheme in Figure III.7.7d, which has been defined as "numerical sensor":

$$v_1 = \{Z_{i,j-2step}, Z_{i,j-step}, Z_{i,j}, Z_{i,j+step}, Z_{i,j+2step}\} \tag{III.7.15a}$$

$$v_2 = \left\{ Z_{i-step,j-2step}, \frac{Z_{i,j-step} + Z_{i-step,j-step}}{2}, Z_{i,j}, \right.$$
$$\left. \frac{Z_{i,j+step} + Z_{i+step,j+step}}{2}, Z_{i+step,j+2step} \right\} \tag{III.7.15b}$$

$$v_3 = \{Z_{i-2step,j-2step}, Z_{i-step,j-step}, Z_{i,j}, Z_{i+step,j+step}, Z_{i+2step,j+2step}\} \tag{III.7.15c}$$

$$v_4 = \left\{ Z_{i-2step,j-step}, \frac{Z_{i-step,j-step} + Z_{i-step,j}}{2}, Z_{i,j}, \right.$$
$$\left. \frac{Z_{i+step,j} + Z_{i+step,j+step}}{2}, Z_{i+2step,j+step} \right\} \tag{III.7.15d}$$

$$v_5 = \{Z_{i-2step,j}, Z_{i-step,j}, Z_{i,j}, Z_{i+step,j}, Z_{i+2step,j}\} \tag{III.7.15e}$$

$$v_6 = \left\{ Z_{i-2step,j+step}, \frac{Z_{i-step,j} + Z_{i-step,j+step}}{2}, Z_{i,j}, \right.$$
$$\left. \frac{Z_{i+step,j} + Z_{i+step,j-step}}{2}, Z_{i+2step,j-step} \right\} \tag{III.7.15f}$$

$$v_7 = \{Z_{i-2step,j+2step}, Z_{i-step,j+step}, Z_{i,j}, Z_{i+step,j-step}, Z_{i+2step,j-2step}\} \tag{III.7.15g}$$

$$v_8 = \left\{ Z_{i-step,j+2step}, \frac{Z_{i-step,j-step} + Z_{i,j+step}}{2}, Z_{i,j}, \right.$$
$$\left. \frac{Z_{i,j-step} + Z_{i+step,j-step}}{2}, Z_{i+step,j-2step} \right\} \tag{III.7.15h}$$

The variable *step* provides the distance between two horizontally or vertically successive nodes of the numerical sensor grid. By increasing the value of *step*, a larger grid is used in order to extract the values forming a neural network test or training pattern.

The digital curvature DC_i of the ith arm of the numerical sensor is computed according to the following formula:

$$DC_i = [v_i[1] + v_i[5] - 2v_i[3]] \cdot \frac{1}{f} \tag{III.7.16}$$

where f is a function assuming the values:

$i = 1$

$\qquad f = (2step)^2$

$i = 2$

$\qquad f = (2step)^2 + step^2$

$i = 3$

$\qquad f = (2step)^2 + (2step)^2$

$i = 4$

$\qquad f = (2step)^2 + step^2$

$i = 5$

$\qquad f = (2step)^2$

$i = 6$

$\qquad f = (2step)^2 + step^2$

$i = 7$

$\qquad f = (2step)^2 + (2step)^2$

$i = 8$

$\qquad f = (2step)^2 + step^2$

In order to extract similar patterns when a surface region has been rotated about z axis, the numerical sensor is allowed to perform rotations of a set of angles corresponding to each sensor arm (vector) (Figure III.7.7e). The adjustment of the sensor with respect to surface rotation is obtained by simply calculating the minimum curvature among the sensor arms, then rotating the sensor in order to bring the minimum curvature arm to the first (horizontal) position, and the other arms accordingly. This is particularly useful when surfaces are characterized by axes of symmetry (e.g., ridges and valleys). A certain amount of invariance under changes of normal direction and plane tilting is provided by simply performing the average of each sensor arm with itself after inverting the order of the elements.

For example, let us consider the 64×64 ridge surface in Figure III.7.8a. By extracting two patterns with a 9×9 mask, the first from the point $(5,5)$ and the second from the point $(32,32)$, we obtain two very similar patterns (Figure III.7.8b).

7.4.3 Surface Models for Training the Neural Tree

Most of the surfaces found in range images can be classified according to the six classes of differential geometry (peak, pit, ridge, valley, saddle, and flat) (Figure III.7.9), represented by the following production rules:

Surface	Rule	
Peak	$\sqrt{R^2 - \left(\left(j - \left(\frac{n}{2}\right)\right)^2 + \left(i - \left(\frac{n}{2}\right)\right)^2\right)}$	$\{i = 1, \ldots, n-1\}, \quad \{j = 1, \ldots, n-1\}$
Pit	$R - \sqrt{R^2 - \left(\left(j - \left(\frac{n}{2}\right)\right)^2 + \left(i - \left(\frac{n}{2}\right)\right)^2\right)}$	$\{i = 1, \ldots, n-1\}, \quad \{j = 1, \ldots, n-1\}$
Ridge	$\sqrt{R^2 - \left(i - \left(\frac{n}{2}\right)\right)^2}$	$\{i = 1, \ldots, n-1\}, \quad \{j = 1, \ldots, n-1\}$
Valley	$R - \sqrt{R^2 - \left(i - \left(\frac{n}{2}\right)\right)^2}$	$\{i = 1, \ldots, n-1\}, \quad \{j = 1, \ldots, n-1\}$
Saddle	$\dfrac{\left(j - \left(\frac{n}{2}\right)\right)^2 R}{\left(\frac{n}{2}\right)^2 R} - \dfrac{\left(i - \left(\frac{n}{2}\right)\right)^2}{\left(\frac{n}{2}\right)^2 R}$	$\{i = 1, \ldots, n-1\}, \quad \{j = 1, \ldots, n-1\}$

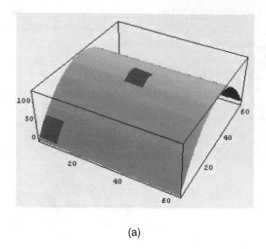

(a)

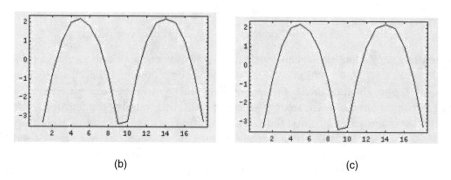

(b)   (c)

FIGURE III.7.8   (a) A 64 × 64 ridge surface. (*b*, *c*) Two patterns extracted with a numerical sensor with $l = 9$, the first from the point (5, 5) and the second one from the point (32, 32).

where $(n - 1) \times (n - 1)$ is the dimension of the matrix representing each considered surface model in a range image and R is the radius of curvature.

To simplify describing these three-dimensional surfaces, we adopt a right-hand coordinate system with the x and y axes making up the horizontal plane and the z axis being the height, with the upward direction being positive. These surfaces were constructed in a 32 × 32 grid of discrete data points, with each point having a height (z value) between 0.0 and 1.0 inclusively.

Sixty-six samples of the synthetic depth maps with 10 different noise levels (from 0–10%) were provided to the NT as a training set. A learning rate $\eta = 0.5$ has been used. Moreover, 30 epochs (i.e., *wait* = 30) and a threshold *toler* = 10^{-6} for the average error $\bar{\delta}$ have been selected. An NT composed of 78 nodes (32 perceptron nodes, 5 splitting nodes, and 41 leaf nodes) with depth $L = 5$ has been obtained.

7.4.4   Experimental Results

7.4.4.1   Synthetic Images

The first experiment considers a 128 × 128 synthetic range image representing five different types of surfaces (i.e., peak, pit, ridge, valley, saddle) posed on a flat region (Figure III.7.10a). This image is characterized by discontinuities between close surfaces. Figure III.7.10b shows the 3D representation

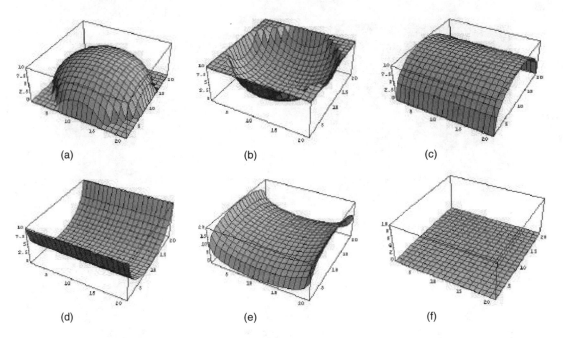

FIGURE III.7.9 Surface models taken from the six classes of the differential geometry (peak, pit, ridge, valley, saddle, and flat).

of the considered image, and Figure III.7.10c shows the classification results. Note that all surfaces are correctly classified, and, in particular, the flat region is detected without misclassification errors. Only a few pixels are wrongly classified in the boundaries regions of close surfaces; this fact can be exploited to identify the edges of close surfaces.

A 128 × 128 synthetic range image representing a continuous surface composed by peaks, pits, and saddles is chosen as a test image (Figure III.7.11a). Figure III.7.11b shows the 3D representation of the considered image, while Figure III.7.11c shows the classification results obtained by a standard NT.[51] Different shades are used to represent different surface patches. Patterns belonging to a small boundary area between peak and saddle surfaces are erroneously assigned to the class ridge, while patterns belonging to a small boundary area between pit and saddle surfaces are erroneously assigned to the valley class. The percentage of misclassification is about 12% of the image points. Figure III.7.11d shows the patterns whose classification has been obtained with a path composed by one or more doubt nodes. Finally, Figure III.7.11e shows the results obtained with the proposed NT. The percentage of misclassification is about 7%. The learning time for the standard NT is about 8 min, and for the proposed NT is about 12 min on a Pentium III at 800 MHz. Classification times are about 5 sec and 9 sec for the standard NT and proposed NT, respectively.

7.4.4.2 Real Images

A real 3D range image representing a complex object (i.e., a bottle) partially submersed on the bottom of the sea is considered (Figure III.7.12a). Figure III.7.12b shows the 3D reconstruction of the image. The scene is composed by different surfaces, i.e., two planar surfaces at approximately the same depth, and two cylindrical surfaces of different dimensions which represent the bottle, divided by significant jump edges. Figure III.7.12c shows the obtained surface classification. The two flat regions are well identified even if some isolated pixels are misclassified. The bottle object is correctly classified by assigning its pixels to the ridge class. However, a small region in its central part is erroneously assigned to the peak class. This is due principally to the presence of noise, which

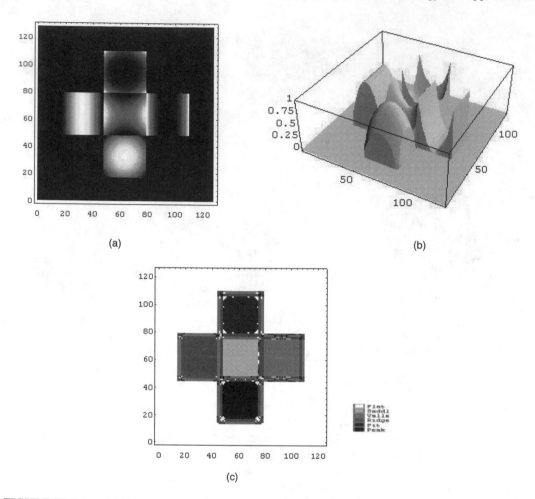

(a)

(b)

(c)

FIGURE III.7.10 (a) 128 × 128 synthetic range image representing five different types of surfaces (i.e., peak, pit, ridge, valley, saddle) posed on a flat region, (b) its 3D representation, and (c) the obtained classification in the absence of noise.

modifies the surface characteristics. Then, the regions between the bottle and the planar surface are correctly classified by the NT: the border is assigned to the ridge class, the slope is assigned to the plane class, and the region nearest to the bottle is assigned to the saddle class. Some pixels are misclassified due to the presence of noise.

7.5 Outdoor Scene Understanding

Figures III.7.13a and b show the general architecture of the advanced neural-based system for outdoor scene understanding. The system is composed of two parts: (1) an off-line module, which is in charge of building up a supervised classifier based on the proposed neural-tree model, and (2) an on-line module, which is dedicated to scene classification. The off-line module (Figure III.7.13a) extracts a large set of patterns from several image prototypes, containing different object models. Let $p(x, y)$ be a pattern, i.e., a feature vector associated with the pixel (x, y) of the considered image. Each pattern should belong to a given object class, and the whole set of pairs, {*pattern, object class*}, makes up the training set (TS). Object models should be taken in different poses to obtain more

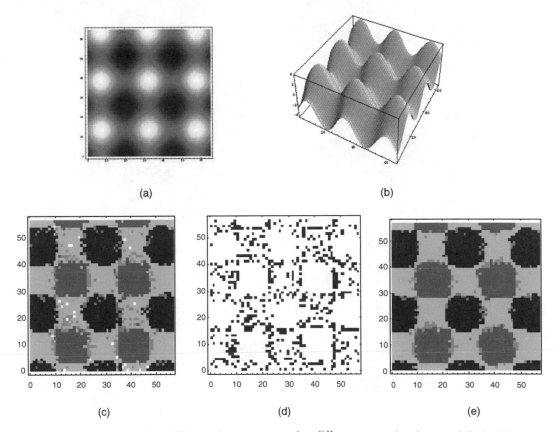

(a)

(b)

(c)   (d)   (e)

FIGURE III.7.11   (a) 128 × 128 range image representing different types of surfaces and (b) its 3D representation. (c) Classification results obtained by a standard NT on a noisy image (with 4% noise level), (d) the patterns whose classification has been obtained with a path composed by one or more doubt nodes, and (e) final results obtained by the proposed NT.

generalized TS.[107,108] A different TS should be created for each different scene. Then, the obtained TS is used to build up the NT (*learning phase*). The on-line module receives in input a visual or an infrared image, $I(x, y)$, whose pixels represent a gray level or a temperature value, respectively. The input image is processed by the feature extraction (FE) module: a pattern $p(x, y)$ is extracted from each pixel (x, y) and then is classified by the NT. The output of the system is represented by an image $C(x, y)$ where each pixel is assigned to one of the considered object classes.

7.5.1   Feature Extraction

Let $p(x, y)$ be a pattern associated with the (x, y) pixel of the input image, $I(x, y)$. A $n \times n$ window $NS(x, y)$,[58] centered in the (x, y) pixel, is used to extract the pattern $p(x, y)$ from the considered image for both learning and classification purposes.

The pattern $p(x, y)$ is composed of two parts: (1) the gray-level values of the pixels belonging to the grid of the window $NS(x, y)$, i.e., $p_1(x, y)$; and (2) some statistical parameters, i.e., $p_2(x, y)$.

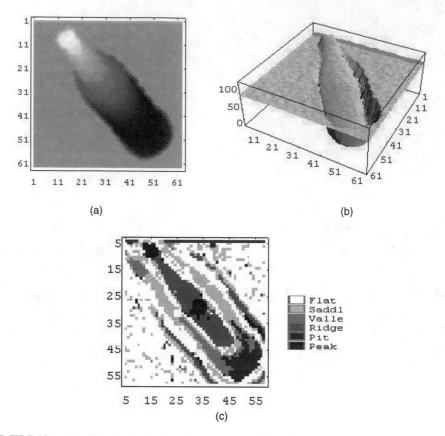

(a)   (b)

(c)

FIGURE III.7.12 (a) 128 × 128 real range image representing a bottle partially submersed on the bottom of the sea, (b) its 3D representation, and (c) the obtained classification by the proposed NT.

In particular, $p_2(x, y)$ is built up by computing the mean value $m(x, y)$, the variance $\sigma(x, y)$, the skewness $S(x, y)$, and the kurtosis $K(x, y)$ parameters on the $n \times n$ window:

$$m(x, y) = \frac{1}{n^2} \sum_{\xi=-\lfloor n/2 \rfloor}^{\lfloor n/2 \rfloor} \sum_{\eta=-\lfloor n/2 \rfloor}^{\lfloor n/2 \rfloor} I(x + \xi \cdot step, y + \eta \cdot step) \qquad \text{(III.7.17a)}$$

$$\sigma(x, y) = \frac{1}{n^2} \sum_{\xi=-\lfloor n/2 \rfloor}^{\lfloor n/2 \rfloor} \sum_{\eta=-\lfloor n/2 \rfloor}^{\lfloor n/2 \rfloor} [I(x + \xi \cdot step, y + \eta \cdot step) - m(x, y)]^2 \qquad \text{(III.7.17b)}$$

$$S(x, y) = \frac{1}{n^2 \sigma^3(x, y)} \sum_{\xi=-\lfloor n/2 \rfloor}^{\lfloor n/2 \rfloor} \sum_{\eta=-\lfloor n/2 \rfloor}^{\lfloor n/2 \rfloor} [I(x + \xi \cdot step, y + \eta \cdot step) - m(x, y)]^3 \qquad \text{(III.7.17c)}$$

$$K(x, y) = \frac{1}{n^2 \sigma^4(x, y)} \sum_{\xi=-\lfloor n/2 \rfloor}^{\lfloor n/2 \rfloor} \sum_{\eta=-\lfloor n/2 \rfloor}^{\lfloor n/2 \rfloor} [I(x + \xi \cdot step, y + \eta \cdot step) - m(x, y)]^4 \qquad \text{(III.7.17d)}$$

where the parameter *step* provides the distance between two horizontal or vertical successive samples. By increasing the value of *step*, a larger grid is used to extract the values forming a training pattern. The value of the variable *step* can be selected according to the noise level affecting the input image, e.g., large values should be used if high noise affects the input image.[57]

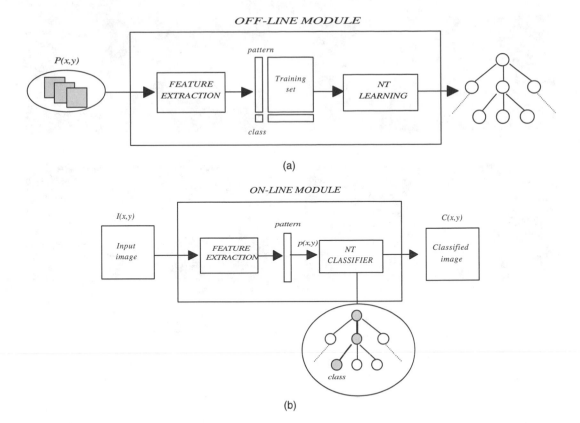

FIGURE III.7.13   General architecture of the classification system for outdoor scene understanding: (a) the off-line module, which extracts from several image prototypes, containing different object models, a large set of patterns; and (b) the on-line module, which classifies unknown images.

7.5.1.1 NT Learning

The obtained TS is used to build up the NT (*learning phase*). Different TS should be created for each different type of scene. Each node of the considered NT takes N input elements (the gray-level image pixels and the four statistical measures) and generates K outputs, one for each class of the problem, e.g., objects composing the considered scene. Generally, the pre-classified image is characterized by some wrong interpretations. Regions belonging to a given object often contain isolated pixels or small areas belonging to other classes. This limitation is due to several factors: (1) the noise affecting the input image, (2) changes in illumination or temperature, (3) weather, (4) shadows, (5) swaying trees, (6) changes in surface reflectivity, and (7) the limited generalization capability of the TS. Moreover, it is worth noting that the main classification errors are localized in the image regions associated with the class of possible mobile objects. In fact, the shape of mobile objects, e.g., vehicles, is generally regular and dependent on their pose in the scene. On the contrary, background objects, e.g., road, vegetation, sky, represent large image regions with irregular shapes independent on the scene viewpoint. Consequently, the classification of background objects can be performed by means of limited TS, while the classification of mobile objects requires large TS containing several samples of each object acquired from different viewpoints.

To this end, a new NT trained with several 2D binary models of the same object seen from different aspect (viewing) angles ϕ, i.e., different viewpoints of the camera with respect to the surface on which the object is placed, has been considered[109] (Figure III.7.14). Each reference model is characterized by a structure and dimensions that are specific for the class to which the object belongs. In general,

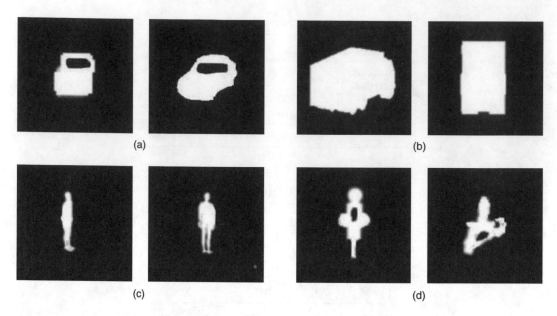

FIGURE III.7.14   Some object models used to build the pattern for learning the NT. The considered object classes are: (a) car, (b) truck, (c) pedestrian, and (d) motorcycle.

the number of 2D models per object depends on the accuracy requested in the aspect angle estimation, i.e., 360 for accuracy of $1°$. Generally, man-made objects have 3D symmetries giving rise to similar shapes when viewed from some aspect angle (ϕ) and its counterpart ($\phi + 180°$). Take the example of some cars seen from the front ($\phi = 0°$) or the rear part ($\phi = 180°$). To this end, the NT is trained with conflicting data, i.e., similar features for two images with aspect angles that are $180°$ apart, causing it to not learn properly. However, the proposed classification strategy with backtracking into the tree allows it to find both aspect angles, ϕ and ($\phi + 180°$), and to select the correct one by the winner-takes-all voting scheme.

7.5.2   Experimental Results

Experiments have been conducted on both visual and infrared images representing complex outdoor scenes, i.e., underwater and road scenes.

7.5.2.1   Underwater Scenes

The first experiment is focused on the classification of underwater images acquired by a visual CCD camera. Classified images should be used to help an autonomous underwater vehicle (AUV) operate in an underwater environment for inspecting a pipeline without human control. The AUV should identify the pipeline structure in each image for two different purposes: (1) visual inspection of the pipeline to identify possible cracks or holes,[110] and (2) pipeline tracking for navigation.[111] The main problem in pipeline object identification is to evaluate typical pipeline characteristics, which may also be very different from one image to another, because of the different material and diameter of the pipeline itself and of the presence of sand and seaweed on it.[110,111] Figure III.7.15a shows an example of a pipeline.

The whole image classification proceeds in two steps:

1. The algorithm automatically scans the image and extracts the patterns by means of a 7×7 mask centered in each pixel.
2. The NT classifies each pattern into two classes: pipeline and seabottom.

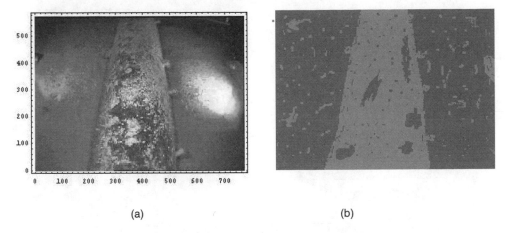

(a)   (b)

FIGURE III.7.15   (a) An underwater image representing a pipeline; (b) final result of the classification process.

A training set composed by 500 underwater images representing a pipeline in different positions and with different seabottom characteristics, e.g., sand, seaweed, etc., has been considered. An NT composed of 75 internal nodes, 4 split nodes, and 168 leaf nodes disposed on $L = 11$ levels has been obtained.

In Figure III.7.15b, the classification result obtained by applying the method on the image in Figure III.7.15a is presented. It is heartening to note that the proposed method is also robust to the presence in the scene of artificial lighting spots due to the illumination produced by the AUV. The image areas characterized by this kind of noise are not recognized as pipeline because of the different texture and shape.

7.5.2.2 Road Scenes

The second experiment refers to the classification of a night road scene from infrared images. These images have been acquired by a THERMOVISION 550 camera on an Italian motorway during the months of October 1997 (at an environment temperature of about 14°). Figures III.7.16a and b show a pair of these infrared images containing multiple vehicles, i.e., cars and small trucks.

Four classes have been selected for the considered scene: (1) vehicles (cars and small trucks), (2) road, (3) vegetation, (4) boundary wall. A TS composed of 2097 patterns (519 for the class vehicles, 528 for the class road, 88 for the class boundary wall, and 962 for the class vegetation) has been used to train the NT. These patterns have been extracted from 50 infrared images of the same scene acquired at different hours during a week. An NT composed of 675 internal nodes, 64 split nodes, and 734 leaf nodes (distributed on $L = 15$ levels) has been obtained. Figures III.7.16c and d show the final classification obtained on the images in Figures III.7.16a and b, respectively.

Table III.7.1 show the percentage of correct pixel classification vs. each object class considered in the experiment.[112,113] It is worth noting that the system is able to reach a percentage of correct classification on the whole scene of about 79% for the image in Figure III.7.16a and of about 75% for the image in Figure III.7.16b.

7.6   Conclusions

In this chapter, an advanced neural model, i.e., a neural tree with learning based on multiple hypotheses, has been applied to the problem of 3D scene understanding.

FIGURE III.7.16   Infrared images representing a motorway scene containing (a) a car and (b) three cars and a small truck, respectively. Classification results obtained on the images in Figures III.7.16a and b, respectively.

TABLE III.7.1   Percentage of Correct Classification vs. Each Object Class for the Images in Figures III.7.16a and b

	Vehicles (%)	Road (%)	Wall (%)	Vegetation (%)	Total (%)
Figure III.7.16a	89.74	77.49	72.56	73.44	78.75
Figure III.7.16b	83.53	84.80	76.44	69.59	74.85

The analysis and interpretation of the content of a 3D scene is a very difficult task. Existing approaches suffer from: (1) high computational requirements; (2) limited domains where the set of objects to be recognized should be predefined; and (3) inadequate testing and performance characterization. Most existing systems have been developed and tested only on synthetic images and/or on indoor scenes.[28,31,33–35] Few works have been considered real scenes where the interpretation task becomes more and more complex.[28,29,32,37–40] The variations in scene illumination, shadows, weather, noise, etc. affect the quality of the acquired images, and therefore the features extracted from them.[41]

In this chapter, images acquired by two different kind of sensors, i.e., range and visual, have been used to interpret real complex scenes, i.e., underwater scenes, road scenes, etc.

Range images have been analyzed by means of a numerical sensor able to extract local surface features at each point of the input image. It has been demonstrated that this procedure is able to extract patterns invariant to rotations around axes parallel to the z axis, translations, variations in curvature, direction of the normal, and scale. Then the proposed NT is applied to classify surfaces according to the six models of the differential geometry. Experiments on synthetic and real 3D range images have demonstrated the efficacy and the performance of the proposed approach.

Optical and infrared images have been considered for outdoor scene understanding. The system is able to recognize both background and mobile objects by overcoming problems of object orientation (rotation), location (translation), and scale (size variation due to viewing distances). Classification of mobile objects into five subsets, i.e., car, van, truck, motorcycle, and pedestrian, yields a success rate greater than 70% on both optical and infrared images.

Acknowledgments

The author would like to thank TECNOMARE, Venezia (IT) and CEOM, Palermo (IT), for providing underwater images, and AUTOVIE VENETE for providing infrared images.

References

1. Hayes-Roth, B., An architecture for adaptive intelligent systems, *Artificial Intelligence*, 72(1–2), 329–365, 1995.
2. Dubois, D., Prade, H., and Yager, R.R., Eds., *Fuzzy Sets for Intelligent Systems*, Morgan Kaufmann, San Mateo, CA, 1993.
3. Musliner, D.J., Durfee, E.H., and Shin, K.G., Circa – A cooperative intelligent real-time control architecture, *IEEE Trans. Systems, Man, Cybern.*, 23(6), 1561 1574, 1993.
4. Groen, F.C.A., Hirose, S., and Thorpe, C.E., Eds., *Intelligent Autonomous Systems*, IOS Press, Amsterdam, 1993.
5. Rodriguez, G., Ed., Intelligent control and adaptive systems, in *Proc. SPIE 1196*, Philadelphia, PA, November 7–8, 1989.
6. Tsoukalas, L.H., Neurofuzzy approaches to anticipation — a new paradigm for intelligent systems, *IEEE Trans. Systems, Man, Cybern., Part B: Cybernetics*, 28(4), 573–582, 1998.
7. Chauvet, P., Lopez-Krahe, J., Taflin, E., and Maltre, H., System for an intelligent office document analysis, recognition and description, *Signal Processing*, 32(1), 161–190, 1993.
8. Lima, P.U. and Saridis, G.N., Intelligent Controllers as Hierarchical Stochastic Automata, *IEEE Trans. Systems, Man, Cybern., Part B: Cybernetics*, 29(2), 151–163, 1999.
9. Casey, R., Ferguson, D., Mohiuddin, K., and Walach, E., Intelligent forms processing system, *Machine Vision and Applications*, 5(3), 143–155, 1992.
10. Roberto, V., Intelligent systems for signal and image understanding, *Signal Processing*, 32(1–2), 1–4, 1993.
11. Zhang, W.R., Nesting, safety, layering, and autonomy — a reorganizable multiagent cerebellar architecture for intelligent control with application in legged robot locomotion and gymnastics, *IEEE Trans. Systems, Man, Cybern., Part B: Cybernetics*, 28(3), 357–375, 1998.
12. Boutilier, C., Shoham, Y., and Wellman, M.P., Economic principles of multi-agent systems, *Artificial Intelligence*, 94(1–2), 1–6, 1997.

13. Rao, H.R., A choice set-sensitive analysis of preference information acquisition about discrete resources, *IEEE Trans. Systems, Man, Cybern.*, 23(4), 1062–1071, 1993.

14. Tsukada, T.K. and Shin, K.G., Distributed tool sharing in flexible manufacturing systems, *IEEE Trans. Robotics and Automation*, 14(3), 379–389, 1998.

15. Kshemkalyani, A.D., Reasoning about causality between distributed nonatomic events, *Artificial Intelligence*, 92(1–2), 301–315, 1997.

16. Vicente, K.J., Christoffersen, K., and Pereklita, A., Supporting operator problem solving through ecological interface design, *IEEE Trans. Systems, Man, Cybern.*, 25(4), 529–545, 1995.

17. Wiebe K.J. and Basu, A., Modelling ecologically specialized biological visual systems, *Pattern Recognition*, 30(10), 1687–1703, 1997.

18. Mohan, A., Papageorgiou, C., and Poggio, T., Example-based object detection in images by components, *IEEE Trans. Pattern Analysis and Machine Intelligence*, 23(4), 349–361, 2001.

19. Segre A. and Elkan, C., A high-performance explanation-based learning algorithm, *Artificial Intelligence*, 69(1–2), 1–50, 1994.

20. Barto, A.G., Bradtke, S.J., and Singh, S.P., Learning to act using real-time dynamic programming, *Artificial Intelligence*, 72(1–2), 81–138, 1995.

21. Bhanu, B. and Poggio, T., Introduction to the special section on learning in computer vision, *IEEE Trans. Pattern Analysis and Machine Intelligence*, 16(9), 865–867, 1994.

22. North, B., Blake, A., Isard, M., and Rittscher, J., Learning and classification of complex dynamics, *IEEE Trans. Pattern Analysis and Machine Intelligence*, 22(9), 1016–1034, 2000.

23. Leckie, C. and Zukerman, I., Inductive learning of search control rules for planning, *Artificial Intelligence*, 101(1–2), 63–98, 1998.

24. Stauffer, C. and Grimson, W.E.L., Learning patterns of activity using real-time tracking, *IEEE Trans. Pattern Analysis and Machine Intelligence*, 22(8), 747–757, 2000.

25. Maio, D. and Rizzi, S., Map learning and clustering in autonomous systems, *IEEE Trans. Pattern Analysis and Machine Intelligence*, 15(12), 1286–1297, 1993.

26. De Jong, G.F., Learning to plan in continuous domains, *Artificial Intelligence*, 65(1), 71–141, 1994.

27. Ishikawa, M., Learning of modular structured networks, *Artificial Intelligence*, 75(1), 51–62, 1994.

28. Grimson, W.E.L. and Huttenlocher, D.P., Special Issue on interpretation of 3D scenes. II. *IEEE Trans. Pattern Analysis and Machine Intelligence*, 14(2), 1992.

29. Azam, M., Potlapalli, H., Janet, J., and Luo, R.C., Outdoor landmark recognition using segmentation, fractal model, and neural networks, in *Proc. DARPA Image Understanding Workshop*, 1996, 189–203.

30. Ayache, N., *Artificial Vision for Mobile Robots: Stereo Vision and Multisensory Perception*, MIT Press, Cambridge, MA, 1991.

31. Grimson, W.E.L., *Object Recognition by Computer*, MIT Press, Cambridge, MA, 1990.

32. Nandhakumar, N. and Aggarwal, J.K., Integrated analysis of thermal and visual images for scene interpretation, *IEEE Trans. Pattern Analysis and Machine Intelligence*, 10(4), 469–481, 1988.

33. Hoover, A., Goldgof, D., and Bowyer, K.W., The space envelope — a representation for 3D scenes, *Computer Vision and Image Understanding*, 69(3), 310–329, 1998.

34. Kim, I.Y. and Yang, H.S., An integrated approach for scene understanding based on Markov Random Field model, *Pattern Recognition*, 28(12), 1887–1897, 1995.

35. Zheng, J.Y. and Tsuji, S., Generating dynamic projection images for scene representation and understanding, *Computer Vision and Image Understanding*, 72(3), 237–256, 1998.

36. Stevens, M.R. and Ross J., Beveridge, Localized scene interpretation from 3D models, range, and optical data, *Computer Vision and Image Understanding*, 80(2), 111–129, 2000.

37. Malik, R. and Whangbo, T., Angle densities and recognition of 3D objects, *IEEE Trans. Pattern Analysis and Machine Intelligence*, 19(1), 52–57, 1997.

38. Mann, R., Jepson, A., and Siskind, J.M., The computational perception of scene dynamics, *Computer Vision and Image Understanding*, 65(2), 113–128, 1997.

39. Drummond, T. and Caelli, T., Learning task-specific object recognition and scene understanding, *Computer Vision and Image sUnderstanding*, 80(3), 315–348, 2000.

40. Bischoff, W.F. and Caelli, T., Scene understanding by rule evaluation, *IEEE Trans. Pattern Analysis and Machine Intelligence*, 19(11), 1284–1288, 1997.

41. Foresti, G.L. Object detection and tracking in time-varying and badly illuminated outdoor environments, *Optical Engineering*, 37(9), 2550–2564, 1998.

42. Foresti, G.L. and Regazzoni, C.S., A hierarchical approach to feature extraction and grouping, *IEEE Trans. Image Processing*, 9(6), 1056–1074, 2000.

43. Dori, D., Liang, Y., Dowell, J., and Chai, I., Sparse-pixel recognition of primitives in engineering drawings, *Machine Vision and Applications*, 6(2–3), 69–82, 1993.

44. Lagunovsky, D. and Ablameyko, S., Straight-line-based primitive extraction in grey-scale object recognition, *Pattern Recognition Letters*, 20(10), 1005–1014, 1999.

45. Yun, H. and Park, K.H., Surface modeling method by polygonal primitives for visualizing three-dimensional volume data, *The Visual Computer*, 8, 246–259, 1992.

46. Borges, D.L. and Fisher, R.B., Class-based recognition of 3D objects represented by volumetric primitives, *Image and Vision Computing*, 15(8), 655–664, 1997.

47. Ripley, B.D., *Pattern Recognition and Neural Networks*, Cambridge University Press, U.K., 1996.

48. Quinlan, J.R., Induction of decision trees, *Machine Learning*, 1, 81–106, 1986.

49. Draper, B.A., Brodley, C.E., and Utgoff, P.E., Goal-directed classification using linear machine decision trees, *IEEE Trans. Pattern Analysis and Machine Intelligence*, 16(9), 888–893, 1994.

50. Baum, E.B., What size net gives valid generalization? *Neural Computation*, 1(1), 151–160, 1989.

51. Utgoff, P.E., Perceptron tree: a case study in hybrid concept representation, in *Proc. 7th Conf. Artificial Intelligence, Saint Paul*, MN, 1988, 601–606.

52. Sirat, J.A. and Nadal, J.P., Neural tree: a new tool for classification, *Network*, 1, 423–438, 1990.

53. Sankar, A. and Mammone, R.J., Neural tree networks, in *Neural Networks: Theory and Application*, Mammone and Zeevi, Eds. Academic Press, 1991, 281–302.

54. Li, T., Tang, Y.Y., and Fang, L.Y., A structure-parameter-adaptive (SPA) neural tree for the recognition of large character set, *Pattern Recognition*, 28(3), 315–329, 1995.

55. Song, H.H. and Lee, S.W., A self-organizing neural tree for large set pattern classification, *IEEE Trans. Neural Networks*, 9(3), 369–380, 1998.

56. Zhang, M. and Fulcher, J., Face recognition using artificial neural networks group-based adaptive tolerance (GAT) trees, *IEEE Trans. Neural Networks*, 7(3), 555–567, 1996.

57. Foresti, G.L. and Pieroni, G.G., 3D object recognition by neural trees, in *Proc. 4th IEEE Int. Conf. Image Processing*, Santa Barbara, CA, October 1997, 408–411.

58. Foresti, G.L., Outdoor scene classification by a neural tree based approach, *Pattern Anal. Appl.*, 2, 129–142, 1999.

59. Hollis, J.E., Brown, D.J., Luckraft, I.C., and Gent, C.R., Feature vectors for road vehicle scene classification, *Neural Networks*, 9(2), 337–344, 1996.

60. Pavlidis, T., Algorithms for shape analysis of contours and waveforms, *IEEE Trans. Pattern Analysis and Machine Intelligence*, 2(4), 301–312, 1980.

61. Matheron, G., *Random Sets and Integral Geometry*, Wiley, New York, 1975.

62. Serra, J., *Image Analysis and Mathematical Morphology*, Academic Press, 1982.

63. Yuille, A., Vincent, L., and Geiger, D., Statistical morphology and Bayesian reconstruction, *J. Mathematical Imaging and Vision*, 1(3), 223–238, 1992.

64. Maragos, P. and Schafer, R.W., Morphological skeleton representation and coding of binary images, *IEEE Trans. Acoustic, Speech and Signal Processing*, 34, 1228–1244, 1986.

65. Anastassopoulos, V. and Venetsanopoulos, A.N., The classification properties of the pecstrum and its use for pattern identification, *Circuits Systems and Signal Process*, 10(3), 293–325, 1991.

66. Maragos, P., Pattern spectrum and multi-shape representation, *IEEE Trans. Pattern Analysis and Machine Intelligence*, 11, 701–716, 1989.

67. Prokop, R.J. and Reeves, A.P., A survey of moment-based techniques for unoccluded object representation and recognition, *Graphical Models and Image Processing*, 54(5), 438–460, 1992.

68. Hu, M.K., Visual pattern recognition by moment invariant, *IEEE Trans. Information Theory*, 8, 179–187, 1962.

69. Maitra, S., Moment invariants, *Proc IEEE*, 67(4), 697–699, 1979.

70. Khotanzad, A. and Hong, Y.H., Invariant image recognition by Zernike moments, *IEEE Trans. Pattern Analysis and Machine Intelligence*, 12(5), 489–497, 1990.

71. Wood, J. Invariant pattern recognition: a review, *Pattern Recognition*, 29(10), 1–17, 1996.

72. Mukundan, R. and Ramakrishnan, K.R., Computation of Legendre and Zernike moments, *Pattern Recognition*, 28(9), 1433–1442, 1995.

73. Liao, S.X. and Pawlak, M., On the accuracy of Zernike moments for image analysis, *IEEE Trans. Pattern Analysis and Machine Intelligence*, 20(12), 1358–1364, 1998.

74. Zhenjiang, M., Zernike moment-based image shape analysis and its application, *Pattern Recognition Letters*, 21(2), 169–177, 2000.

75. Kim, W.Y. and Kim, Y.S., A region-based shape descriptor using Zernike moments, *Signal Processing: Image Communication*, 16(1–2), 95–102, 2000.

76. Sethi, I.K., Entropy nets: from decision trees to neural networks, *Proc. IEEE*, 78(10), 1605–1613, 1990.

77. Haykin, S., *Neural Networks: A Comprehensive Foundation*, IEEE Press, MA, 1994.

78. Pontil, M. and Verri, A., Support vector machines for 3D object recognition, *IEEE Trans. Pattern Analysis and Machine Intelligence*, 20(6), 637–646, 1998.

79. Lehtokangas, M., Cascade-correlation learning for classification, *IEEE Trans. Neural Networks*, 11(3), 795–798, 2000.

80. Sethi, I.K., Entropy nets: from decision trees to neural networks, *Proc. IEEE*, 78(10), 1605–1613, 1990.

81. Golea, M. and Marchand, M., A growth algorithm for neural network and decision trees, *Europhysics Letters*, 12(3), 205–210, 1990.

82. Sankar, A. and Mammone, R.J., Growing and pruning neural tree networks, *IEEE Trans. Computers*, 42(3), 291–299, 1993.

83. D'Alché-Buc, F., Zwierski, D., and Nadal, J.P., Trio learning: a new strategy for building hybrid neural trees, *Int. J. Neural System*, 5(4), 259–274, 1994.

84. Rahim, M.G., A neural tree network for phoneme classification with experiments on the TIMIT database, in *Proc. IEEE Int. Conf. Acoustic, Speech, Signal Processing*, San Francisco, CA, March 23–26, 1992, 345–348.

85. Farrell, K.R., Mammone, R.J., and Assaleh, K.T., Speaker recognition using neural networks and conventional classifiers, *IEEE Trans. Acoutic, Speech, and Signal Processing*, 2, 194–205, 1994.

86. Behnke, S. and Karayiannis, B., Competitive neural trees for pattern classification, *IEEE Trans. Neural Networks*, 9(6), 1352–1369, 1998.

87. Song, H.H. and Lee, S.W., A self-organizing neural tree for large-set classification, *IEEE Trans. Neural Networks*, 9(3), 369–380, 1998.

88. Rontogiannis, A. and Dimopoulos, N.J., A probabilistic approach for reducing the search cost in binary decision trees, *IEEE Trans. System, Man, Cybern*, 25(2), 362–370, 1995.

89. Besl, P.J. and Jain, R.C., Segmentation through variable order surface fitting, *IEEE Trans. Pattern Analysis and Machine Intelligence*, 10(2), 167–192, 1988.

90. Pieroni, G.G. and Triphaty, S.P., A multi-resolution approach to segmenting surfaces, in issue of *Machine Vision*, Springer Verlag, 1989.

91. Boyter, B.A., Three-dimensional matching using range data, *Proc. 1st Conf. Artificial Intelligence Applications*, 211–216, 1984.

92. Oshima, M. and Shirai, Y., Object recognition using three-dimensional information, *IEEE Trans. Pattern Analysis and Machine Intelligence*, 5(4), 353–361, 1983.

93. Handerson, T.C., Efficient 3D object representation for industrial vision systems, *IEEE Trans. Pattern Analysis and Machine Intelligence*, 5(6), 609–617, 1983.

94. Again, G.J. and Binford, T.O., Computer description of curved objects, *Proc. 3rd Int. Jt. Conf. Artificial Intelligence*, 629–640, 1973.

95. Hebert, M. and Ponce, J. A new method for segmenting 3D scenes into primitives, *Proc. 6th Int. Conf. Pattern Recognition*, 836–838, 1982.

96. Oshima, M. and Shirai, Y., Object recognition using three-dimensional information, *IEEE Trans. Pattern Analysis and Machine Intelligence*, 5(4), 353–361, 1983.

97. Sethi, I.K. and Jayaramamurthy, S.N., Surface classification using characteristics contours, *Proc. 7th Int. Conf. Pattern Recognition*, 438–440, 1984.

98. Trucco, E. and Fisher, R.B., Experiments in curvature-based segmentation of range data, *IEEE Trans. Pattern Analysis and Machine Intelligence*, 17(2), 177–182, 1995.

99. Canny, J., A computational approach to edge detection, *IEEE Trans. Pattern Analysis and Machine Intelligence*, 8(6), 679–698, 1986.

100. Inokuchi, S., Nita, T., and Matsuday, F., A 3D edge region operator for range pictures, *Proc. 6th Int. Conf. Pattern Recognition*, 918–920, 1982.

101. Fan, T.G., Medioni, G., and Nevetia, R., Description of surfaces from range data using curvature properties, *Proc. Int. Conf. Computer Vision and Pattern Recognition*, 86–91, 1986.

102. Brady, M., Ponce, J., Yuille, A., and Asada, H., Describing surfaces, *Computer Vision, Graphics and Image Processing*, 32, 1–28, 1985.

103. Lee, D. and Pavlidis, T., Edge detection through residual analysis, *Proc. Int. Conf. Computer Vision and Pattern Recognition*, 215–222, 1988.

104. Yokoya, N. and Levine, M., Range image segmentation based on differential geometry: a hybrid approach, *IEEE Trans. Pattern Analysis and Machine Intelligence*, 11, 643–649, 1989.

105. Jain, A.K. and Nadabar, S., MRF model-based segmentation of range images, *Proc. 3rd Int. Conf. Computer Vison*, Osaka, Japan, 667–671, 1990.

106. Ghosal, S. and Mehrotra, R., Range surface characterization and segmentation using neural networks, *Pattern Recognition*, 28(5), 711–727, 1995.

107. Koller, D. Danilidis, K., and Nagel, H.H., Model-based object tracking in monocular image sequences of road traffic, *Int. J. Computer Vision*, 10(2), 257–281, 1993.

108. Khotanzad, A. and Liou, J.J., Recognition and pose estimation of unoccluded three-dimensional objects from a two-dimensional perspective view by banks of neural networks, *IEEE Transaction on Neural Networks*, 7(4), 897–905, 1996.

109. Foresti, G.L. and Regazzoni, C.S., A real-time model based method for 3D object orientation estimation in outdoor scenes, *IEEE Signal Processing Letters*, 4(9), 248–251, 1997.

110. Foresti, G.L., Gentili S., and Zampato, M., Autonomous underwater vehicle guidance by integrating neural networks and geometric reasoning, *International Journal of Imaging Systems and Technology*, 10, 385–396, 1999.

111. Foresti, G.L., Visual inspection of sea-bottom structures by an autonomous underwater vehicle, *IEEE Transaction on System, Man and Cybernetics: Part B*, 31(5), October 2001.

112. Rachkovskij, D.A. and Kussul, E.M., Datagen – a generator of datasets for evaluation of classification algorithms, *Pattern Recognition Letters*, 17(9), 537–544, 1998.

113. Zhang, Y.J., Evaluation and comparison of different segmentation algorithms, *Pattern Recognition Letters*, 18(10), 963–974, 1997.

8

Shape Representation and Automatic Model Inference in Character Recognition Systems

Hirobumi Nishida
Ricoh Software Research Center

8.1 Introduction

Handwritten character recognition has found many applications in office automation, and has stimulated both basic research and development of practical systems.[4] Compared with machine-printed character recognition, the prime difficulty in research and development of handwritten character recognition systems lies in the variety of shape deformation. Human beings can recognize deformed characters easily, but the mechanism of character recognition of human beings is unknown. Furthermore, character shapes are defined artificially with little logic.

In the *pattern matching* approach to character recognition, theoretical studies have been conducted based on mathematical statistics and functional analysis.[5] However, pattern matching techniques have some difficulties in handwritten character recognition because those methods are inherently position dependent, and shape deformation of handwritten characters is often too complex to be handled with simple normalization procedures. On the other hand, the *structural* approach[13] is powerful for handwritten character recognition, and therefore many practical systems have been developed based on structural pattern recognition techniques. Much effort has been devoted to developing practical systems, whereas few theoretical studies have been conducted for character recognition based on structural methods. The development of handwritten character recognition systems depends heavily on manually designed and *ad hoc* rules. Furthermore, most of the methods are so specific to problems, i.e., type of the character set such as numerals, that it is costly to extend the system, e.g., from numerals to Roman alphabets, Katakana, etc. The development of handwritten character recognition systems requires much manpower and time for manual adjustment of heuristic rules, and therefore, there has been a criticism that character recognition is an art rather than a sound engineering technology.

In particular, construction of class descriptions (prototypes) is an essential step in any model-based machine vision systems. A substantive criticism against the structural approach is the difficulty of automatic inference of class descriptions. In many recognition systems based on structural methods, much time and manpower are usually spent in manual construction of class descriptions. Another criticism is that a large number of class descriptions are required for taking account of shape variations. During the development process of recognition systems, the number of class descriptions constantly increases for dealing with a variety of possible cases. Maintenance of such recognition systems is difficult because an intractably large number of class descriptions are incorporated into the systems.

We have found that in order to break through the difficulties, it is important to study primary features of handwritten characters, mathematical description of character shapes in terms of the primary features, and mathematical models of shape transformation of handwritten characters. In particular, throughout more than three decades of research, it has been found that shape description with straight lines, arcs, and corners is not appropriate to handwritten characters, but some *quasi-topological* features such as convexity and concavity are more flexible and powerful. *Directional features* such as upward, downward, leftward, and rightward are essential. *Singularities* such as branch points and crossings are also appealing features.

Furthermore, the problem of automatic, inductive construction of class descriptions from the data can be considered as constructing *inductively, from the data set*, some shape descriptions that tolerate certain types of *shape transformations*. We define and analyze explicitly shape transformations to be applied to patterns in terms of some particular scheme of shape analysis and description, and we obtain systematic, complete, *a priori*, high-level knowledge about effects of all possible deformations caused by the transformations. Then the problem of automatic construction of class descriptions can be stated as generalizing shape descriptions that can be transformed to each other by particular types of shape transformations. The generalization process is controlled by high-level models of shape transformations, and is supported and guaranteed by high-level knowledge.

In this approach, the design of algorithms for model construction proceeds by the following steps:

1. *Specification of the scheme for shape analysis and description, and the scheme for class descriptions incorporating the shape transformations*: For dealing with the shape transformations rigorously and systematically, some mathematical language is required for shape analysis and description. Based on the shape description, we can define the scheme for class descriptions.

2. *Specification of the types of shape transformations to deal with*: In general, the more types of shape transformations that are incorporated, the fewer class descriptions are required. As a first step, we need to clearly specify the types of shape transformations for consideration without ambiguity so that they can be treated mathematically and systematically.

3. *Systematic, complete knowledge of the effects of the transformations based on the exhaustive analysis of the a priori effects of the transformations*: To obtain high-level knowledge controlling and supporting the generalization process, we need to analyze the *a priori* effects of transformations completely and systematically based on the specifications of Steps 1 and 2. Furthermore, the analysis must be carried out mathematically without depending on heuristics, and should lead to a small, tractable number of distinct cases. Once we have a complete, *a priori* knowledge of all possible cases, we can analyze all instances satisfying certain realistic conditions in a unified, systematic way without resorting to heuristics.

4. *Design of algorithms for generalizing shape descriptions which can be transformed to each other by the specified types of shape transformations*: Based on the high-level knowledge introduced in Step 3, we can design algorithms for model construction. Generalization is performed by comparing two structural descriptions and finding out relevant component correspondence by consistent-labeling algorithms[2] along with the high-level models of the shape transformations.

In this chapter, based on the approach described above, we describe a mathematical model of shape description[8,11] and an approach to automatic construction of structural models incorporating certain types of transformations commonly occurring in unconstrained handwritten characters.[9,12] Of course, handwritten characters cannot be modeled with mathematical tools alone, and domain-specific knowledge is required, because character shapes have been defined artificially, historically with cultural backgrounds, and a character is not an image of a physical object. However, the approach described in this chapter is more systematic and extendable than the existing methods, and character recognition systems can be developed efficiently.

8.2   Structural Representation of Character Shapes

In this section, we describe the scheme for shape analysis and description, and the scheme for class descriptions.

8.2.1   Structural Analysis of Curves

We describe a method for structural analysis and representation of curves based on quantized directional features and quasi-convexity/concavity. In handwriting, since different strokes are written separately, each of them should have independent information. The first step in analyzing handwritten character shapes is to decompose the line picture into independent components (strokes). Each stroke is analyzed separately by integrating low-level structural features into high-level features in a systematic way. Then, characters are described through a few components with this rich and independent information.

Figure III.8.1 is a flowchart of the structural analysis of curves. In Figure III.8.1, items in rectangular boxes are objects on each representation level, and items in ellipses are operations on the objects. Arrows in Figure III.8.1 mean that the higher-level representation is given by applying some

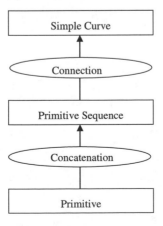

FIGURE III.8.1   Structural analysis of curves. The items in boxes are the objects on each representation level, and the ones in ellipses are the operations on the objects.

operations to objects on the lower level. First, curve primitives are defined. Binary operations on the primitives are defined by classifying the concatenations of primitives according to the direction of convexity. We show that the operators have some explicit properties. Next, the primitive sequences are generated by linking the concatenations of primitives. A label is given to each primitive sequence according to the properties of the primitives and their concatenations forming the sequence. A primitive sequence can be connected with others by sharing its first or last primitive. In other words, two binary operations are introduced to primitive sequences. Furthermore, we show that the operations can be applied only if the labels of the two primitive sequences satisfy the specific relations. Finally, the structure of a curve is described by a string of primitive sequences.

8.2.1.1 Singular Point Decomposition

Curves in the real world are not ideal in the mathematical sense in that they have thickness. The first step of structural analysis of curves is to extract lines in the mathematical sense from line drawings. Many algorithms for pixelwise thinning have been proposed,[3] but they have intrinsic limitation due to local processing. Therefore, their output is hard to use for structural analysis of curves directly, because the figures of branch points and corners are distorted and counterintuitive. On the other hand, some non-pixelwise thinning methods have been developed.[14] They give a rough description and features of lines, but a more adequate and fine representation is necessary for a structural description of curves.

To resolve this problem, new non-pixelwise thinning methods have been developed based on shape decomposition techniques.[18] The principle is the following. First, the image is decomposed into simple/regular parts (lines) and complex/singular parts (branching and crossing).[17] Next, the simple/regular parts are analyzed and described as thin lines. Finally, the residual parts (complex/singular parts) are analyzed in conjunction with neighboring simple/regular parts, and they are described as junctions of lines or endpoints.

The output from the thinning is a piecewise linear curve expressed in terms of the undirected graph $G = (V, E)$, where V is the set of vertices and E is the set of edges. Furthermore, we assume that the graph $G = (V, E)$ is a simple graph (a graph which contains neither a self-loop nor a multiple edge). The coordinate function:

$$coord : V \to R^2$$

is defined for each element v in V, where *coord* (v) corresponds to the coordinates of v.

The vertex in V whose order (the number of edges incident to the vertex) is equal to or more than 3 is called a *singular point* of G. For instance, the vertices p and q in Figure III.8.2 are singular points, whose orders are 4 for p and 3 for q. For each singular point v, we introduce n new vertices v_i such that $coord(v_i) = coord(v)$, $i = 1, 2, \ldots, n$ (n is the order of v).

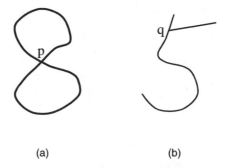

(a)   (b)

FIGURE III.8.2   Thinned line pictures. p and q are singular points of orders 4 and 3, respectively.

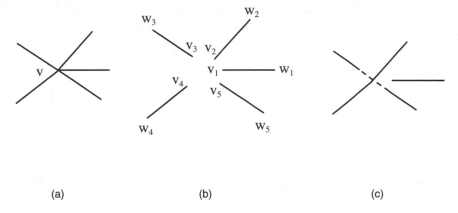

(a)   (b)   (c)

FIGURE III.8.3 Decomposition of a singular point of order 5.

Next, for n vertices $w_i (i = 1, 2, \ldots, n)$ which are adjacent to v, we add virtual edges $(v_i, w_i) (i = 1, 2, \ldots, n)$, and remove the vertex v. Though the vertices $v_i (i = 1, 2, \ldots, n)$ have the same coordinates as v, they are regarded as different from each other. After applying this operation, the graph G is decomposed into connected components, each of which is a simple arc. Figure III.8.3(a) depicts a singular point of the order 5, and Figure III.8.3(b) illustrates the result of the operation mentioned above.

Let $coord(v)$ be (x_0, y_0) and $coord(w_i)$ be (x_i, y_i). Now, we compute $S(i, j) (i < j, i, j = 1, 2, \ldots, n)$:

$$S(i, j) = \frac{\overrightarrow{q_i} \cdot \overrightarrow{q_j}}{|\overrightarrow{q_i}| \cdot |\overrightarrow{q_j}|}, \tag{III.8.1}$$

where

$$\overrightarrow{q_k} = (x_k - x_0, y_k - y_0), \ k = 1, 2, \ldots, n.$$

Next, we find the sequence:

$$S(i_1, j_1) \leq S(i_2, j_2) \leq \cdots \leq S(i_m, j_m) \tag{III.8.2}$$

such that:

$$S(i_1, j_1) = \min\{S(i, j) : i, j \in I, i < j\},$$

$$S(i_k, j_k) = \min \left\{ S(i, j) : i, j \in I - \bigcup_{\lambda=1}^{k-1} \{i_\lambda, j_\lambda\}, i < j \right\}$$

for $k = 2, 3, \ldots, m$, where $I = \{1, 2, \ldots, n\}$ and $m = \lfloor n/2 \rfloor$ ($\lfloor r \rfloor$ is the greatest integer that is equal to or smaller than r). Then, we connect a pair of edges (v_{i_k}, w_{i_k}) and (v_{j_k}, w_{j_k}) by regarding the two vertices v_{i_k} and $v_{j_k} (k = 1, 2, \ldots, m)$ as identical.

A series of the above operations is called the *singular point decomposition*. It decomposes the vertex v into $\lceil n/2 \rceil$ ($\lceil r \rceil$ is the smallest integer that is equal to or greater than r) vertices, and generates $\lceil n/2 \rceil$ pairs of edges incident to v. Furthermore, the graph obtained by applying the singular point decomposition to the graph G is called the *stroke graph* of G. Each connected component of the stroke graph is called a *stroke*, which is topologically equivalent to either a simple arc or a simple closed curve. Therefore, the curve is decomposed into unicursal components by taking account of the smoothness of directional changes at the singular points.

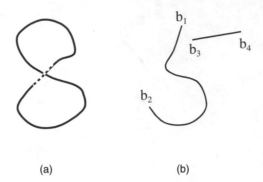

(a)   (b)

FIGURE III.8.4   Singular point decomposition for line pictures shown in Figure III.8.2.

Figure III.8.3(c) illustrates the decomposition of the singular point shown in Figure III.8.3(a). Since $n = 5$ in this case, three pairs of edges, $\{1, 3\}, \{2, 4\}, \{5, \phi\}$, are connected by the procedure mentioned above.

Definition of $S(\cdot, \cdot)$ depends on the application and complexity of a shape. We have given the simplest one suitable for simple shapes such as numerals and Roman alphabets. When local features around the singular point are complex, e.g., for some l:

$$S(i_l, j_l) \approx S(i_l, k), \quad j_l \neq k$$

or:

$$S(i_l, j_l) \approx S(k, j_l), \quad i_l \neq k$$

the decision should be made in terms of global features of adjacent curves (sequences of edges), rather than the local rules for adjacent edges. The definition of $S(\cdot, \cdot)$ can be replaced by the absolute value of the curvature at the singular point estimated from curve fitting, e.g., cubic B-splines or Bernstein-Bezier curve fitting.

The graphs shown in Figure III.8.4 are obtained by applying the singular point decomposition to the graphs in Figure III.8.2. The graph in Figure III.8.2(a) is transformed into the stroke graph which consists of a simple loop, and the one in Figure III.8.2(b) into the stroke graph of two simple arcs ($b1, b2$) and ($b3, b4$).

8.2.1.2 Primitives

On a two-dimensional plane, we define an orthonormal coordinate system $\sum = (O; \mathbf{e}_1, \mathbf{e}_2)$ by taking two unit vectors $\mathbf{e}_1 = (1, 0)$ and $\mathbf{e}_2 = (0, 1)$ together with the origin O, where the counterclockwise angle from $\mathbf{e}_1$ to $\mathbf{e}_2$ is $\pi/2$. Furthermore, for a natural number N, we consider a group of N lines $\{L_i : i = 0, 1, \ldots, N - 1\}$ passing through the origin, where the clockwise angle from the vector $\mathbf{e}_2$ to $L_i (i = 0, 1, \ldots, N - 1)$ is $\psi_i = \frac{i}{N}\pi$. By taking two unit vectors $\mathbf{u}_{i,0} = (\sin \psi_i, \cos \psi_i)$ and $\mathbf{u}_{i,1} = -\mathbf{u}_{i,0}$ on $L_i (i = 0, 1, \ldots, N - 1)$, $2N$ directions are introduced by this line group on a two-dimensional plane. For instance, when $N = 2, 4, 6$, directional codes are introduced as shown in Figure III.8.5.

For the point $P(x, y)$, we define the *height* $h_i(x, y)$ along the line $L_i (i = 0, 1, \ldots, N - 1)$ as the projection of P onto L_i, i.e.:

$$h_i(x, y) = x \sin \psi_i + y \cos \psi_i \tag{III.8.3}$$

We define a *primitive* of curves as a curve whose height is always either nonincreasing or nondecreasing along all the lines $L_i (i = 0, 1, \ldots, N - 1)$ as one traverses the curve. If a primitive p is perpendicular to some $L_i (i = 0, 1, \ldots, N - 1)$, the primitive p is called *singular*.

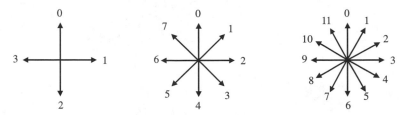

FIGURE III.8.5   Directional codes for $N = 2$ (left), $N = 4$ (middle), and $N = 6$ (right).

For the nonsingular primitive p, the two end points $H(x_h, y_h)$ and $T(x_t, y_t)(y_h > y_t)$ are called *head* and *tail*, respectively, of the primitive p. Furthermore, a nonsingular primitive p is called a *primitive of type d* if $h_i(x_h, y_h) > h_i(x_t, y_t)$ for $i = 0, \ldots, N - d - 1$, and $h_i(x_h, y_h) < h_i(x_t, y_t)$ for $i = N-d, \ldots, N-1$. Figure III.8.6 illustrates instances of primitives of type $d (d = 0, \ldots, N-1)$ when $N = 2, 4$. Figure III.8.7 shows decomposition of strokes into primitives when $N = 2$ (four directions).

8.2.1.3   Concatenation of Primitives

We introduce the concatenation operators of the primitives. We assume that a primitive can be concatenated to others only at its end points.

Definition 1: $\Delta(a, b)$, where a and b are primitives. Suppose that two primitives a and b are concatenated. Let P be the joint of a and b, and let P_a and P_b be points sufficiently close to P such that P_a and P_b are contained in only a and only b, respectively. $\Delta(a, b)$ is defined as:

$$\Delta(a, b) = \text{sgn} \begin{vmatrix} x_a - x_p & x_b - x_p \\ y_a - y_p & y_b - y_p \end{vmatrix} \tag{III.8.4}$$

$$\text{sgn } x = \begin{cases} -1 & \text{if } x < 0 \\ 0 & \text{if } x = 0 \\ 1 & \text{if } x > 0 \end{cases}$$

where (x_p, y_p), (x_a, y_a), and (x_b, y_b) are the coordinates of P, P_a, and P_b, respectively.

In the following, for two primitives a and b, we introduce concatenation operations along $L_i (i = 0, 1, \ldots, N - 1)$:

$$concat\ (i; a.b) \in \{i, N + i, -1\} \tag{III.8.5}$$

Equation III.8.5 is defined by classifying concatenations of primitives according to the local configuration around their joints. Let P be the joint of a and b.

Rule

1. *concat (i; a.b)* $= i$ if $\Delta(a, b) = 1$ and the height of the joint P is maximal around P along the line L_i.
2. *concat (i; a.b)* $= N + i$ if $\Delta(a, b) = 1$ and the height of the joint P is minimal around P along the line L_i.
3. *concat (i; a.b)* $= -1$ if $\Delta(a, b) = 1$ and the height of the curve created by the concatenation of the primitives a and b is always either nonincreasing or nondecreasing along the line L_i as one traverses it.

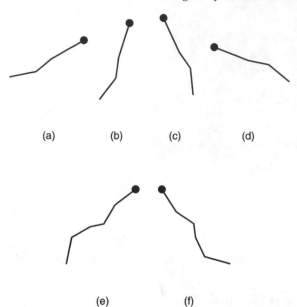

(a)   (b)   (c)   (d)

(e)   (f)

FIGURE III.8.6 Instances of primitives. When $N = 4$, (a), (b), (c), and (d) are instances of primitives of type 0, 1, 2, and 3, respectively. When $N = 2$, (a), (b), and (e) are those of type 0, and the others are of type 1. Heads of primitives are denoted by filled circles.

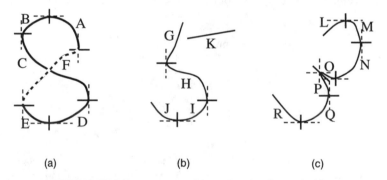

(a)   (b)   (c)

FIGURE III.8.7 Decomposition of strokes into primitives.

The direction of convexity around a joint of two primitives has correspondence to elements of the set $\{0, 1, \ldots, 2N - 1\}$. For $i = 0, 1, \ldots, N - 1$, the direction of $\mathbf{u}_{i,0}$ corresponds to the integer i, and that of $\mathbf{u}_{i,1}$ corresponds to $N + i$.

It can be shown that by collecting the elements such that *concat* $(i; a.b) \geq 0$ from $\{concat\ (i; a.b) :$ $i = 0, 1, \ldots, N - 1\}$, we can create the sequence, for some $l \geq 1$, $(j(1), j(2), \ldots, j(l))$ such that:

$$j(k + 1) = (j(k) + 1)\%(2N), \quad k = 1, 2, \ldots, l - 1, \tag{III.8.6}$$

where $a\%b(b > 0)$ is the residue when a is divided by b, and $0 \leq a\%b < b$. Therefore, we write a concatenation of primitives a and b as:

$$a \xrightarrow{\ j(1), j(l)\ } b \tag{III.8.7}$$

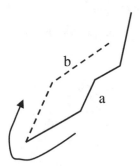

FIGURE III.8.8   Concatenation of two primitives. Two primitives a and b are concatenated such that we turn to the right (clockwise) around the joint of a and b when traversing the curve from a to b.

The arrow "→" means that the primitive a is concatenated to b so that we turn to the right (clockwise) around their joint when traversing them from a to b (Figure III.8.8), and "$j(1), j(l)$" denotes the direction codes of the convexity formed by a and b. For instance, the following concatenations are obtained from primitives shown in Figure III.8.7:

$$A \xrightarrow{1,1} F, \quad E \xrightarrow{3,3} F, \quad D \xrightarrow{2,2} E, \quad C \xrightarrow{1,1} D, \quad C \xrightarrow{3,3} B,$$
$$B \xrightarrow{0,0} A, \quad H \xrightarrow{3,3} G, \quad H \xrightarrow{1,1} I, \quad I \xrightarrow{2,2} J, \quad L \xrightarrow{0,0} M,$$
$$M \xrightarrow{1,1} N, \quad N \xrightarrow{2,2} O, \quad P \xrightarrow{3,0} O, \quad P \xrightarrow{1,1} Q, \quad Q \xrightarrow{2,2} R.$$

Note that two codes are assigned to the concatenation of O and P.

Definition 2: For $d \in \{0, \ldots, N-1\}$, $Q \in \{head, tail\}$, and $c \in \{-1, 1\}$, a function $dir(d, Q, c)$ is defined as follows:

$$dir(d, head, 1) = (2N - d)\%(2N)$$
$$dir(d, tail, 1) = N - d$$
$$dir(d, head, -1) = N - 1 - d$$
$$dir(d, tail, -1) = 2N - 1 - d$$

Proposition 1: The concatenation operator has the following properties:

1. If $a \xrightarrow{j_0, j_1} b$, then $0 \le (j_1 - j_0)\%2N \le 2N - 1$.

2. For the nonsingular primitive a of type d, if $a \xrightarrow{j_0, j_1} b$, and a and b are concatenated at the head of a, then $j_0 = dir(d, head, 1)$.

3. For the nonsingular primitive a of type d, if $a \xrightarrow{j_0, j_1} b$, and a and b are concatenated at the tail of a, then $j_0 = dir(d, tail, 1)$.

4. For the nonsingular primitive b of type d, if $a \xrightarrow{j_0, j_1} b$, and a and b are concatenated at the head of b, then $j_1 = dir(d, head, -1)$.

5. For the nonsingular primitive b of type d, if $a \xrightarrow{j_0, j_1} b$, and a and b are concatenated at the tail of b, then $j_1 = dir(d, tail, -1)$.

6. If $a \xrightarrow{j_0, j_1} b, b \xrightarrow{j_2, j_3} c$, and the primitive b is nonsingular, then $(j_2 - j_1)\%(2N) = 1$.

7. If $a \xrightarrow{j_0, j_1} b$ and $a \xrightarrow{j_2, j_3} c$, then $j_0 \equiv j_2 + N \pmod{2N}$.

8. If $b \xrightarrow{j_0, j_1} a$ and $c \xrightarrow{j_2, j_3} a$, then $j_1 \equiv j_3 + N \pmod{2N}$.

8.2.2   Primitive Sequences of Curves

By linking the concatenations (III.8.7) of primitives, the *primitive sequence* is constructed:

$$a_0 \xrightarrow{j(1,0),\, j(1,1)} a_1 \xrightarrow{j(2,0),\, j(2,1)} \cdots \xrightarrow{j(n,0),\, j(n,1)} a_n \qquad \text{(III.8.8)}$$

The label of the primitive sequence, *PS-label* for short, $\langle r, d \rangle$ is given to the sequence III.8.8 composed of $n + 1$ primitives by the following formulae:

$$d = j(1, 0) \qquad \text{(III.8.9)}$$

$$r = \sum_{i=1}^{n} \{(j(n, 1) - j(n, 0))\%(2N)\} + \sum_{i=1}^{n-1} \{(j(n + 1, 0) - j(i, 1))\%(2N)\} + 2 \qquad \text{(III.8.10)}$$

r and d represent the angular span and the direction of the first primitive quantized in $2N$ directions. Furthermore, the two end points of the primitive sequence are called *h-point* (on a_0) and *t-point* (on a_n).

When the sequence III.8.8 is cyclic, PS-label $\langle 0, 0 \rangle$ is given to the primitive sequence. For instance, circles, ellipses, and convex polygons have the PS-label $\langle 0, 0 \rangle$. The PS-label $\langle 1, d \rangle (d = 0, \ldots, N-1)$ is given to the primitive sequence composed of just one nonsingular primitive p of type $d(d = 0, \ldots, N - 1)$. Figure III.8.9 shows instances of curves with the PS-label $\langle 5, 2 \rangle$.

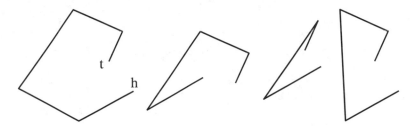

FIGURE III.8.9   Instances of curves with PS-label $\langle 5, 2 \rangle$.

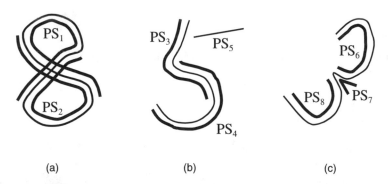

(a)   (b)   (c)

FIGURE III.8.10   Decomposition of curves into primitive sequences.

For instance, the following primitive sequences are generated as shown in Figure III.8.10:

$$PS_1 : C \xrightarrow{3,3} B \xrightarrow{0,0} A \xrightarrow{1,1} F \ \langle 4, 3 \rangle,$$

$$PS_2 : C \xrightarrow{1,1} D \xrightarrow{2,2} E \xrightarrow{3,3} F \ \langle 4, 1 \rangle,$$

$$PS_3 : H \xrightarrow{3,3} G \langle 2, 3 \rangle,$$

$$PS_4 : H \xrightarrow{1,1} I \xrightarrow{2,2} J \langle 3, 1 \rangle,$$

$$PS_5 : \langle 1, 0 \rangle,$$

$$PS_6 : L \xrightarrow{0,0} M \xrightarrow{1,1} N \xrightarrow{2,2} O \langle 4, 0 \rangle,$$

$$PS_7 : P \xrightarrow{3,0} O \langle 3, 3 \rangle,$$

$$PS_8 : P \xrightarrow{1,1} Q \xrightarrow{2,2} R \langle 3, 1 \rangle.$$

8.2.3 Connection of Primitive Sequences

Two primitive sequences are connected to one another by sharing the first primitive (**h-connection**) or the last primitive (**t-connection**). For two adjacent primitive sequences (e_0 and e_1), the connections are denoted by $e_0 \xrightarrow{h} e_1$ or $e_0 \xrightarrow{t} e_1$. For the primitive sequences illustrated in Figure III.8.10, their connections are described as follows:

$$PS_1 \xrightarrow{t} PS_2 \xrightarrow{h} PS_1, \qquad PS_3 \xrightarrow{h} PS_4, \qquad PS_6 \xrightarrow{t} PS_7 \xrightarrow{h} PS_8.$$

Proposition 2: Let $\langle r_0, d_0 \rangle$ and $\langle r_1, d_1 \rangle$ be PS-labels of e_0 and e_1, respectively.

1. If $e_0 \xrightarrow{h} e_1$, then $d_0 - d_1 \equiv N \pmod{2N}$.
2. If $e_0 \xrightarrow{t} e_1$, then $r_0 + d_0 \equiv r_1 + d_1 + N \pmod{2N}$.
3. If $e_0 \xrightarrow{h} e_1$ and $e_0 \xrightarrow{t} e_1 (r_0 \geq r_1 \geq 2)$, then $r_0 \geq 3, r_0 \equiv r_1 \pmod{2N}$, and $d_0 - d_1 \equiv N \pmod{2N}$.

8.2.4 Structural Description of Curves

We have constructed the primitive sequences and their connections for a simple (open) arc or a simple closed curve. Now, we can describe the structure of the curve by a string of PS-labels and connections. For instance, structures of curves shown in Figure III.8.6 is described as follows:

$$\langle 4, 3 \rangle \xrightarrow{t} \langle 4, 1 \rangle \xrightarrow{h} \langle 4, 3 \rangle^*, \quad \langle 2, 3 \rangle \xrightarrow{h} \langle 3, 1 \rangle, \quad \langle 1, 0 \rangle, \quad \langle 4, 0 \rangle \xrightarrow{t} \langle 3, 3 \rangle \xrightarrow{h} \langle 3, 1 \rangle.$$

The mark "*" means that the string is cyclic, and that the last element is identical to the first one.

We mention some properties of closed curves in terms of the PS-labels. We assume that the closed curve C is composed of n primitive sequences $e_1, e_2, \ldots, e_n$ (n is an even number) such that:

$$e_1 \xrightarrow{h} e_2 \xrightarrow{t} e_3 \xrightarrow{h} \cdots \xrightarrow{h} e_n \xrightarrow{t} e_1^*. \tag{III.8.11}$$

Let the PS-label of e_i be $\langle r_i, d_i \rangle$. Now, we give the total amount T of the directional change along the curve C in terms of the PS-labels.

Proposition 3: The total amount T of the directional change along the curve C is given by:

$$|T| = \frac{\pi}{N} \left| \sum_{i=1}^{n} (-1)^i r_i \right| \tag{III.8.12}$$

Proposition 3 suggests that r_i represents the rotation number along the primitive sequence e_i. Furthermore, $T = \pm 2m\pi$ (m is an integer) for closed curves, and $T = \pm 2\pi$ for simple closed curves in particular. Thus, we obtain the following proposition:

Proposition 4: If the curve C is a closed curve, then:

$$\sum_{i=1}^{n}(-1)^{i}r_{i} \equiv 0(mod\ 2N).\tag{III.8.13}$$

In particular, the curve C is a simple closed curve, then:

$$\sum_{i=1}^{n}(-1)^{i}r_{i} = \pm 2N\tag{III.8.14}$$

Finally, we mention the relationship between the PS-label set and singular points on a closed curve.
Proposition 5: If:

$$\sum_{i=1}^{n}(-1)^{i}p_{i} \neq \pm 2N,$$

then there is a singular point on the closed curve C.

8.2.5 Class Description

We describe a shape class of a character with a set of *primitive sequences* forming the character, together with rules for stroke concatenation and linkage, as shown in Figure III.8.11. Since the representation ignores metric information such as size and position, some patterns representing different characters may have the identical structure. For instance, as shown in Figures III.8.12 and III.8.13, some instances of numeral "6" have the same structure as some of "0." Therefore, simple, additional parameters, such as size and position of primitive sequences, are also used to enrich the shape description.

The formal description of a class P is as follows: $P = (M_{P}, \pi_{P}, T_{P}, \Gamma_{P}, \Sigma_{P})$. M_{P} is a number of primitive sequences. $\pi_{P}(k)(k = 1, \ldots, M_{P})$ is a set of eligible PS-labels of the primitive sequence k. T_{P} is a set of possible connections or concatenations of end points of primitive sequences, and it is a set of eligible operations of stroke concatenation or linkage. For $c = (i, p; j, q) \in T_{p}$, where $1 \leq i, j \leq M_{P}$, and $p, q \in \{$"h-point", "t-point", "head", "tail"$\}$, this means that the point p of the primitive sequence i and the point q of j may be connected or concatenated by the stroke concatenation or linkage. Γ_{P} is a set of h/t-connections among primitive sequences. Σ_{P} is

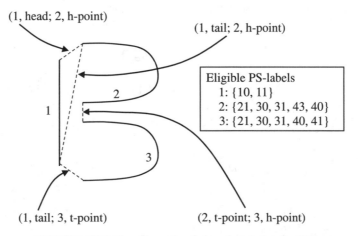

FIGURE III.8.11   Example of class description for "B."

FIGURE III.8.12   Examples of numeral 6.

FIGURE III.8.13   Examples of numeral 0.

statistics (means and standard deviations) for additional parameters (size, position, etc.) of primitive sequences.[10]

For example, when $N = 2$, a class of "B" is described as follows: $B = (3, \pi_B, T_B, \phi, \Sigma_B)$, where $\pi_B(1) = \{10, 11\}$, $\pi_B(2) = \{21, 30, 31, 43, 40\}$, $\pi_B(3) = \{21, 30, 31, 40, 41\}$ ($\langle r, d \rangle$ is denoted by an integer $10 \times r + d$). $T_B = \{(1, \text{"head"}; 2, \text{"}h\text{-point"}), (1, \text{"tail"}; 2, \text{"}h\text{-point"}), (1, \text{"tail"}; 3, \text{"}t\text{-point"}), (2, \text{"}t\text{-point"}; 3, \text{"}h\text{-point"})\}$ (Σ_B is omitted here.) This class is illustrated in Figure III.8.11, where the three primitive sequences are denoted by solid lines and the four stroke connections are denoted by dotted lines.

Based on this class description and the shape analysis, shape matching is performed in the following way.[7] Some local features (*primitives* and *concatenation* of two adjacent primitives) are extracted from the input. In the structural matching, by some combinatorial search, primitives are grouped into primitive sequences so that the groups of primitives are similar to corresponding primitive sequences in the class description and the primitives align within a primitive sequence. Some primitives may not be grouped to any primitive sequences in the class description, and such primitives are checked if it can be explained as either noise or stroke concatenation/linkage in terms of the list of eligible operations registered in the class description.

An alternative approach to shape matching is to describe the input pattern in the bottom-up manner using the shape description, and to apply some simple matching method to the class descriptions and the structural description of the input pattern. Although the computational cost of this approach is quite low, we have to prepare a very large number of class descriptions taking account of all possible cases due to random variations as well as structural deformations. Therefore, this approach is not advisable, and we need to explore some methods for combinatorial matching with *reasonable cost*.[16] The above-mentioned algorithm for shape matching[7,10] along with the class descriptions is an instance of such methods.

8.3   Structural Transformation of Shapes

In this section, we specify the types of shape transformations commonly occurring in unconstrained handwritten characters. In general, structural shape transformations can be classified into two categories: continuous and discontinuous. A continuous transformation is a shape transformation that does not change the global structure of the shape. Furthermore, a small amount of a continuous transformation should also cause a small amount of change of shape features. Continuous transformations are, in general, mathematically tractable. In algebraic topology, continuous shape transformation is discussed based on the equivalence relation *homotopic*[6] defined on a set of continuous functions. On the other hand, discontinuous transformations are catastrophic, and change the global structure and features of the shape. Discontinuous transformations are quite intractable, and there had been few systematic, mathematical tools until René Thom proposed the theories on catastrophe[15] and structural stability.[19] Therefore, practitioners in computer vision and pattern recognition have been obliged to resort to heuristic methods for coping with discontinuous transformations, which are intrinsically intractable.

8.3.1 Continuous Transformation of Shapes

A primitive sequence is characterized by PS-label $\langle r, d \rangle$ which represents the angular span and the direction. If we rotate a curve with PS-label $\langle r, d \rangle$ slightly, then the angular span r is sometimes changed to $r \pm 1$ or the direction to $d \pm 1$. Therefore, we consider the continuous transformation of a curve with respect to angular span (r) and direction (d). To discuss how continuous transformation of a curve is represented in terms of PS-label, we examine the relation between the PS-labels $\langle r, d \rangle$ and $\langle r \pm 1, d \pm 1 \rangle$ with respect to rotation of the curve. Then we generalize the relation to a partial order of the PS-label set:

$$\{\langle r, d \rangle : r = 2, 3, \ldots ; d = 0, 1, \ldots, 2N - 1\}$$

to give an explicit representation of continuous transformation of curves.

Let a primitive sequence with the label $\langle n, d \rangle$ be:

$$a_0 \xrightarrow{j_0, j_0} a_1 \xrightarrow{j_1, j_1} \cdots \xrightarrow{j_{n-3}, j_{n-3}} a_{n-2} \xrightarrow{j_{n-2}, j_{n-2}} a_{n-1}, \qquad \text{(III.8.15)}$$

where we assume that a_i is nonsingular for simplicity, and $j_k = d + k \pmod{2N}, k = 0, 1, \ldots, n-2$. To create a primitive sequence with the label $\langle n + 1, {}^* \rangle$ from sequence III.8.15, we add one primitive in front of a_0 or after a_{n-1}:

$$a_{-1} \xrightarrow{j_{-1}, j_{-1}} a_0 \xrightarrow{j_0, j_0} a_1 \xrightarrow{j_1, j_1} \cdots \xrightarrow{j_{n-3}, j_{n-3}} a_{n-2} \xrightarrow{j_{n-2}, j_{n-2}} a_{n-1},$$

$$a_0 \xrightarrow{j_0, j_0} a_1 \xrightarrow{j_1, j_1} \cdots \xrightarrow{j_{n-3}, j_{n-3}} a_{n-2} \xrightarrow{j_{n-2}, j_{n-2}} a_{n-1} \xrightarrow{j_{n-1}, j_{n-1}} a_n,$$

where $j_{-1} = d - 1 \pmod{2N}$ and $j_{n-1} = d + n - 1 \pmod{2N}$. Therefore, we can expand the primitive sequence with the label $\langle r, d \rangle (r \geq 2)$ to $\langle r + 1, d - 1 (\text{mod } 2N) \rangle$ or $\langle r + 1, d \rangle$ by adding one primitive to the head or the tail of the sequence, respectively. Similarly, we can reduce the primitive sequence with the label $\langle r, d \rangle (r \geq 3)$ to $\langle r - 1, d + 1 (\text{mod } 2N) \rangle$ or $\langle r - 1, d \rangle$ by removing the first or the last primitive of the sequence, respectively. From the above discussion, we can define a partial order in the PS-label set in order to represent the transformation of curve structure according to rotation or curls of curves.

Definition 3: The partial order "$\subseteq_P$" in the PS-label set

$$\{\langle r, d \rangle : r = 2, 3, \ldots ; d = 0, 1, \ldots, 2N - 1\}$$

is defined as follows:

1. $\langle r, d \rangle \subseteq_P \langle r, d \rangle$
2. $\langle r, d_0 \rangle \subseteq_P \langle r + n, d_n \rangle$ $(n \geq 1)$ if and only if there is a sequence:

$$(\langle r, d_0 \rangle, \langle r + 1, d_1 \rangle, \ldots, \langle r + n - 1, d_n - 1 \rangle, \langle r + n, d_n \rangle)$$

such that $d_i = d_{i+1}$ or $d_i = d_{i+1} + 1 (\text{mod } 2N)$ for $i = 0, 1, \ldots, n - 1$.

Hasse diagrams are useful for graphic representation of order relations.

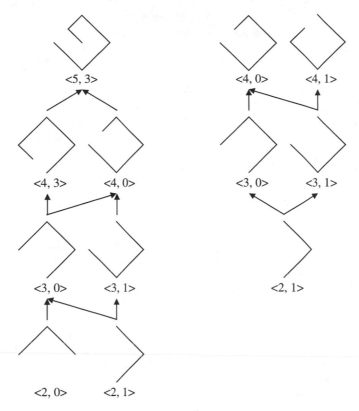

FIGURE III.8.14 The order relation " $\subseteq_P$" in the PS-label set.

Definition 4: For a partially ordered set, a *Hasse diagram* is a graphic representation of all ordered pairs of the partial order with the following properties:

1. A partial order $a \leq b$ is represented by an arc pointing upward from a to b.
2. There is an arc from a to b only if there is no other element c such that $a \leq c$ or $c \leq b$. Therefore, an arc is not present in a Hasse diagram if it is implied by the transitivity of the relation.

Figure III.8.14 shows some examples of Hasse diagrams of the PS-labels in terms of the partial order " $\subseteq_P$" when $N = 2$. For instance, the curve with PS-label $\langle 3, 1 \rangle$ is transformed into $\langle 4, 0 \rangle$ with an additional rotation, and vice versa by removing a rotation. The partial order represents such transformation with simple arithmetics on integers.

In class descriptions, it is natural to require that elements in $\pi(k)(1 \leq k \leq N_P)$ be transformed continuously to each other. Based on the above discussions, we can introduce the structure of $\pi(k)$ in terms of " $\subseteq_P \cdot$" For simplicity, we only mention the case of $N = 2$ below.

PSL-Condition:

(P-0) $\langle 0, 0 \rangle, \in \pi(k)$ if and only if $\pi(k) = \{\langle 0, 0 \rangle\}.\langle 1, {}^* \rangle \in \pi(k)$ if and only if $\pi(k) \subseteq \{\langle 1, j \rangle : j \in \{0, 1, 2, 3\}\}$.

(P-1) Let $\mu(k) = \min\{r : \langle r, d \rangle \in \pi(k)\}$ and $M(k) = \max\{r : \langle r, d \rangle \in \pi(k)\}$. For any $\langle r, m \rangle \in \pi(k)(2 \leq \mu(k) \leq r < M(k))$, there exists $\langle r + 1, n \rangle \in \pi(k)$ such that $\langle r, m \rangle \subseteq_P \langle r + 1, n \rangle$, i.e., $m = n$ or $m = n + 1 \pmod 4$. For any $\langle r, n \rangle \in \pi(k)(2 < \mu(k) \leq r \leq M(k))$, there exists $\langle r - 1, m \rangle \in \pi(k)$ such that $\langle r - 1, m \rangle \subseteq_P \langle r, n \rangle$.

(P-2) Let $D(k, r) = \{\langle r, d \rangle : \langle r, d \rangle \in \pi(k)\}$. For any $r (2 \leq \mu(k) \leq r \leq M(k))$, $\#D(k, r) \leq 2$. If $\#D(k, r) = 2$, the two elements $\langle r, d_1 \rangle$ and $\langle r, d_2 \rangle$ satisfy $d_1 = d_2 \pm 1 \pmod 4$.

(P-0) is clear, because the primitive sequence with the label $\langle 0, 0 \rangle$ or $\langle 1, * \rangle$ is a special one. It is required that elements in $\pi(k)$ be transformed continuously to each other. (P-1) is derived from this requirement in terms of "$\subseteq_P$." On Hasse diagrams such as Figure III.8.14, there must be a path between any pair $\langle r_0, d_0 \rangle$ and $\langle r_1, d_1 \rangle$ $(r_o \neq r_1)$ of elements of $\pi(k)$. (P-2) mentions that a primitive sequence with the label $\langle r (\geq 2), d \rangle$ can be transformed to $\langle r, d \pm 1 \rangle$ by small rotation. PS-label sets $\pi_B(1)$, $\pi_B(2)$, and $\pi_B(3)$ illustrated in Figure III.8.11 satisfy **PSL-Condition**.

8.3.2   Discontinuous Transformations

Among various types of 2D patterns, handwritten characters are quite unique in their shape deformation. In particular, discontinuous transformations commonly occurring in handwriting can be classified as follows:

T1. Concatenating two end points of curve components by moving them slightly (stroke concatenation). End points of strokes coincide and the end points consequently disappear (Figures III.8.15(a) and (b)).

T2. Connecting two end points of curve components with an additional curve (Figures III.8.15(c), (d), and (e)).

These transformations are quite simple, but change the global features and structures of the shape significantly. Furthermore, it is difficult to systematically analyze the *a priori* effects of these transformations, and therefore we are obliged to depend on heuristics to handle them.

There have been some studies conducted in order to accommodate systematically such transformations into recognition systems. Wakahara et al.[20] have conducted a series of researches on on-line handwriting recognition, and they have found it necessary to incorporate rules of stroke linkage into shape prototypes for recognition of unconstrained handwritten characters. Camillerap et al.[1] incorporated explicit models of distortions into cursive handwriting recognition systems based on the observation that a cusp is often transformed to a loop or a hump, and vice versa. Nishida[7] generalized these approaches and proposed a robust shape-matching algorithm that tolerates some structural deformations caused by stroke connection in handwriting, stroke breaking, and a variety of stroke order, direction, and number.

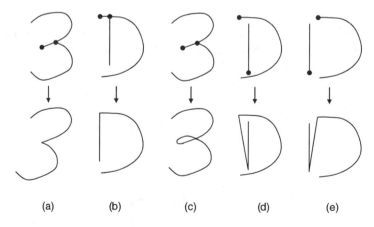

(a)   (b)   (c)   (d)   (e)

FIGURE III.8.15   Discontinuous structural transformations of curves. In (a) and (b), the two curves are moved slightly so that they are concatenated at the end points (denoted by filled circles). In (c), (d), and (e), the end points (denoted by filled circles) are connected with additional curves.

In this section, we obtain systematic, complete, *a priori* knowledge of the effects of the transformations based on their exhaustive analysis. For simplicity, we only mention main results when $N = 2$.[9] Complete results for general cases can be found in Nishida.[8]

8.3.2.1 Structural Transformation by Concatenating End Points

We analyze the structural transformation caused by each operation of T1, i.e., concatenating two end points of primitive sequences by moving them slightly. Suppose that there are two primitive sequences p_0 and p_1 with PS-labels $\langle r_0, d_0 \rangle$ and $\langle r_1, d_1 \rangle$ $(r_0, r_1 \geq 2)$, respectively. Analysis for other cases, namely (1) $r_0 \geq 2$ and $r_1 = 1$, (2) $r_0 = r_1 = 1$, and (3) concatenating h-point and t-point of one primitive sequence, is given in Nishida.[8,9]

Theorem 1: If t-point of the primitive sequence p_0 and h-point of p_1 are concatenated, then the local structure of the curve is transformed as follows:

- If $k \equiv (d_1 - r_0 - d_0)\%4 = 0, 1, 3$, then the primitive sequences p_0 and p_1 can be merged into one primitive sequence P_0 with PS-label $P_0 : \langle r_0 + r_1 - 2 + (k + 2)\%4, d_0 \rangle$. ($a\%b, b > 0$ is the residue when a is divided by b and $0 \leq a\%b < b$.)
- If $k = 1, 2$, then a new primitive sequence can be created and the local structure of the curve can be transformed to $P_0 \overset{t}{\_} P_1 \overset{h}{\_} P_2$, where PS-labels of the primitive sequences are $P_0 : \langle r_0, d_0 \rangle$, $P_1 : \langle 2 + (-k - 2)\%4, (d_1 + 2)\%4 \rangle$, and $P_2 : \langle r_1, d_1 \rangle$. The primitive sequences P_0 and P_2 correspond to p_0 and p_1, respectively.

Table III.8.1 summarizes Theorem 1. When $k = 1$, the curve structure can be transformed in two ways, according to the configuration of two primitive sequences. For instance, if PS-labels of p_0 (left of Figure III.8.16(a)) and p_1 (right) are $\langle 2, 1 \rangle$ and $\langle 2, 0 \rangle (k = (0 - 2 - 1)\%4 = (-3)\%4 = 1)$, the local structure of the curve can be either one primitive sequence P_0 with PS-label $\langle 5, 1 \rangle$ (Figure III.8.16(b)), or $P_0 \overset{t}{\_} P_1 \overset{h}{\_} P_2$ with PS-labels $P_0 : \langle 2, 1 \rangle$, $P_1 : \langle 3, 2 \rangle$, and $P_2 : \langle 2, 0 \rangle$ (Figure III.8.16(c)).

TABLE III.8.1 Structural Transformation when Concatenating t-point of Primitive Sequence p_0 and h-point of p_1

Condition	M	Structure	PS-label
$k = 0, 1, 3$	1	P_0	$P_0 : \langle r_0 + r_1 - 2 + (k + 2)\%4, d_0 \rangle$
$k = 1, 2$	3	$P_0 \overset{t}{\_} P_1 \overset{h}{\_} P_2$	$P_o : \langle r_0, d_0 \rangle, P_1 : \langle 2 + (-k - 2)\%4, (d_1 + 2)\%4 \rangle, P_2 : \langle r_1, d_1 \rangle$

Note: M is the number of primitive sequences after the two primitive sequences are concatenated. $k = (d_1 - r_0 - d_0)\%4$.

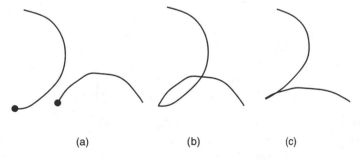

(a)   (b)   (c)

FIGURE III.8.16 When we concatenate the two end points denoted by filled circles in (a), there are two possible structures (b) and (c) according to the transformation rules.

TABLE III.8.2 Structural Transformation when Concatenating h-point of Primitive Sequence P_0 and h-point of p_1

Condition	M	Structure	PS-label
$k = 2, 3, 0$	2	$P_0 \overset{h}{=} P_1$	$P_0 : \langle r_0, d_0 \rangle, P_1 : \langle r_1 + (k + 2)\%4, (d_0 + 2)\%4 \rangle$
$k = 0, 1, 2$	2	$P_0 \overset{h}{=} P_1$	$P_0 : \langle r_0 + (2 - k)\%4, (d_1 + 2)\%4 \rangle, P_1 : \langle r_1, d_1 \rangle$

Note: M is the number of primitive sequences after the two primitive sequences are concatenated. $k = (d_1 - d_0)\%4$.

TABLE III.8.3 Structural Transformation when Concatenating t-point of Primitive Sequence p_0 and t-point of p_1

Condition	M	Structure	PS-label
$k = 0, 1, 2$	2	$P_0 \overset{t}{=} P_1$	$P_0 : \langle r_0, d_0 \rangle, P_1 : \langle r_1 + (2 - k)\%4, d_1 \rangle$
$k = 0, 2, 3$	2	$P_0 \overset{t}{=} P_1$	$P_0 : \langle r_0 + (k + 2)\%4, d_0 \rangle P_1 : \langle r_1, d_1 \rangle$

Note: M is the number of primitive sequences after the two primitive sequences are concatenated. $k = (r_1 + d_1 - r_0 - d_0)\%4$.

TABLE III.8.4 Structural Transformation when Connecting t-point of Primitive Sequence p_0 and h-point of p_1 by an Additional Primitive

Condition	M	Structure	PS-label
$k = 2, 3$	1	P_0	$P_0 : \langle r_0 + r_1 - 2 + (k + 2)\%4, d_0 \rangle$
$k = 0, 1, 2, 3$	1	P_0	$P_0 : \langle r_0 + r_1 + k, d_0 \rangle$
$k = 0, 1, 2, 3$	3	$P_0 \overset{t}{=} P_1 \overset{h}{=} P_2$	$P_0 : \langle r_0, d_0 \rangle, P_1 : \langle 2 + (-k - 2)\%4, (d_1 + 2)\%4 \rangle, P_2 : \langle r_1, d_1 \rangle$
$k = 0, 1, 2, 3$	3	$P_0 \overset{t}{=} P_1 \overset{h}{=} P_2$	$P_0 : \langle r_0 + (k + l + 2)\%4, d_0 \rangle, P_1 : \langle 2 + l, (d_1 + 2)\%4 \rangle, P_2 : \langle r_1, d_1 \rangle$ $(l = 0, 1; (k + l + 2)\%4 = 0, 1, 2)$
$k = 0, 1, 2, 3$	3	$P_0 \overset{t}{=} P_1 \overset{h}{=} P_2$	$P_0 : \langle r_0, d_0 \rangle, P_1 : \langle 2 + l, (r_0 + d_0 - l)\%4 \rangle,$ $P_2 : \langle r_1 + (k + l + 2)\%4, (r_0 + d_0 - l - 2)\%4 \rangle$ $(l = 0, 1; (k + l + 2)\%4 = 0, 1, 2)$

Note: M is the number of primitive sequences after the two primitive sequences are connected. $k = (d_1 - r_0 - d_0)\%4$.

Theorem 2: If h-point of the primitive sequence p_0 and h-point of p_1 are concatenated, then the local structure of the curve is transformed as shown in Table III.8.2.

Theorem 3: If t-point of the primitive sequence p_0 and t-point of p_1 are concatenated, then the local structure of the curve is transformed as shown in Table III.8.3.

In Tables III.8.1 through III.8.3, M is the number of primitive sequences after the two primitive sequences are concatenated, and P_0 and P_{M-1} on the new curve correspond to p_0 and p_1 on the original curve. When $M \geq 3$, primitive sequences P_1 through P_{M-2} do not exist on the original curve, and they are introduced by the operation of concatenating the two end points.

8.3.2.2 Structural Transformation by Connecting End Points

We analyze the structural transformation caused by each operation of T2, i.e., connecting two end points of primitive sequences with an additional curve. Suppose that there are two primitive sequences p_0 and p_1 with PS-labels $\langle r_0, d_0 \rangle$ and $\langle r_1, d_1 \rangle (r_0, r_1 \geq 2)$, respectively. Analysis for other cases is given in.[8,9]

Theorem 4: If t-point of the primitive sequence p_0 and h-point of p_1 are connected by an additional primitive, then the local structure of the curve is transformed as shown in Table III.8.4.

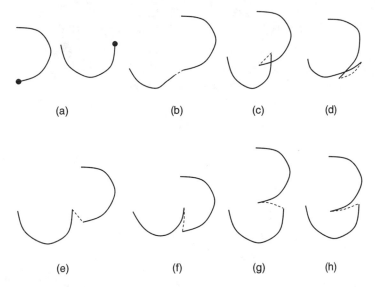

FIGURE III.8.17   When we connect the two end points denoted by filled circles in (a) with an additional primitive, there are seven possible structures (b) through (h) according to the transformation rules. The additional primitives are denoted by dotted lines.

TABLE III.8.5   Structural Transformation when Connecting h-point of Primitive Sequence p_0 and h-point of p_1 by an Additional Primitive

Condition	M	Structure	PS-label
$k = 2, 3$	2	$P_0 \xrightarrow{h} P_1$	$P_0 : \langle r_0, d_0 \rangle, P_1 : \langle r_1 + (k+2)\%4, (d_0 + 2)\%4 \rangle$
$k = 0, 1, 2$	2	$P_0 \xrightarrow{h} P_1$	$P_0 : \langle r_0, d_0 \rangle, P_1 : \langle r_1 + k + 2, (d_0 + 2)\%4 \rangle$
$k = 0, 1, 2$	2	$P_0 \xrightarrow{h} P_1$	$P_0 : \langle r_0 + (2-k)\%4, (d_1 + 2)\%4 \rangle, P_1 : \langle r_1, d_1 \rangle$
$k = 2, 3$	2	$P_0 \xrightarrow{h} P_1$	$P_0 : \langle r_0 + 6 - k, (d_1 + 2)\%4 \rangle, P_1 : \langle r_1, d_1 \rangle$
$k = 0, 1, 2, 3$	2	$P_0 \xrightarrow{h} P_1$	$P_0 : \langle r_0 + l, (d_0 - l)\%4 \rangle, P_1 : \langle r_1 + (k+l-2), (d_0 - l + 2)\%4 \rangle$
			$(l = 0, 1, 2; (k + l - 2)\%4 = 0, 1, 2)$
			$P_0 : \langle r_0, d_0 \rangle, P_1 : \langle 2 + l, (d_0 + 2)\%4 \rangle,$
$k = 0, 1, 2$	4	$P_0 \xrightarrow{h} P_1 \xrightarrow{l} P_2 \xrightarrow{h} P_3$	$P_2 : \langle 2 + (l - k + 2), (d_1 + 2)\%4 \rangle, P_3 : \langle r_1, d_1 \rangle,$
			$(l = 0, 1; (l - k + 2)\%4 = 0, 1)$

Note: M is the number of primitive sequences after the two primitive sequences are connected. $k = (d_1 - d_0)\%4$.

When end points of primitive sequences are connected by an additional primitive, there are more possible structures for the new curve than when two end points are concatenated. When $k = 3$, the curve structure can be transformed in seven ways according to the configuration of two primitive sequences. For instance, if PS-labels of p_0 (left of Figure III.8.17(a)) and p_1 (right) are $\langle 2, 1 \rangle$ and $\langle 2, 2 \rangle (k = (2 - 2 - 1)\%4 = (-1)\%4 = 3$, the local structure of the curve can be one of the following:

1. One primitive sequence P_0 with PS-label $\langle 3, 1 \rangle$ (Figure III.8.17(b))
2. One primitive sequence P_0 with PS-label $\langle 7, 1 \rangle$ (Figure III.8.17(c))
3. $P_0 \xrightarrow{l} P_1 \xrightarrow{h} P_2$ with PS-labels $P_0 : \langle 2, 1 \rangle, P_1 : \langle 5, 0 \rangle,$ and $P_2 : \langle 2, 2 \rangle$ (Figure III.8.17(d))
4. $P_0 \xrightarrow{l} P_1 \xrightarrow{h} P_2$ with PS-labels $P_0 : \langle 3, 1 \rangle, P_1 : \langle 2, 0 \rangle,$ and $P_2 : \langle 2, 2 \rangle$ (Figure III.8.17(e))
5. $P_0 \xrightarrow{l} P_1 \xrightarrow{h} P_2$ with PS-labels $P_0 : \langle 4, 1 \rangle, P_1 : \langle 3, 0 \rangle,$ and $P_2 : \langle 2, 2 \rangle$ (Figure III.8.17(f))

6. $P_0 \xrightarrow{t} P_1 \xrightarrow{h} P_2$ with PS-labels $P_0 : \langle 2, 1 \rangle$, $P_1 : \langle 2, 3 \rangle$, and $P_2 : \langle 3, 1 \rangle$ (Figure III.8.17(g))

7. $P_0 \xrightarrow{t} P_1 \xrightarrow{h} P_2$ with PS-labels $P_0 : \langle 2, 1 \rangle$, $P_1 : \langle 3, 2 \rangle$, and $P_2 : \langle 4, 0 \rangle$ (Figure III.8.17(h))

Theorem 5: If h-point of the primitive sequence p_0 and h-point of p_1 are connected by an additional primitive, then the local structure of the curve is transformed as shown in Table III.8.5.

Theorem 6: If t-point of the primitive sequence p_0 and t-point of p_1 are connected by an additional primitive, then the local structure of the curve is transformed as shown in Table III.8.6.

TABLE III.8.6   Structural Transformation when Connecting t-point of Primitive Sequence p_0 and t-point of p_1 by an Additional Primitive

Condition	M	Structure	PS-label
$k = 2, 3$	2	$P_0 \xrightarrow{t} P_1$	$P_0 : \langle r_0, d_0 \rangle$, $P_1 : \langle r_1 + 6 - k, d_1 \rangle$
$k = 0, 1, 2$	2	$P_0 \xrightarrow{t} P_1$	$P_0 : \langle r_0, d_0 \rangle$, $P_1 : \langle r_1 + (2 - k)\%4, d_1 \rangle$
$k = 0, 1, 2$	2	$P_0 \xrightarrow{t} P_1$	$P_0 : \langle r_0 + k + 2, d_0 \rangle$, $P_1 : \langle r_1, d_1 \rangle$
$k = 2, 3$	2	$P_0 \xrightarrow{t} P_1$	$P_0 : \langle r_0 + (k - 2)\%4, d_0 \rangle$, $P_1 : \langle r_1, d_1 \rangle$
$k = 0, 1, 2, 3$	2	$P_0 \xrightarrow{t} P_1$	$P_0 : \langle r_0 + l, d_0 \rangle$, $P_1 : \langle r_1 + (l - k + 2), d_1 \rangle$
			$(l = 0, 1, 2; (l - k + 2)\%4 = 0, 1, 2)$
			$P_0 : \langle r_0, d_0 \rangle$, $P_1 : \langle 2 + l, (d_0 + r_0 - l)\%4 \rangle$,
$k = 1, 2, 3$	4	$P_0 \xrightarrow{t} P_1 \xrightarrow{h} P_2 \xrightarrow{t} P_3$	$P_2 : \langle 2 + (l + k - 2), (d_0 + r_0 - l + 2)\%4 \rangle$, $P_3 : \langle r_1, d_1 \rangle$,
			$(l = 0, 1; (l + k - 2)\%4 = 0, 1)$

Note: M is the number of primitive sequences after the two primitive sequences are connected. $k = (r_1 + d_1 - r_0 - d_0)\%4$.

In Tables III.8.4 through III.8.6, M is the number of primitive sequences after the two primitive sequences are connected with an additional primitive, and P_0 and P_{M-1} on the new curve correspond to p_0 and p_1 on the original curve.

8.4   Automatic Model Inference

In this section, based on the analysis, we describe the algorithm for automatic construction of structural models incorporating discontinuous transformations. We state the main problem to solve and describe the algorithms, along with their outline.

8.4.1   Problem Definition

We state clearly the problem to solve in this section. Given the dataset for a particular character (e.g., "B"), suppose that each sample is analyzed and described, and that the set of structural descriptions is obtained. The structural descriptions are grouped together by structural pattern clustering in terms of continuous transformation, so that elements belonging to the same group can be transformed continuously to each other within the group (subclass). A pair of class descriptions C and C' are merged together if there is a one-to-one mapping f from primitive sequences of C to those of C' such that:

- For each primitive sequence p of the class C and the corresponding primitive sequence $f(p)$ of the class C', the union of the PS-label sets $\pi_c(p)$ and $\pi_{c'}(f(p))$ satisfies **PSL-Condition**.
- Connections of primitive sequences in the class C are preserved in C' under the mapping f.
- Connections of primitive sequences in the class C' are preserved in C under the mapping f^{-1}.

The inductive learning is performed by merging recursively a pair of classes that satisfy the conditions for generalization until no pair of classes satisfies the conditions anymore. We also utilize statistics

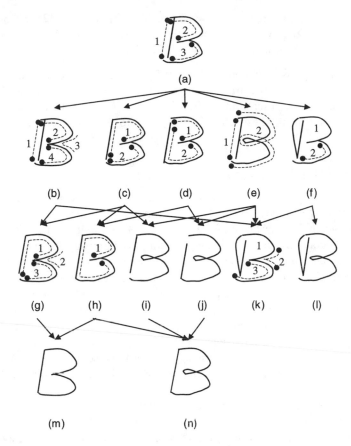

FIGURE III.8.18 A transition diagram of shape transformation when we apply just one operation of T1 or T2 at one time recursively to the original shape illustrated by (a). Dashed/dotted lines denote curve components (primitive sequences) along with their indices.

of shape parameters attached to primitive sequences as secondary information. If the grouping conditions are satisfied, the new statistics (mean values and standard deviations) are computed for these parameters by combining statistics of each component of the class C and its corresponding component in the class C'. The Mahalanobis distance (discriminant efficiency) is also computed between the pair of mean vectors. If several mappings f are found, the one is selected that minimizes the distance. Furthermore, statistical constraints are put on the standard deviation of each parameter and the Mahalanobis distance. If some of the statistical constraints are violated, the pair is not grouped together. Note that we use the structural description as dominant information and the statistics as secondary.

Each of Figures III.8.18(a) through (n) represents a subclass created by this structural clustering procedure. Now, the main problem is to create such a class description shown in Figure III.8.11 from the set of subclass descriptions as shown in Figures III.8.18(a) through (n).

For instance, when $N = 2$, the five subclasses corresponding to Figures III.8.6(a), (b), (c), (e), and (h) are described as follows ($T_a = T_b = T_c = T_e = T_h = \phi$):

a. $B_a = (3, \pi_a, T_a, \phi, \Sigma_a)$, where $\pi_a(1) = \{10, 11\}, \pi_a(2) = \{21, 30, 31, 43, 40\}$, $\pi_a(3) = \{21, 30, 31, 40, 41\}$. (PS-label $\langle r, d \rangle$ is denoted by an integer $10 \times r + d$ and Σ_a is omitted here.)

b. $B_b = (4, \pi_b, T_b, \Gamma_b, \Sigma_b)$, where $\pi_b(1) = \{10, 11\}, \pi_b(2) = \{21, 30, 31, 43, 40\}$, $\pi_b(3) = \{23, 32, 33, 42\}, \pi_b(4) = \{21, 30, 31, 40, 41\}, \Gamma_b = \{2 \stackrel{t}{\rightharpoonup} 3 \stackrel{h}{\rightharpoonup} 4\}$.

c. $B_c = (2, \pi_c, T_c, \phi, \Sigma_c)$, where $\pi_c(1) = \{30, 43, 40, 53\}$, $\pi_c(2) = \{21, 30, 31, 40, 41, 51\}$.

e. $B_e = (2, \pi_e, T_e, \phi, \Sigma_e)$, where $\pi_e(1) = \{10, 11\}$, $\pi_e(2) = \{61, 70, 71, 83, 80, 93\}$.

h. $B_h = (1, \pi_h, T_h, \phi, \Sigma_h)$, where $\pi_h(1) = \{61, 70, 71, 80\}$.

The descriptions of these classes are quite different from each other because they are transformed to each other by the discontinuous transformations T1 and T2. However, all these classes can be represented by the class description shown in Figure III.8.11. In this section, we present an algorithm for generalizing these class descriptions to a *super-class* description as shown in Figure III.8.11.

8.4.2   Outline of the Algorithm

Figure III.8.18 shows a transition diagram for shape transformation when we apply just one operation of T1 or T2 registered in T_B of the class for "B" at one time recursively to the original shape illustrated Figure III.8.18(a), which is an instance of the class for "B" without any transformations T1 or T2 applied. From this diagram, it is found that the problem can be translated into the problem of finding out, inductively from the dataset, such initial shape classes as Figure III.8.18(a) and transformation rules associated with directed links in the diagram, and therefore we need to address the problem of automatic construction of such a transition diagram as Figure III.8.18 from the given dataset. In order to construct such a transition diagram, for each subclass C, we need to find another subclass C_0 such that C_0 can be transformed to C by applying just one operation of T1 or T2 to C_0. Furthermore, in order to find C_0, we need to find a relevant component correspondence between two shape classes represented in the form of structural descriptions. Usually, the discontinuous shape transformations change topology and global features, and therefore it is difficult to find a component correspondence between two class descriptions.

However, in our case, based on the complete, systematic, *a priori* knowledge about the effects of the transformations T1 and T2, we can design an algorithm for finding component correspondence between two deformed patterns and can introduce a generalization rule for model construction, inferring that a subclass can be transformed to another subclass by a particular operation. For instance, we can find that the subclass B_a is transformed to the other subclass B_b based on the *a priori* knowledge, given in Table III.8.1, that by concatenating t-point of a primitive sequence with PS-label $\langle 3, 0 \rangle$ (primitive sequence 2 of Figure III.8.18(a)) and h-point of a primitive sequence with PS-label $\langle 3, 1 \rangle$ (primitive sequence 3), the local structure composed of the two primitive sequences can be changed to $\langle 3, 0 \rangle \xrightarrow{t} \langle 2, 3 \rangle \xrightarrow{h} \langle 3, 1 \rangle$ (corresponding to primitive sequences 2, 3, and 4 of Figure III.8.18(b), respectively). Therefore, transition diagrams can be created by comparing two structural descriptions and finding out relevant component correspondence through consistent labeling algorithms[2] along with the high-level models of the shape transformations.

After the transition diagram is completed, the subclasses are generalized into the class corresponding to the root nodes of the diagrams by traversing them from leaves to the root, and finally the class as shown in Figure III.8.11 is created automatically from the dataset. All the shape subclasses illustrated in Figure III.8.18 can be represented by this single class.

8.4.3   Algorithm

Suppose that N_c subclasses for a particular character (for instance, "B") have been created from the given dataset. In this subsection, for the set S composed of N_c subclasses, we describe the algorithm for automatic construction of the transition diagram such as Figure III.8.18 and structural models (classes) as illustrated in Figure III.8.11. Note that T_c (the set of eligible operations in each class C) is initially empty.

The algorithm for constructing transition diagrams is shown in Figure III.8.19. This algorithm is operated for the set S. The procedure `ps_config` checks whether component (primitive sequence)

Algorithm: construction of transition diagram D
begin
  transition diagram $D \leftarrow$ empty;
  for each $C \in \{0, 1, \ldots, N_C - 1\}$ **do**
    begin
      $C_0 \leftarrow ABSENT$;
      $d_0 \leftarrow \infty$;
      $class\,[C]\,.matched \leftarrow 0$;
      $class\,[C]\,.dist \leftarrow \infty$;
      for each $C' \in \{0, 1, \ldots, N_C - 1\}$ **do**
        if $C' \neq C$ **then**
          begin
            $M_{C'} \leftarrow$ number of primitive sequences of class C';
            for each $i \in \{0, 1, \ldots, M_{C'} - 1\}$ **do**
              for each $j \in \{0, 1, \ldots, M_{C'} - 1\}$ **do**
                ps_config(class[C], class[C'], i, j);
            if $class\,[C]\,.matched$ & $class\,[C]\,.dist < d_0$ &
              $class\,[C]\,.dist < thr$ **then**
            begin
              $C_0 \leftarrow C'$;
              $d_0 \leftarrow class\,[C]\,.dist$;
            end
          end
      if C_0 is not $ABSENT$ **then**
        add arc $C_0 \rightarrow C$ to transition diagram D;
    end
end

FIGURE III.8.19   Algorithm for constructing transition diagrams.

correspondence can be found between the two subclasses C and C' when we assume that the subclass C' is possibly transformed to the subclass C by applying just one operation of T1 or T2 to primitive sequences i and j of the subclass C'. If some component correspondence can be found, then the field *matched* of the subclass C is set to 1 inside the procedure ps_config (Figure III.8.20). Furthermore, based on the component correspondence, statistical distance between the two subclasses C and C' is computed for some additional geometrical parameters on primitive sequences,[10] and the distance is put to the field *dist* of the subclass C. Among the subclasses, along with operations of T1 or T2 for which some component correspondences are found, the one that minimizes statistical distance is selected as C_0. When each arc $C_0 \rightarrow C$ is created, the corresponding operation of T1 or T2 is added to T_{c_0} (the set of eligible operations of T1 and T2 of the subclass C_0). We illustrate these algorithms with examples.

We consider the procedure called ps_config(class(b),class(a),2,3), where $M_a = 3$ and $M_b = 4$. When the second rule listed in Table III.8.1 is applied, the local structure composed of primitive sequences 2 and 3 of the subclass (a) is transformed to three primitive sequences P_0, P_1, and P_2 that are connected as $P_0 \overset{l}{-} P_1 \overset{h}{-} P_2$. We find out a relevant subset $\{Q_0, Q_1, Q_2\}$ of primitive sequences of the subclass (b). For $\langle 2, 1 \rangle \in \pi_a(2)$ and $\langle 2, 1 \rangle \in \pi_a(3)$, the PS-labels for P_0, P_1, and P_2 are $\langle 2, 1 \rangle \in \pi_b(2)$, $\langle 2, 3 \rangle \in \pi_b(2)$, $\langle 2, 1 \rangle \in \pi_a(4)$ according to the transformation rule. Furthermore, primitive sequences 2, 3, and 4 of the subclass (b) are connected as $2 \overset{l}{-} 3 \overset{h}{-} 4$. Therefore, the two primitive sequences 2 and 3 in the subclass (a) can be transformed to the three primitive sequences 2, 3, and 4 in the subclass (b), i.e., $Q_0 = 2$, $Q_1 = 3$, and $Q_2 = 4$. We check the rest of primitive sequences: 1 in the subclass (a) and 1 in (b). PS-label sets of both primitive sequences are equal, and therefore a mapping $f : \{1, 2, 3\} \rightarrow \{1, 2, 3, 4\} - \{3\}$ is found as $f(1) = 1$, $f(2) = 2$, and $f(3) = 4$. Next, we consider the procedure call ps_config(class(e),class(a),2,3), where $M_a = 3$ and

Algorithm: ps_config(class C, class C', i, j)
for each transformation rule in Theorems 1–6

- Suppose that by applying the transformation rule under consideration, two primitive sequences i and j with PS-labels $\langle r_i, d_i \rangle$ and $\langle r_j, d_j \rangle$ are transformed to K primitive sequences $P_0, \ldots, P_{K-1}$ with PS-labels $\langle R_k(r_i, d_i, r_j, d_j), D_k(r_i, d_i, r_j, d_j) \rangle$ ($k = 0, \ldots, K-1$) that are connected as $P_0 \overset{c_1}{\ldots} \overset{c_{K-1}}{} P_{K-1}$ ($c_k \in \{h, t\}$, $k = 1, \ldots, K-1$).

- Find a subset $\{Q_0, \ldots, Q_{K-1}\}$ of primitive sequences of class C such that the local structure composed of i and j can be transformed to the structure composed of this subset on C. The subset must satisfy the following conditions: (a) $Q_0, \ldots, Q_{K-1}$ are connected to each other as $Q_0 \overset{c_1}{\ldots} \overset{c_{K-1}}{} Q_{K-1}$ as specified by the transformation rule, and (b) there are some pair $\langle r_i, d_i \rangle \in \pi_{C'}(i)$ and $\langle r_j, d_j \rangle \in \pi_{C'}(j)$ such that $\langle R_k(r_i, d_i, r_j, d_j), D_k(r_i, d_i, r_j, d_j) \rangle \in \pi_C(Q_k)$ for $k = 0, \ldots, K-1$.

- **for each** found subset $\{Q_0, \ldots, Q_{K-1}\}$

  - Find a mapping $f : \{1, \ldots, M_{C'}\} \to \{1, \ldots, M_C\} - \{Q_1, \ldots, Q_{K-2}\}$ ($f(i) = Q_0$, $f(j) = Q_{K-1}$) that satisfies the generalization conditions specified in Section 8.4.1.

  - **for each** found f

    * *class* $[C]$*.matched* $\leftarrow 1$.
    * $d \leftarrow$ statistical distance with respect to f.
    * **if** $d <$ class$[C]$.dist **then**
      · *class* $[C]$*.dist* $\leftarrow d$.
      · Record the operation (concatenation/connection and its type), mapping f, indices i, j, K, and $\{Q_0, \ldots, Q_{K-1}\}$ in *class* $[C]$.

FIGURE III.8.20   Algorithm for finding component correspondence.

$M_e = 2$. When the second rule in Table III.8.4 is applied, the local structure composed of primitive sequences 2 and 3 of the subclass (a) is transformed to one primitive sequence P_0. We find out a relevant subset $\{Q_0\}$ of primitive sequences of the subclass (b). For $\langle 2, 1 \rangle \in \pi_a(2)$ and $\langle 2, 1 \rangle \pi_a(3)$, the PS-labels for P_0 is $\langle 6, 1 \rangle \in \pi_b(2)$ according to the transformation rule (Table III.8.4). Therefore, the two primitive sequences 2 and 3 in the subclass (a) can be transformed to the primitive sequences 2 in the subclass (e), i.e., $Q_0 = 2$. We check the rest of primitive sequences: 1 in the subclass (a) and 1 in (e). PS-label sets of both primitive sequences are equal, and therefore a mapping $f : \{1, 2, 3\} \to \{1, 2\}$ is found as $f(1) = 1$ and $f(2) = f(3) = 2$.

The transition diagram created by the above procedure is different from the one shown in Figure III.8.18. Since only one parent node C_0 is determined for each C and there is only one edge coming into C, the diagram is a tree (without loops). When there are two or more connected components in the diagram, the number of connected components is equal to the number of classes after the model construction procedure is operated. Figure III.8.21 shows the algorithm for generalizing subclasses into the root class in each tree in the transition diagram D.

We illustrate this algorithm for the path $(a) \to (c) \to (h)$ in Figure III.8.18. The subclass (h) corresponds to a leaf node in the transition diagram, and there is an arc $(c) \to (h)$ along with $N = 1$, $Q_0 = 1$, $f(1) = f(2) = 1$, and $t = (1, "h\text{-point}"; 2, "t\text{-point}")$. Then, in the subclass (c), T_c is updated to $\{(1, "h\text{-point}"; 2, "t\text{-point}")\}$, and the arc $(c) \to (h)$ and the subclass (h) are deleted. The subclass (c) now corresponds to a leaf node in the transition diagram, and there is an arc $(a) \to (c)$ along with $N = 1$, $Q_0 = 1$, $f(1) = f(2) = 1$, $f(3) = 2$, and $t = (1, "head"; 2,$

Algorithm Generalization(D)

while there is some arc in D

- **for each** leaf subclass C in D

  - $C_0 \leftarrow$ subclass that has an arc $C_0 \rightarrow C$.
  - Retrieve K, Q_i $(i = 0, \ldots, K-1)$, f, and the applied transformation t from subclass C. (They are recorded in subclass C in procedure ps_config.)
  - For each primitive sequence $k \in \{1, \ldots, M_{C_0}\}$, if $f(k) \neq Q_0$ and $f(k) \neq Q_{K-1}$ then the PS-label set $\pi_{C_0}(k)$ is replaced with $\pi_{C_0}(k) \cup \pi_C(f(k))$.
  - The statistics Σ_{C_0} of parameters (means and standard deviations) are updated for each primitive sequence based on mapping f.
  - Add elements in Γ_C to Γ_{C_0} (the set of possible connections or concatenations of end points of primitive sequences).
  - Add transformation t to Γ_{C_0}.
  - Delete arc $C_0 \rightarrow C$ and subclass C.

FIGURE III.8.21   Generalization algorithm.

"h-point"). Then, in the subclass (a), $\pi_a(3)$ is updated to $\{21, 30, 31, 40, 41, 51\}$ by taking the union of $\pi_a(3)$ and $\pi_c(2)$. T_a is first replaced with the union of T_a and T_c, and it is updated to $\{(1, \text{"tail"}; 3, \text{"}t\text{-point"})\}$. Note that indices of primitives and locations of points need to be changed based on the mapping f and the applied transformation. Furthermore, t is added to T_a, and T_a is updated to $\{(1, \text{"tail"}; 3, \text{"}t\text{-point"}), (1, \text{"head"}; 2, \text{"}h\text{-point"})\}$. The arc $(a) \rightarrow (c)$ and the subclass (c) are deleted. Therefore, the three subclasses (a), (c), and (h) are generalized to the class $B_a = (3, \pi_a, T_a, \phi, \Sigma_a)$, where $\pi_a(1) = \{10, 11\}$, $\pi_a(2) = \{21, 30, 31, 43, 40\}$, $\pi_a(3) = \{21, 30, 31, 40, 41, 51\}$, $\{(1, \text{"tail"}; 3, \text{"}t\text{-point"}), (1, \text{"head"}; 2, \text{"}h\text{-point"})\}$. By applying this algorithm to the transition diagram, the class illustrated in Figure III.8.11 is created.

8.5   Discussion

Handwritten character recognition has been a main research subject in pattern recognition. The prime difficulty in research and development of the technology lies in the variety of shape deformation. Various methods such as contour analysis and background analysis have been proposed.[5] Since most of the practical methods utilize some kinds of distance measures or matching methods on the feature space in which the features are position dependent, the normalization on the image is indispensable for coordinate transformation. In other words, image normalization has been a main method to cope with shape deformation. However, it is difficult to know the extent of displacement of the image with respect to the standard coordinate system before recognizing it. It implies that malfunction of normalization may distort the shape and result in rejection or substitution error. Furthermore, model construction often relies on hand-tuning in structural methods.

We describe the shape of characters by two types of features, i.e., symbolic, qualitative, and discrete features (directional features and quasi-convexity/concavity), and statistical, quantitative, and continuous features (geometrical parameters such as position and size of primitive sequences). The former is regarded as dominant information, while the latter is secondary information attached to structural components. The structural description is size and translation invariant, and therefore normalization of image is not necessary for global feature extraction. Furthermore, statistical constraints for deformation can be put on the structural description in terms of the geometrical parameters, and the categories with the same structure can be discriminated by statistical analysis of geometrical parameters. The structural description and parameterization are simple and agree with human intuition. Therefore, we can construct hybrid recognition systems[10] in cooperation with statistical methods based on the structural description and the geometrical parameters.

Since different strokes were written independently, they should have independent information and their transformations should be independent of each other. Our recognition method is close to on-line character recognition techniques, although primitive sequences have nothing to do with time sequences of handwriting.

Furthermore, a systematic and rigorous description scheme of shapes is a basis for systematic clustering of character shapes. Automatic construction of structural models has been one of the most difficult problems in pattern recognition. As no "perfect" recognition system exists now, it is too idealistic and impractical to look for perfect learning procedures. Therefore, models should be simply structured and easy for humans to understand intuitively so that models can be edited manually. Our approach satisfies this requirement and we would also stress that the compact, flexible, and rigorous description scheme makes it possible. As Winston mentions,[21] *learning from examples strongly depends on good descriptions*. We have developed a new method for clustering character shapes in terms of structural features, and applied it to automatic model inference.

References

1. Camillerap, J., Lorette, G., Menier, G., Oulhadj, H., and Pettier, J.C., Off-line and on-line methods for cursive handwriting recognition, in *From Pixels to Features: Frontiers in Handwriting Recognition*, Impedovo, S. and Simon, J.-C., Eds., Elsevier, Amsterdam, 1992, 273–287.
2. Haralick, R.M. and Shapiro, L.G., The consistent labeling problem: part 1, *IEEE Trans. Pattern Analysis and Machine Intelligence*, 1, 173–184, 1979.
3. Lam, L., Lee, S.W., and Suen, C.Y., Thinning methodologies: a comprehensive survey, *IEEE Trans. Pattern Analysis and Machine Intelligence*, 14, 879–885, 1992.
4. Mori, S., Suen, C.Y., and Yamamoto, K., Historical review of OCR research and development, *Proc. IEEE*, 80, 1029–1058, 1992.
5. Mori, S., Nishida, H., and Yamada, H., *Optical Character Recognition*, Wiley, New York, 1999.
6. Nash, C. and Sen, S., *Topology and Geometry for Physicists*, Academic Press, London, 1983.
7. Nishida, H., Model-based shape matching with structural feature grouping, *IEEE Trans. Pattern Analysis and Machine Intelligence*, 17, 315–320, 1995.
8. Nishida, H., A structural model of curve deformation by discontinuous transformations, *Graphical Models and Image Processing*, 58, 164–179, 1996.
9. Nishida, H., Automatic construction of structural models incorporating discontinuous transformations, *IEEE Trans. Pattern Analysis and Machine Intelligence*, 18, 400–411, 1996.
10. Nishida, H., Shape recognition by integrating structural descriptions and geometrical/statistical transforms, *Computer Vision and Image Understanding*, 64, 248–262, 1996.
11. Nishida. H. and Mori, S., Algebraic description of curve structure, *IEEE Trans. Pattern Analysis and Machine Intelligence*, 14, 516–533, 1992.
12. Nishida. H. and Mori, S., An algebraic approach to automatic construction of structural models, *IEEE Trans. Pattern Analysis and Machine Intelligence*, 15, 1298–1311, 1993.
13. Pavlidis, T., *Structural Pattern Recognition*, Springer-Verlag, New York, 1977.
14. Pavlidis, T., A vectorizer and feature extractor for document recognition, *Computer Vision, Graphics, and Image Processing*, 35, 111–127, 1986.
15. Poston, T. and Stewart, I., *Catastrophe Theory and Its Applications*, Pitman Publishing, London, 1978.

16. Rocha, J. and Pavlidis, T., A shape analysis model with applications to a character recognition system, *IEEE Trans. Pattern Analysis and Machine Intelligence*, 16, 393–404, 1994.

17. Simon, J.-C. and Baret, O., Regularities and singularities in line pictures, in *Structured Document Image Analysis*, Baird, H.S., Bunke, H., and Yamamoto, K., Eds., Springer-Verlag, Heidelberg, 1992, 261–280.

18. Suzuki, T. and Mori, S., Structural description of line images by the cross section sequence graph, *Int. J. Pattern Recognition and Artificial Intelligence*, 7, 1055–1076, 1993.

19. Thom, R., *Stabilité Structurelle et Morphogénése*, InterEditions, Paris, 1977.

20. Wakahara, T., Murase, H., and Odaka, K., On-line handwriting recognition, *Proc. IEEE*, 80, 1181–1194, 1992.

21. Winston, P.H., Learning structural descriptions from examples, in *The Psychology of Computer Vision*, Winston, P.H., Ed., McGraw-Hill, New York, 1975, chap. 5.

9

Novel Noise Modeling Using AI Techniques

Abdul Wahab
Nanyang Technological University

Chai Quek
Nanyang Technological University

Abstract. This chapter discusses the specific application of expert systems in modeling noise and signal spectral amplitude; in particular, two novel AI modeling approaches are used as modeling tools. They are hybrid fuzzy neural techniques and associative memories. Speech and image signals are often corrupted by additive noise and most of these perturbations are nonlinear. Thus, estimating and modeling them for signal enhancement is always a problem. Various modeling algorithms, especially those operating in the frequency domain, allow a very good understanding of the power spectral density (PSD) of the signal to be modeled. However, most of these spectral modeling techniques assume that the perturbation is either a Gaussian- or Gamma-type distribution, which may not be true in most practical cases. In addition,

0-8493-1121-7/03/$0.00+$1.50

information about the past is missing. However, it plays an important role in estimating the perturbation signals for nonlinear time variant (NLTV) systems.

The amplitude spectral estimator (ASE) for signal enhancement is used as an example in this chapter to show how modeling the perturbation with different AI techniques provides better performance for specific applications. The classical approaches toward noise modeling using Gaussian and Gamma distribution are unable to provide an accurate estimation of the noise and produce musical and clicking artifacts that are unacceptable for most applications. Post-processing of signals corrupted by the musical noise is very costly.

Alternative AI approaches to the modeling of noise and signal spectra is investigated in this chapter. Extensive comparison is made using the new AI approaches for the modeling of the perturbation; e.g., hybrid fuzzy neural techniques and associative memories. In hybrid fuzzy neural techniques, noise estimates are predicted on the basis of past information, while in associative memories the noise references are used to point to the closest associative memory location. This chapter discusses novel examples of the application of the pseudo output product (POP) hybrid fuzzy neural techniques and the associative memories using the cerebellar model articulation controller (CMAC) for noise modeling.

9.1   Introduction

The advancement of communication and network technology has allowed signals to be transmitted and received at a much lower bandwidth than its original baseband requirement. Image and speech signals are first transformed into digital sequences and transmitted via the Internet to be displayed or heard by the receiver using either a telephone modem, a high-speed cable modem, or a mobile telephony. In an ideal environment where the signal is noise-free, no other processing is required. Thus, compressing and expanding the image and speech data sequences will be straightforward. However, in most cases, the image and speech signals are corrupted by noise either from the environment or due to multiple path effect. A very good example of signal that is corrupted by both the environment and the multiple path situations is hands-free telephony in a vehicular environment.

9.1.1   Hands-Free Telephony

There is an urgent need for hands-free mobile telephony to be used in a vehicular environment, where drivers need not use an earplug or mouthpiece as it is today. This hands-free mobile telephone system should also allow telephone conversation involving passengers and drivers of other vehicles (teleconferencing) with minimum interference and allowing the far-end speaker to recognize each individual speaker and the speech. The hands-free mobile telephone system must remove all the perturbations that are present in the environment before any coding for compression can be undertaken prior to transmission. Some of these ills include man-made noise: the engine noise, the tire and road noise, and noise from nature, e.g., the wind and rain.[1,2]

Acoustic echo is generated due to the coupling between the loudspeaker and the microphone. In a hands-free environment, the speaker is not in physical contact with the terminal. The microphone picks up the far-end speech after traveling through the communication channel and, consequently, echo is heard at the far end. The length of the acoustic echo path depends on the chamber. For a saloon car with a sampling frequency of 8000 samples, the echo path would be on the order of hundreds of samples, while for a conference room this could have an acoustic echo path in the thousands of samples. Both can result in an empty room effect and an echo which is audible. Therefore, the computational complexity of the acoustic echo canceler may be very high and depends critically on the echo cancellation algorithm.

A tremendous amount of work has been done toward canceling these degradations (both the echo and noise) to improve the corrupted speech signal. Most of this algorithm uses the minimum mean

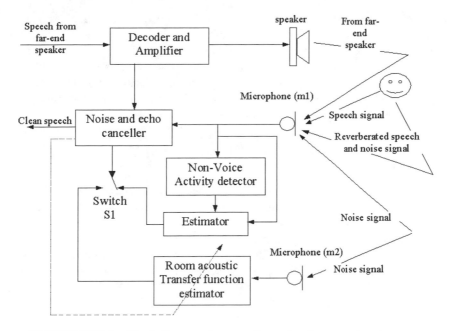

FIGURE III.9.1 Block diagram of a general speech enhancement system.

square error (MSE) algorithm to cancel the unwanted noise and echo signal.[3–7] All these techniques adopt the Weiner filter approach to cancel the unwanted signal from the corrupted signal and assume a reasonably good model of the actual unwanted signal. The estimate of the unwanted signal is derived from either a separate microphone or from a single microphone using the non-voice activity detector (NVAD) to determine the noise (unwanted) signals. The noise estimate, sometimes referred to as the reference signal, cannot be correlated to the speech signal to be recovered, as this will cause the recovered speech to be canceled as well. In addition, there are similar approaches for generating the Weiner filter directly using the neural network approach.[8] However, this requires a heavy computational load due to the structure of the neural network proposed. Basically, two main approaches are possible for recovering the clean speech from a corrupted speech; i.e., by modeling the acoustic transfer function of the room that created the perturbation, or by modeling the entire perturbation as a lump circuit as shown by Figure III.9.1.

The speech enhancement system consists of a two-microphone system, where one is used to capture the noise signal to be used as reference signal, and the other microphone captures the corrupted speech from the near-end speaker. At the same time, the signal coming from the far-end speaker is fed back into the microphone as echoes. The noise and echo canceler module of Figure III.9.1 cancels these noise and echo signals before transmitting to the far-end speaker. Microphone m2 is placed in such a position that only noise (or reference) signal is captured, while microphone m1 will capture all the other signals, including the speech, coming from the near-end speaker and noise coming from the environment and the vehicle. Switch S1 is required to switch the system between using dual microphones vs. a single-microphone system. In a dual-microphone system, both microphones (m1 and m2) will be used, with m1 only capturing the reference signal and m2 the corrupted speech. In the case of a single-microphone system, the non-voice activity detector (NVAD) is required to determine if a particular frame of data captured consists of voice or purely noise, so that a noise spectral estimate can be stored and used for future references. The single-microphone system normally creates musical artifacts due to the inaccuracy in estimating the noise estimate; but for positive signal-to-noise ratio, the processed speech is still comprehensible.

In this chapter, we introduce the amplitude spectral estimator in detail to allow an appreciation of the problems and concerns in its use. Various techniques based on the classical approach will also be introduced to demonstrate the complexity of the post-processing required to ensure a reasonably clean speech signal that could be recovered and be comprehensible by the far-end speaker.

Subsequently, an introduction to the cerebellar model articulation controller (CMAC) and a novel fuzzy neural network (POPFNN-CRI) in noise modeling, and some experimental results showing the noise-modeling base on various input algorithms will be presented.

9.2   The Amplitude Spectral Estimator (ASE)

In general, it is significantly easier to enhance speech from a signal degraded by additive noise by estimating the spectral amplitude associated with the original speech than to estimate both the amplitude and phase angle of the speech. It is the short-time spectral amplitude rather than phase that is important for speech intelligibility and quality. Consider a stationary random signal $s(n)$ that has been degraded by uncorrelated additive noise $d(n)$ with a known power spectral density. The power spectral density (PSD) of the signal can be easily estimated through spectral subtraction as shown by Equation III.9.1:

$$y(n) = s(n) + d(n)$$
$$P_y(\omega) = P_s(\omega) + P_d(\omega) \tag{III.9.1}$$

where $P_y(\omega)$ represents the PSD of $y(n)$ the corrupted speech signal, $P_s(\omega)$ represents the PSD of $s(n)$ the stationary random signal, and $P_d(\omega)$ represents the PSD of $d(n)$ the uncorrelated additive noise.

Assuming that all the signals are windowed and the original speech signals $s(n)$ are processed in a short-time basis, Equation III.9.1 can be re-expressed as the power spectrum given by Equation III.9.2:

$$|Y(\omega)|^2 = |S(\omega) + D(\omega)|^2$$
$$|Y(\omega)|^2 = |S(\omega)|^2 + |D(\omega)|^2 + S(\omega) \cdot D^*(\omega) + S^*(\omega) \cdot D(\omega) \tag{III.9.2}$$

where $Y(\omega)$ is the Fourier transform of the windowed corrupted speech signal $y(n)$, $S(\omega)$ is the Fourier transform of the windowed original speech signal $s(n)$, $D(\omega)$ is the Fourier transform of the windowed noise signal $d(n)$, $D^*(\omega)$ is the complex conjugate of $D(\omega)$, and $S^*(\omega)$ is the complex conjugate of $S(\omega)$.

Since the original speech and the noise signals are uncorrelated and the noise signal has zero mean, the cross terms between the original speech signal and the noise signal $S(\omega) \cdot D^*(\omega)$ and $S^*(\omega) \cdot D(\omega)$ are zero. Thus, the power spectrum of the corrupted speech $|Y(\omega)|^2$ in Equation III.9.2 can be expressed as the sum of the original speech power spectrum $|S(\omega)|^2$ and the noise power spectrum $|D(\omega)|^2$.

The objective of the speech enhancement is to estimate the $|\hat{S}(\omega)|$ of $|S(\omega)|$ and from Equation III.9.2 we note that by applying the minimum mean square error (MMSE), the speech signal can be estimated as given in Equation III.9.3:

$$|\hat{S}(\omega)|^2 = |Y(\omega)|^2 - E[|D(\omega)|^2] \tag{III.9.3}$$

where $E[\cdot]$ denotes the expectation or ensembled average, $|Y(\omega)|^2$ is the power spectrum of the corrupted speech captured by microphone m1, and $|D(\omega)|^2$ is the noise power spectrum.

From Figure III.9.1, with microphone m2 capturing the noise (reference) signal, the cleaned speech of Equation III.9.3 can be implemented using the dual-microphone method. In the case of the single-microphone system, the noise power spectrum can be estimated during speech pauses.

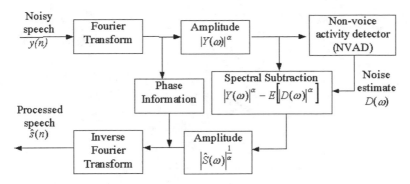

FIGURE III.9.2 Block diagram of the generalized amplitude spectral subtractor for speech enhancement.

9.2.1 Generalized Form of the ASE

The general form of Equation III.9.3 is expressed in Equation III.9.4:

$$\hat{s}(n) = F^{-1}[(|Y(\omega)|^{\alpha} - \beta \cdot E[|D(\omega)|^{\alpha}])^{\frac{1}{\alpha}}] \qquad (\text{III.9.4})$$

where α and β are the two parameters considered for generalization.[9,10]

Figure III.9.2 shows the block diagram of the spectral subtraction method described by Equation III.9.4. The input noisy speech is first transformed into frequency domain using Fourier transformation. Since phase information does not affect hearing quality, only the amplitude of the corrupted speech signal will be processed. At each short time of the corrupted speech segment, the processed speech is extracted from the corrupted speech power spectrum by subtracting the noise spectral density estimated using the NVAD (Figure III.9.2). The negative value of the right-hand side of Equation III.9.4 can be dealt with by taking either a half-wave or a full-wave rectification. At the output, the phase information from the corrupted speech together with the amplitude of the processed speech will be passed through the inverse Fourier transform to recover the original speech in the time domain. Another variation to Equation III.9.4 for speech enhancement[11,12] uses the maximum likelihood instead of the MMSE method in relation to the amplitude spectral subtraction. Other frequency-domain transforms approach were also possible, based on the discrete cosine or sine transform.

9.2.2 Weiner Filter Approach

Different techniques discussed thus far in estimating the spectral amplitude of the speech attempt to enhance the speech-to-noise (S/N) ratio without affecting the high S/N ratio and attenuating the low S/N ratio. In the Weiner filter approach, speech enhancement is achieved by enhancing the S/N ratio. Consider the spectral speech estimate $\hat{S}(\omega)$ in the form of a zero-phase frequency response $H(\omega)$ applied to the corrupted speech input $Y(\omega)$ given by Equation III.9.5:

$$H(\omega) = \frac{\hat{S}(\omega)}{Y(\omega)} \qquad (\text{III.9.5})$$

Substituting Equation III.9.4 into Equation III.9.5 and letting $\alpha = 2$, the zero-phase frequency response can then be re-expressed as Equation III.9.6:

$$H(\omega) = \left(\frac{|Y(\omega)|^2 - \beta \cdot E[|D(\omega)|^2]}{|Y(\omega)|^2} \right)^{\frac{1}{2}} \qquad (\text{III.9.6})$$

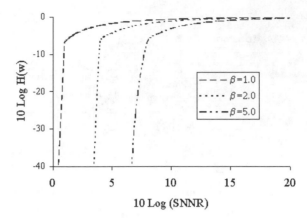

FIGURE III.9.3   Attenuation characteristics for negative β greater than 1.

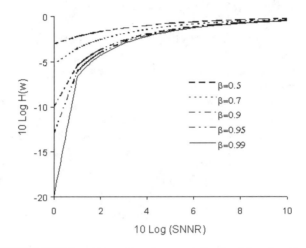

FIGURE III.9.4   Attenuation characteristics for β less than 1.

This simplifies to Equation III.9.7:

$$H(\omega) = \left(\frac{SNNR - \beta}{SNNR}\right)^{\frac{1}{2}} \tag{III.9.7}$$

where

$$SNNR = \frac{|Y(\omega)|^2}{E[|D(\omega)|^2]} \tag{III.9.8}$$

The SNNR expressed in Equations III.9.7 and III.9.8 is the signal-plus-noise-to-noise power ratio at each frequency (ω) of the short-time speech segment. Only positive values of Equation III.9.8 will be considered. Equation III.9.7 forms the Weiner filter in the speech enhancement process, and by applying Equation III.9.5 after deriving the Weiner filter $H(\omega)$ from Equation III.9.7, the processed speech can easily be determined using the inverse Fourier transform. Figures III.9.3 and III.9.4 show the attenuation curve of the zero-phase frequency response with respect to the SNNR. A lower SNNR will yield a higher attenuation between the input and output. For β greater than one (Figure III.9.3), the processed speech is being attenuated for more than 40 dB if the SNNR is less than 2 dB. But for corrupted speech signals with negative SNNR, a β value of less than one will be required (Figure III.9.4).

Thus, the choice of β from Figures III.9.3 and III.9.4 of Equations III.9.6, III.9.7, and III.9.8 will depend on the SNNR of the input signal. If SNNR is high, then $\beta > 1$ will suffice; but if the SNNR of the corrupted signal is low, the speech enhancement can only be effective with values of β less than 1. Much work has been done to extend the idea of using the MMSE short-time amplitude spectral estimator approach in speech enhancement.[13–15] Ephraim and Malah's (1984) original paper proposed the use of noise and speech variances from the speech frame and the corrupted speech input, respectively, to estimate the signal-to-noise ratio by assuming that the noise is of a white Gaussian nature and the cleaned speech to be a zero-mean white Gaussian process. Ephraim and Malah[13] used the SNNR based on the variance estimate to derive the Weiner filter transfer function. The problem with the use of variances as the noise estimate is that one would need to compute the NVAD. In order to have a better estimate of the noise and the speech variances, one would need to process the corrupted speech a number of times similarly with the noise during the non-voice activity. Extending Equations III.9.6, III.9.7, and III.9.8 on the basis of uncorrelated stationary random process using their respective power spectral density given by Equation III.9.1, the linear estimator of $s(n)$ that minimizes the mean square error can be obtained by filtering $y(n)$ with the noncausal Wiener filter frequency response. This is given by Equation III.9.9:

$$H(\omega) = \frac{P_s(\omega)}{P_s(\omega) + P_d(\omega)} = \frac{E[|S(\omega)|^2]}{E[|S(\omega)|^2] + E[|D(\omega)|^2]} \tag{III.9.9}$$

Beaugeant (1998), Pascal (1996), and Jeannès (1996) proposed the use of the power spectral density (PSD) of the corrupted speech,[14–16] the noise during the non-voice and the estimated speech, to derive the Wiener filter and this is described in Equation III.9.10:

$$H(\omega) = \frac{P_s(\omega)}{P_s(\omega) + P_d(\omega)} = \frac{1}{1 + \frac{1}{SNR_{Total}}} \tag{III.9.10}$$

where

$$SNR_{Total} = \frac{P_s(\omega)}{P_D(\omega)}$$

The SNR_{Total} contains both the *a priori* and the *a posteriori* signal-to-noise ratios.

In an analysis of Equation III.9.1 between the interval $[0, \ T]$, the kth spectral components of the input signal, $s(t)$, the noise, $d(t)$, and the corrupted signal, $y(t)$, can be represented by $S_k \overset{\Delta}{=} |S_k| < \alpha_k$, D_k, and $Y_k \overset{\Delta}{=} |Y_k| < \vartheta_k$, respectively. On the other hand, the observed signal $\{y(t), 0 \leq t \leq T\}$ can be rewritten in terms of its spectral components Y_k given by Equation III.9.11:

$$y(t) = \lim_{K \to \infty} \sum_{k=-K}^{K} Y_k \exp\left(j\frac{2\pi}{T}kt\right) dt \quad k = 0, \pm 1, \pm 2, \ldots \tag{III.9.11}$$

Based on the Gaussian statistical model for spectral components, Equation III.9.11 converges to $y(t)$ for every $t \in [0, T]$. This means that the minimum mean square error (MMSE) estimation problem can be reduced to the estimation of $\hat{A}_k$. Furthermore, since the spectral components are assumed to be statistically independent, the amplitude spectral estimator only needs Y_k, as shown by Equation III.9.12:

$$|\hat{S}_k| = E[|S_k||Y_k], \hat{A}_k \approx \frac{\xi_k}{1 + \xi_k} \cdot \gamma_k \cdot Y_k$$

$$\xi_k = \frac{E[|\hat{S}_k|^2]}{E[D_k^2]} \approx \frac{\text{var}(|\hat{S}_k|)}{\text{var}(D_k)}$$

$$\gamma_k = \frac{E[Y_k^2]}{E[D_k^2]} \approx \frac{E[(|S_k| + D_k)^2]}{\text{var}(D_k)} \approx \frac{E[|S_k|^2]}{\text{var}(D_k)} + 1 \qquad \text{(III.9.12)}$$

$$G_{MMSE} = \frac{\hat{S}_k}{Y_k} = \frac{\xi_k}{1 + \xi_k} \cdot \gamma_k$$

where $\hat{S}_k$ = estimate of the original speech based on the Bayer's theorem, $E[\cdot]$ = expectation or ensembled average, ξ_k = *a priori* signal-to-noise ratio, γ_k = *a posteriori* signal-to-noise ratio, D_k = noise amplitude, and var$(\cdot)$ = variance.

Notice from Equation III.9.12 the gain of the system (G_{MMSE}) is defined equivalently as $H(\omega)$ in Equations III.9.5, III.9.9, and III.9.10. In Equation III.9.12 the noise was estimated using the variances of the lumped noise signal as indicated by Ephraim and Malah.[13]

9.2.3   Speech Enhancement Using the Spectral Subtraction Based on SNR

Given both the *a priori* SNR and the a *posteriori* SNR from Equation III.9.12 of the short-time Fourier transform, the total signal-to-noise ratio (SNR_{Total}) can be generalized as shown by Equation III.9.13:

$$SNR_{post} = \left| \frac{|Y(k, \omega)|^2}{E[|D(\omega)|^2]} - 1 \right|$$

$$SNR_{priori} = \frac{|H(k - \lambda, \omega)Y(k - \lambda, \omega)|^2}{E[|D(\omega)|^2]} = \frac{|\hat{S}(k - \lambda, \omega)|^2}{E[|D(\omega)|^2]} \qquad \text{(III.9.13)}$$

$$SNR_{Total} = SNR_{post}(1 - \beta) + SNR_{priori}\beta$$

where β can take a value from 0 to 1, $k = 1, 2, 3, 4 \ldots$, and λ is the number of overlapping samples.

Substituting Equation III.9.13 into Equation III.9.10, the speech enhancement system described by Figure III.9.5 can be realized. The speech enhancement is only carried out once every frame rather than every sample. An overlap filter is necessary to avoid any kind of musical tone or clicking noise side effects, especially during the silent period.

Notice the similarity of Figures III.9.5 and III.9.2 by which both process the corrupted speech in frequency domain and the phase information is left untouched. The spectral subtraction in Figure III.9.5 is then carried out by the Weiner filter.[17] The *a posteriori* signal-to-noise ratio is derived from the short-time Fourier transform of the corrupted input and the noise of microphones m1 and m2, respectively. The *a priori* signal-to-noise ratio is derived from the delayed version of the short-time Fourier transform of the estimated original speech and the noise spectrum. Once the

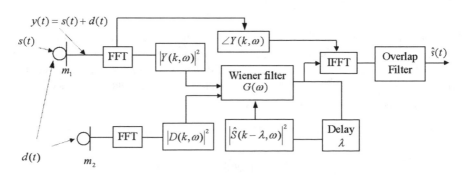

FIGURE III.9.5   Block diagram of the noise canceler using the ASE algorithm.

SNR_{Total} has been calculated based on Equation III.9.13 and substituting into Equation III.9.10, the Weiner filter can subsequently be derived. The estimated original speech is subsequently passed through an overlap filter to remove unwanted musical artifacts.

9.2.4   The Overlap Filter

The processed data sequence $s(t)$ is given by:

$$s(t) = \cdots + [s_{(n-1)N+1+\lambda}, s_{(n-1)N+2+\lambda}, \ldots, s_{nN}, s_{nN+1}, \ldots, s_{nN+\lambda}]^T$$

$$+ \cdots + [s_{nN+1}, \ldots, s_{(n+1)N}]^T + \cdots + [s_{(n+1)N-\lambda+1}, \ldots s_{(n+1)N},$$

$$s_{(n+1)N+1}, \ldots, s_{(n+2)N-\lambda}]^T + \cdots \qquad (\text{III}.9.14)$$

where N = number of samples per block of frame that is being processed, n = frame number, λ = number of samples that is overlapped between the present and the previous frame, and T = transpose of the matrix.

The first set of data in Equation III.9.14 represents the previous frame of data followed by the present frame, and lastly the next frame of data. It can be seen from Equation III.9.14 that when there is no overlap (i.e., $\lambda = 0$) the string of data $s(t)$ is just a data sequence with N number of samples per frame of data. There are many different ways to handle the overlap data sequence and one of the simplest forms is the averaging method. In the averaging method the overlap sequences from the previous and the present frames are smoothed out just by taking the average of the two overlap frames.

Consider the sequence of data that overlap in previous frame $s_{N-\lambda}^{f-1}, s_{N-\lambda+1}^{f-1}, \ldots, s_N^{f-1}$ and the present frame $s_{N-\lambda}^f, s_{N-\lambda+1}^f, \ldots, s_N^f$. Equation III.9.14 can be rewritten within the overlap data samples as shown by Equation III.9.15:

$$[s_{N-\lambda}^{f-1}, s_{N-\lambda+1}^{f-1}, \ldots, s_N^{f-1}] \cdot M1 + [s_{N-\lambda}^f, s_{N-\lambda+1}^f, \ldots, s_N^f].M2 \qquad (\text{III}.9.15)$$

where $M1$ and $M2$ are $\lambda \times \lambda$ diagonal matrices defined by:

$$M1 = \begin{bmatrix} m_{11} & 0 & \cdots & 0 \\ 0 & m_{22} & \cdots & 0 \\ \vdots & \cdots & \ddots & \vdots \\ 0 & 0 & \cdots & m_{\lambda\lambda} \end{bmatrix} \text{ and } M2 = \begin{bmatrix} 1-m_{11} & 0 & \cdots & 0 \\ 0 & 1-m_{22} & \cdots & 0 \\ \vdots & \cdots & \ddots & \vdots \\ 0 & 0 & \cdots & 1-m_{\lambda\lambda} \end{bmatrix}$$

$$(\text{III}.9.16)$$

where m_{xx} defines different shapes of smoothing functions and can only take value from 0 to 1.

Equations III.9.17, III.9.18, and III.9.19 describe the different shapes for the smoothing algorithms. To achieve a triangular function, m_{xx} is defined by Equation III.9.17:

$$m_{xx} = 1 - \frac{x}{\lambda}, \text{ for } x = 1, \ldots, \lambda \qquad (\text{III}.9.17)$$

The cosine smoothing function is described by m_{xx} of Equation III.9.18:

$$m_{xx} = \frac{1 + \cos(\frac{x}{\lambda}\pi)}{2}, \text{ for } x = 1, \ldots, \lambda \qquad (\text{III}.9.18)$$

and for exponential decay, m_{xx} is defined by Equation III.9.19:

$$m_{xx} = \frac{e^{-(\frac{x}{\lambda}\pi)} - e^{(-1)}}{1 - e^{(-1)}}, \text{ for } x = 1, \ldots, \lambda \qquad (\text{III}.9.19)$$

From Equation III.9.13 the computation for the SNR requires only the ratio of the amplitude spectrum. Any reasonable form of the frequency translation should be able to provide a sufficiently good representation of the power spectrum ratio for both the *a posteriori* and *a priori* SNR. Experiments were carried out using three different transforms; i.e., the fast Fourier transform (FFT), discrete cosine transform (DCT), and discrete sine transform (DST). In the case of the DCT or the DST, the phase angle is not required and the sampling frequency will need to be doubled.

9.3   Experimental Results

Actual field recordings on a moving vehicle were superimposed with a prerecorded speech to produce a corrupted observation with an average SNR in the range 5 to –20 dB. Experiments were carried out to study the noise cancellation effect by using:

- The LMS algorithm with automatic step-size adjustment to ensure convergence
- The amplitude spectral estimator based on the PSD and the noise variances estimate

An experiment was conducted to study the performance of the amplitude spectral estimator (ASE) and its comparison with the least mean square (LMS) algorithm. The performance of the ASE algorithm was compared with the LMS algorithm using simulated data and prerecorded data from a vehicle, respectively.

9.3.1   ASE and LMS Performance Comparison

Prerecorded speech and noise data were added to form a corrupted speech with SNR of -10 dB. The two-microphones algorithm was employed as shown by Figure III.9.5. The corrupted speech simulates a signal captured by microphone m1 and the noise-only by microphone m2. The performance measure is based on the signal-to-noise ratio of the processed speech as compared to that of the corrupted speech. The signal-to-noise ratio of the corrupted speech in microphone m1 is also varied to study the performance differences between the ASE and LMS algorithms. Microphone m2 also uses the prerecorded noise data, but this time is added with the same original speech of lower amplitude. Two different experiments were conducted with microphone m2 having SNR of -19 and -22 dB.

For the LMS algorithm, taps size was 512 with an initial step-size of 0.5. For the amplitude spectral estimator (ASE), the number of taps was set at 512 with 25% overlap and α was set at 0.85. Table III.9.1 shows that for both the LMS and ASE algorithms the speech enhancement performance is very good when the corrupted speech is greater than -5 dB, Notice also from Table III.9.1 that when

TABLE III.9.1   Performance Comparison between the ASE and LMS Algorithms

Algorithm	Reference Input SNR	Corrupted Speech SNR			
		-15 dB	-10 dB	-5 dB	0 dB
		Output Processed Speech SNR (dB)			
ASE	-22 dB	+10	+10	+10	+10
	-19 dB	+7	+10	+10	+10
LMS	-22 dB	-9	+6	+8	+10
	-19 dB	-3	+4	+7	+10

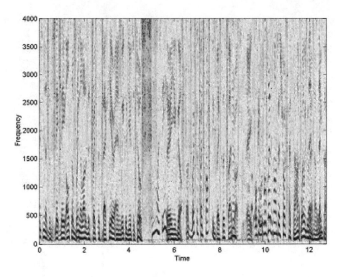

FIGURE III.9.6   3D spectrogram plot of the original speech signal.

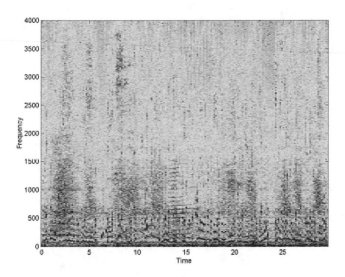

FIGURE III.9.7   3D spectrogram plot of the corrupted signal.

the corrupted speech SNR is less than $-15\,$dB, the LMS algorithm fails to remove the unwanted noise signals, while the ASE algorithm is still able to cancel most of the noise. In addition, the performance of the ASE improves if the reference signal in microphone m2 is only a pure noise and the signal is not correlated to the original speech.

Figures III.9.6 and III.9.7 show the spectrogram plots of the original speech and corrupted speech with SNR $= -10$ dB, respectively. It can be seen that most of the noise occurs at lower frequencies of up to 1000 Hz. For frequencies above 1500 Hz, the speech signal that is corrupted with very large noise amplitude is still observable. In the case of the corrupted speech processed using the LMS algorithm, the result of the processed speech spectrogram (Figure III.9.8) still shows some amount of noise data. In addition, the LMS algorithm also produces echo and can be observed from the spectrogram plot of Figure III.9.8 as repeated lines close to each other. In the case of the ASE algorithm (Figure III.9.9), most of the noise data has been cancelled. The original

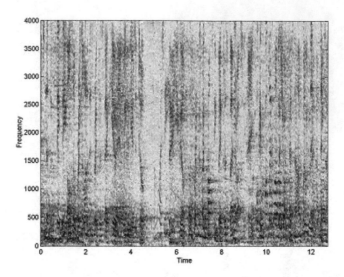

FIGURE III.9.8   3D spectrogram plot of enhanced speech using the LMS algorithm.

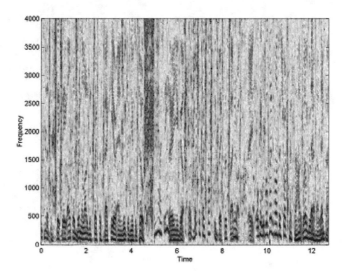

FIGURE III.9.9   3D spectrogram plot of enhanced speech using the ASE algorithm.

speech recovered, which is very close to the spectrogram of the original speech, is shown in Figure III.9.6.

When the SNR of the observation was forced to -20 dB, i.e., in a very noisy vehicle chamber, the LMS algorithm failed to converge. But the performances based on ASE were reasonably good and yielded speech signal still recognizable as shown by the 3D plot of the corrupted and processed speech periodogram (Figures III.9.10, III.9.11, and III.9.12). In Figure III.9.10, the noise amplitude is sufficiently high such that all other speech signal was suppressed. The LMS algorithm processed the corrupted speech and managed to remove the low-frequency noise, but at the same time other noise signals were also generated (Figure III.9.11). In Figure III.9.12 the ASE algorithm suppressed most of the noise, but produced some other musical artifacts. The processed speech for the LMS is not recognizable while that of the ASE algorithm is still recognizable.

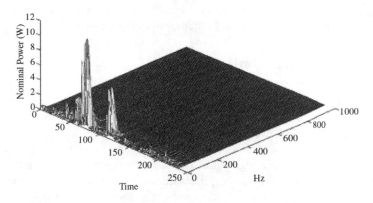

FIGURE III.9.10   3D periodogram plot of the corrupted speech with SNR $= -20$ dB for frequencies between 1 and 1000 Hz.

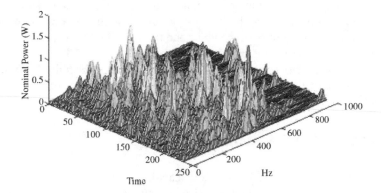

FIGURE III.9.11   3D periodogram plot of the processed speech using the LMS algorithm for frequencies between 1 and 1000 Hz.

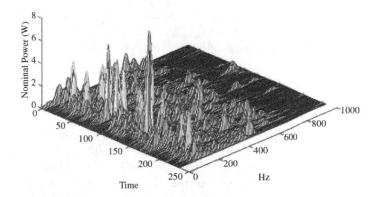

FIGURE III.9.12   3D periodogram plot of the processed speech using the ASE algorithm for frequencies between 1 and 1000 Hz.

   It can be seen from the plots of Figure III.9.10 that the noise signal is far too dominant and the speech signal cannot be traced in the 3D periodogram plots. In the case of the LMS algorithm (Figure III.9.11), the noise level was very significant and as a result, other degrading artifacts were audible at the output of the Wiener filter. But the enhanced speech at the output of the ASE was able to produce a cleaner speech that was easily recognizable (Figure III.9.12).

9.4   ASE Performance Limitation

The ASE speech enhancement scheme has been shown to be able to enhance speech or any other signal sequences that have been corrupted by additive noise. The performance of the ASE algorithm is comparable to that of the LMS algorithm at high SNR of greater than -2 dB, but for low SNR of less than -10 dB the ASE performs better. In addition, the echo effect that occurs in the LMS algorithm is not present in the ASE algorithm. The noise distribution is assumed to be Gaussian and the speech enhancement needs only to be processed once per block of data sequence. In order to remove the musical artifact and other clicking noise, the smoothing overlap filter was introduced. An overlapping sequence of less than 60% should be sufficient to remove the unwanted artifacts.

However, one major problem still exists in estimating the noise. Noise data from one frame to another does change and the distribution may not be Gaussian. The accuracy of the non-voice activity detector for a noisy environment may be in question, thus resulting in a wrong noise estimate that may have the noise distribution of both the noise and the speech signal itself.

Our findings from the experiments clearly indicate that the ASE-DCT algorithm is a viable technique for speech enhancement. The variance technique to estimate the noise PSD shows robustness of the system performance, but musical noise effect exists due to the inaccuracy of the signal-to-noise ratio estimates.

9.5   Cerebellar Model Articulation Control

Figure III.9.13 shows the CMAC block diagram. This is similar to the specialized on-line learning control architecture. The CMAC memory consists of a two-dimensional array that stores the value of $x_n(kT)$ as the content of an element in the array with coordinates i, j.[8] The coordinates i, j are derived by quantizing the reference input $y_{ref}(kT)$ and plant output $y_p(kT - T)$. The quantization process is described in Equation III.9.20, where k represents a discrete step and T represents the sampling period. During the initial operation, the plant derives almost all its control input from the classical controller, while the CMAC memory is initialized to zero. During each subsequent control step, the classical control actuation signal $x_c(kT)$ is used to build the CMAC characteristic surface:

$$Q(y(kT)) = \left\lfloor \frac{n(y(kT) - y_{min})}{y_{max} - y_{min}} \right\rfloor \tag{III.9.20}$$

where y_{max} = maximum value of y, y_{min} = minimum value of y, and n = resolution of CMAC memory.

The CMAC memory can also be visualized as a neural network consisting of a cluster of two-dimensional *self-organizing neural networks* (SOFM). However, instead of a random initialization of the neural net weights, they are fixed such that they form a two-dimensional grid (Figure III.9.14).

The winning neuron in the CMAC memory at time step k is identified as the neuron with weights $Q(y_{ref}(kT))$ and $Q(y_p(kT - T))$ given the inputs $y_{ref}(kT)$ and $y_p(kT - T)$. The weights are effectively the coordinates i, j of the location of the neuron in the SOFM. The output of the winning neuron can be directly obtained from the weight $w_{i,j}$ of the output neuron.

CMAC learning is a competitive learning process that is similar to the Kohonen and the SOFM learning rules. However, since the weights of the cluster of neurons that represent indices to the CMAC memory are fixed, learning only occurs in the output neuron. The CMAC learning rule is based on the Grossberg competitive learning rule and is applied only to the output layer. No

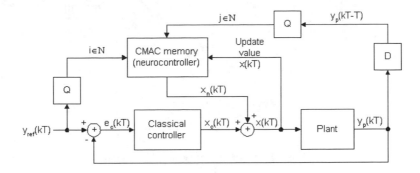

FIGURE III.9.13 CMAC block diagram.

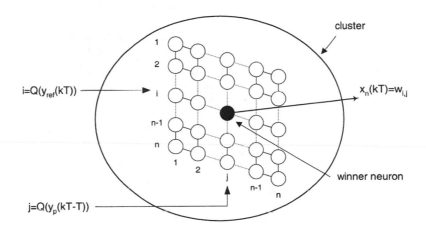

FIGURE III.9.14 Architecture of CMAC memory.

competitive Kohonen learning rule is applied to the input layer. The CMAC learning rule can therefore be represented by Equation III.9.21:[18]

$$i = Q(y_{\mathrm{ref}}(kT)), \ j = Q(y_p(kT - T)); \ i, j \in N$$

$$w_{i,j}^{(k+1)} = w_{i,j}^{(k)} + \lambda(x(kT) - w_{i,j}^{(k)}) \tag{III.9.21}$$

where λ = learning constant, $x(kT)$ = plant input at discrete step k, $y_{\mathrm{ref}}(kT)$ = reference input at discrete step k, $y_p(kT - T)$ = plant output at discrete step $k-1$, $w_{i,j}^{(k)}$ = contents of CMAC cell with coordinates $i, \ j$ at discrete step k, and $Q(\cdot)$ = the quantization function defined in Equation III.9.20.

9.5.1 Modified CMAC

It is difficult to train the CMAC memory because the plant characteristic surface has to be learned while a classical model controller controls the plant. For a particular control setting, the plant output typically follows a certain trajectory. Hence, only the weights of the output neurons (array cells) visited by the path of this trajectory are updated. Therefore, the training of the CMAC has to be carefully planned. This poses a problem in many control applications that require on-line learning in which the control rules are not readily available. The *Modified CMAC* (MCMAC) architecture has been proposed[19] to overcome this limitation by using the plant closed loop error $e_c(kT)$ and plant output $y_p(kT)$ as the models in the training. This allows

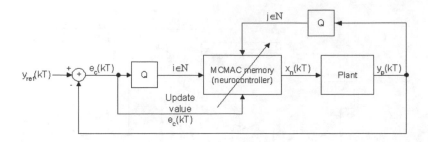

FIGURE III.9.15   MCMAC block diagram.

on-line training as well as ease in the planning of the training trajectory. The training path can now be more directly controlled using the reference plant input $y_{ref}(kT)$. Figure III.9.15 shows the MCMAC block diagram. This architecture is similar to the CMAC, except that the quantized closed loop error $e_c(kT) = y_{ref}(kT) - y_p(kT)$ and plant output $y_p(kT)$ are the indices to the 2D MCMAC memory array instead of the original $y_{ref}(kT)$ and $y_p(kT-T)$ in CMAC. The MCMAC learning rule is given in Equations III.9.22 through III.9.25:

$$m = Q(e_c(kT)), n = Q(y_p(kT)); m, n \in N$$

$$W_{m,n}^{(k+1)} = W_{m,n}^{(k)} + \lambda e_c(kT) \tag{III.9.22}$$

$$w_{m,n}^{(k+1)} = w_{m,n}^{(k)} + h_{m,n}(\lambda(1-\alpha)e_c(kT) + \alpha \Delta w_{m,n}^{(k)})$$

$$\text{for } |m - i| \leq N, |n - j| \leq N; i, j \in N \tag{III.9.23}$$

$$h_{m,n} = e^{(\frac{-|r_{m,n}-r_{i,j}|^2}{2\sigma^2})} \tag{III.9.24}$$

$$\Delta w_{m,n}^{(k+1)} = w_{m,n}^{(k+1)} - w_{m,n}^{(k)} \tag{III.9.25}$$

where λ = learning constant, α = momentum constant, $y_{ref}(kT)$ = reference input at discrete step k, $y_p(kT)$ = plant output at discrete step k, $e_c(kT)$ = closed loop error $y_{ref}(kT) - y_p(kT)$ at discrete step k, N = neighborhood constant, $|r_{m,n} - r_{i,j}|$ = distance between cell with coordinates m, n and cell with coordinates i, j, σ = radius of the Gaussian distribution, $h_{m,n}$ = neighborhood function for neuron with coordinates m, n, $w_{m,n}^{(k)}$ = contents of MCMAC cell with coordinates m, n at discrete step k, and $Q(\cdot)$ = the quantization function defined in Equation III.9.20.

Both CMAC and MCMAC attempt to model the plant's characteristics on the basis of the input indices. The difference between the two is that the former learns using the plant input $x(kT)$ and output $y_p(kT)$, while the latter learns using the closed loop error $e_c(kT)$ and plant output $y_p(kT)$. Hence, MCMAC does not require an inverse model of the plant. In addition, MCMAC does not require a classical controller to be operational during the learning phase. It uses the closed loop error and the plant output as the model. This has an added advantage in the determination of the training trajectory through the control of the reference input $y_{ref}(kT)$.

Although the modification to CMAC in MCMAC has removed the need for a classical controller, the training of MCMAC still requires careful planning such that all the cells in the MCMAC memory are visited. The contents of the MCMAC memory represent the plant characteristic to be controlled by the neurocontroller. The MCMAC memory can be visualized using a 3D characteristic surface. The axes of this MCMAC contour surface consist of the cell indices m, n (namely $Q(e_c(kT))$ and $Q(y_p(kT))$) and the content value of each cell. This is subsequently used to analyze the training rate of both the MCMAC learning rule and the improved learning rule.

9.6 MCMAC for Generic Amplitude Spectral Estimator

As presented earlier, the amplitude spectral estimator (ASE) employs the short-time Fourier transform (STFT) method to estimate the power spectral density for both the reference input and the noisy speech. Due to the back-to-back configuration of the stereo microphone pair (Figure III.9.5), the channel facing the speaker is expected to contain mostly the noisy speech, whereas the opposite channel will be primarily the reference signal and the secondary echo reflected from the windshield. Understandably, there will be some portion of the reference signal in the front channel as well, due to reflection from the chamber walls, the primary acoustic echo.

We also note that the process of corrupting the speech data in a vehicular environment is not linear or just purely additive in nature. In most cases the relationship between the original speech and the noise is convolution instead of addition where the corrupted speech can be represented by $y(t) = s(t)^*d(t)$, where $s(t)$ and $d(t)$ are the original speech and the noise signal, respectively, and $*$ indicates convolution. Since the system is not linear, and Fourier analysis and filtering cannot be applied directly, a simple LMS or the ASE algorithm discussed earlier cannot be used. Therefore, we expect that a higher-order CMAC with a nonlinear basis function is a viable solution to both the ambient convolutive noise term inherent to the environment and the traditional additive noise term. Experimental results, which we present in the next section, demonstrate not only the effectiveness of the MCMAC system when coupled with an ASE in the treatment of the convolutive nature of the corrupted speech, but also the robustness of the process as a viable candidate for deployment in next-generation vehicular communication systems.

The SNR_{priori} and the $SNR_{posteriori}$ values were computed from signal variances for the address indices of the MCMAC memory. In the training mode, Equations III.9.12 and III.9.13 were used to derive the weights for the Weiner filter. These weights were then stored into the MCMAC memory. In the recall mode, all the memory elements, pointed by the SNR_{priori} and the $SNR_{posteriori}$ as address indices, are added together with respect to the neighborhood basis function.

In Figure III.9.16 we show the proposed speech enhancement system using an MCMAC, in which the Weiner filter is constantly updated.[20] The variance of the signal-to-noise ratio is used as the

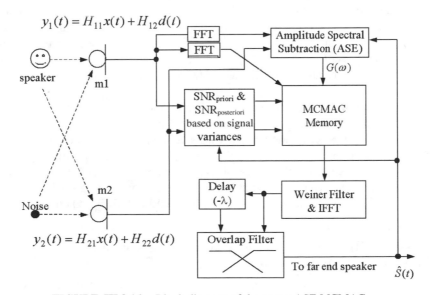

FIGURE III.9.16 Block diagram of the stereo ASE-MCMAC.

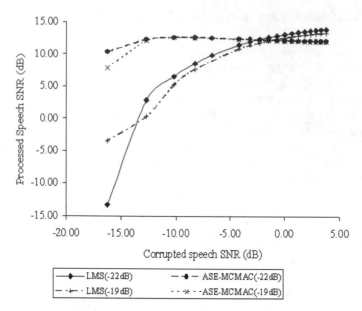

FIGURE III.9.17   Plot of the processed speech SNR vs. corrupted speech SNR for ASE-MCMAC and LMS algorithm using stereo microphone.

address index to the MCMAC memory and the third dimension of the MCMAC memory is the frequency point.

9.6.1   Unsupervised Learning Stage of the ASE-MCMAC

The ASE algorithm as configured in Figure III.9.16 operates in the frequency domain since we need to compute the spectra used in SNR values of Equations III.9.10 to III.9.13. The profile of the Weiner filter weights is obtained from Equations III.9.22 through III.9.25 and stored in the MCMAC memory. These new or updated values from the MCMAC memory are read and used as the Weiner filter weights in estimating the de-noised speech without any musical artifacts. This enhanced speech is subsequently used to calculate SNR values needed in the ASE module and the SNR variance module. This closed loop nature of the ASE-MCMAC allows the system to employ an unsupervised learning.

In the recall mode, however, the information needed for the computation of address indices are all in time domain, except for the processing of the corrupted signal by the Weiner filter; the latter needs to be undertaken in the frequency domain. Therefore, a number of FFTs are omitted to shorten the processing time. Additional savings are achieved from the frame-based updating of the Wiener filter coefficients, as opposed to the sample-by-sample computations used in the LMS-based schemes. Our frame-by-frame approach is consistent with the subsequent speech coding stage, which is normally an analysis-by-synthesis procedure including LPC, RTP, CELP, MELP, and others.

9.6.2   Result in Training and Recalling the Noise Spectral for Speech Enhancement

A two-microphone system was used to measure the performance of the ASE–MCMAC algorithm in comparison with the LMS algorithm. As shown by Figure III.9.16, both inputs to microphones m1 and m2 were simulated using prerecorded speech and noise signal from a vehicular environment. The

SNR of both microphone m1 and m2 were varied, and results of the processed speech using the ASE–MCMAC and the LMS algorithm are shown in Figure III.9.17. The performance of ASE–MCMAC is similar to that of the ASE algorithm. At low input SNR the LMS algorithm fails to remove the noise data, but the ASE–MCMAC algorithm performs very well provided the SNR of the reference input into microphone m2 is less than the SNR of microphone m1. A noise-only signal at m2 will be ideal, but in practice this is not the case. The advantage of the ASE–MCMAC over the classical ASE approach is that once the noise spectral has been trained, only a recall process

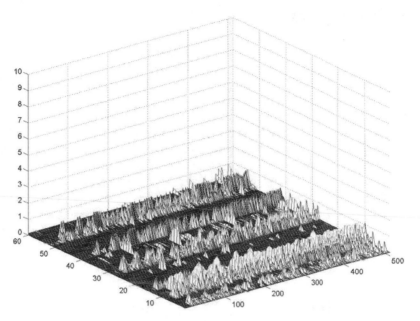

FIGURE III.9.18   CMAC profile for the Weiner filter.

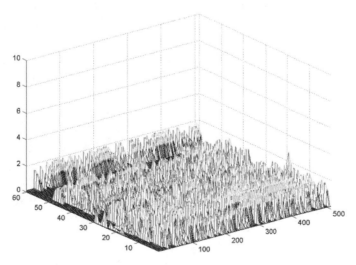

FIGURE III.9.19   MCMAC profile of the Weiner filter.

will be required to process the corrupted speech. In addition, during training the corrupted speech is also processed at the same time.

Figures III.9.18 and, III.9.19 show the plot of the Weiner filter profile using both the classical CMAC and the modified CMAC algorithm, respectively. Notice from Figure III.9.18 that only some of the CMAC memory was trained, while for the MCMAC case almost all the CMAC memory were trained during the learning cycle. Thus, no further training is required for a fully trained CMAC memory unless the profile of the Weiner filters changes drastically.

9.7 The Hybrid Fuzzy Neural Method

Fuzzy neural networks are hybrid systems that possess the advantages of both neural networks and fuzzy systems. The integration of fuzzy systems and neural networks combines the human inference style and natural language description of fuzzy systems with the learning and parallel processing of neural networks. There are numerous approaches to integrate fuzzy systems and neural networks.[21–28] An extensive bibliography on fuzzy neural network can be found in Reference 29.

Zadeh suggested the use of fuzzy logic as a framework to manage vagueness and uncertainty.[30] When modeling vagueness, fuzzy predicates without well-defined boundaries concerning the set of objects may be applied. The rationale for using fuzzy logic is that the denotations of vague predicates are fuzzy sets rather than probability distributions. In many situations, vagueness and uncertainty are simultaneously presented since any precise or imprecise fact may be uncertain as well. Fuzzy set and possibility theories provide a unified framework to deal with vagueness and uncertainty.[31] In this section, the pattern of approximate reasoning in pseudo outer-product fuzzy neural network using the compositional rule of inference and a Singleton fuzzifier called *POPFNN-CRI(S)* is described in detail.

9.7.1 The POPFNN-CRI(S)

The proposed Singleton pseudo outer-product based fuzzy neural network (POPFNN-CRI(S)) is developed on the basis of possibility theory and the fuzzy compositional rule of inference. The structure of POPFNN-CRI(S) resembles the neural-network-like structure of POPFNN-TVR,[19] but has strict correspondence to the inference steps in fuzzy rule-based systems that employ the compositional rule of inference using standard fuzzy operators and Singleton fuzzifiers.

9.7.1.1 Architecture of POPFNN-CRI(S)

The proposed POPFNN-CRI(S) architecture for a multi-input multi-output (MIMO) system is a five-layer neural network as shown in Figure III.9.20. For simplicity, only the interconnections for the output y_m are shown.

Each layer in POPFNN-CRI(S) performs a specific fuzzy operation. The inputs and outputs of the POPFNN-CRI(S) are represented as nonfuzzy vector $\mathbf{X}^T = [x_1, x_2, \ldots x_i, \ldots x_{n1}]$ and nonfuzzy vector $\mathbf{Y}^T = [y_1, y_2, \ldots y_l, \ldots y_{n5}]$, respectively. Fuzzification of the input data and defuzzification of the output data are respectively performed by the condition and output linguistic layers, while the fuzzy inference is collectively performed by the rule-base and consequence layers. The number of neurons in the condition and rule-base layers are defined in Equations III.9.26–III.9.28, respectively. A detailed description of the functionality of each layer is given as follows:

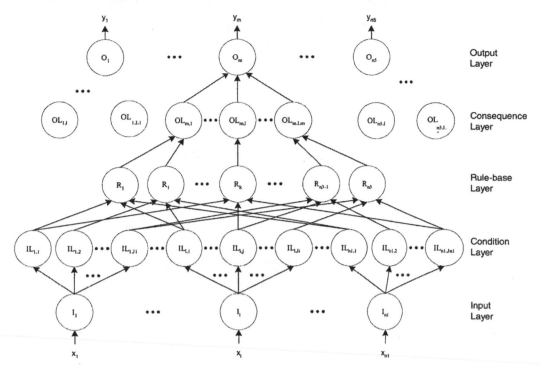

FIGURE III.9.20 Structure of POPFNN-CRI(S).

$$n2 = \sum_{i=1}^{n1} Ji \tag{III.9.26}$$

$$n4 = \sum_{m=1}^{n5} Lm \tag{III.9.27}$$

$$n3 = \left[\prod_{i=1}^{n1} Ji \right] \times \left[\prod_{m=1}^{n5} Lm \right] \tag{III.9.28}$$

where Ji = number of linguistic labels for the ith input, Lm = number of linguistic labels for the mth output, $n1$ = number of inputs, $n2$ = number of neurons in the condition layer, $n3$ = number of rules or rule-base neurons, $n4$ = number of linguistic labels for the output, and $n5$ = number of outputs.

9.7.1.2 Input Linguistic Layer

Neurons in the input linguistic layer are called input linguistic nodes. Each input linguistic node I_i represents an input linguistic variable of the corresponding nonfuzzy input x_i. Each node transmits

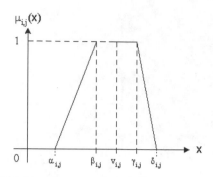

FIGURE III.9.21   Trapezoidal-shaped membership function.

the nonfuzzy input directly to the condition layer. The net input and output of an input linguistic node are defined in Equation III.9.29:

$$\text{Net input: } f_i^I = x_i$$

$$\text{Net output: } o_i^I = f_i^I \tag{III.9.29}$$

where x_i = value of the ith input.

9.7.1.3   Condition Layer

Neurons in the condition layer are called input-label nodes. Each input-label node $IL_{i,j}$ represents the jth linguistic label of the ith linguistic node from the input layer. The input-label nodes constitute the antecedent of the fuzzy rules. Each node is represented by a trapezoidal membership function $\mu_{i,j}(x)$ described by a fuzzy interval formed by four parameters $(\alpha_{i,j}, \beta_{i,j}, \gamma_{i,j}, \delta_{i,j})$ and a centroid $v_{i,j}$ as shown in Figure III.9.21. This fuzzy interval is also known as a trapezoidal fuzzy number.[26]

The subinterval $[\beta_{i,j}, \gamma_{i,j}]$, where $\mu_{i,j}(x) = 1$, is called the kernel of the fuzzy interval, and the subinterval $[\alpha_{i,j}, \delta_{i,j}]$ is called the support. When the kernel reduces to a single point, the fuzzy interval becomes a triangular fuzzy number.[26] The semantic interpretation of this representation is as follows:

1. The linguistic term is totally compatible for values of x between $\beta_{i,j}$, $\gamma_{i,j}$.
2. The linguistic term is incompatible with values smaller than $\alpha_{i,j}$ or greater than $\delta_{i,j}$.
3. Between $\alpha_{i,j}$ and $\beta_{i,j}$, the degree of compatibility increases linearly.
4. Between $\gamma_{i,j}$ and $\delta_{i,j}$, the degree of compatibility decreases linearly.

The net input and net output of an input-label node are given in Equation III.9.30:

$$\text{Net input: } f_{i,j}^{II} = o_i^I$$

$$\text{Net output: } \alpha_{i,j}^{II} = \begin{cases} 0 & \text{if } f_{i,j}^{II} < \alpha_{i,j}^{II} \text{ or } f_{i,j}s^{II} > \delta_{i,j}^{II} \\ \frac{\alpha_{i,j}^{II} - f_{i,j}^{II}}{\alpha_{i,j}^{II} - \beta_{i,j}^{II}} & \text{if } \alpha_{i,j}^{II} \le f_{i,j}^{II} \le \beta_{i,j}^{II} \\ 1 & \text{if } \beta_{i,j}^{II} \le f_{i,j}^{II} \le \gamma_{i,j}^{II} \\ \frac{\delta_{i,j}^{II} - f_{i,j}^{II}}{\delta_{i,j}^{II} - \gamma_{i,j}^{II}} & \text{if } \gamma_{i,j}^{II} \le f_{i,j}^{II} \le \delta_{i,j}^{II} \end{cases} \tag{III.9.30}$$

Where $[\alpha_{i,j}^{II}, \delta_{i,j}^{II}]$ = kernel of the fuzzy interval for the jth linguistic label of the ith input, $[\beta_{i,j}^{II}, \gamma_{i,j}^{II}]$ = support of the fuzzy interval for the jth linguistic label of the ith input, and o_i^I = output of ith input node.

9.7.1.4 Rule-Base Layer

Neurons in the rule-base layer are called rule nodes. Each rule node R_k represents a fuzzy if-then rule. The net input and output of a rule node are given in Equation III.9.31:

$$\text{Net input: } f_k^{III} = \min(o_{i,j}^{II})$$

$$\text{Net output: } o_k^{III} = f_k^{III} \tag{III.9.31}$$

where $o_{i,j}^{II}$ = output of the input-label node $IL_{i,j}$ that forms the antecedent conditions for the ith input to the kth fuzzy rule R_k.

9.7.1.5 Consequence Layer

Neurons in the consequence layer are called output-label nodes. The output-label node $OL_{m,l}$ represents the lth linguistic labels of the output y_m. The net input and output of the output-label node $OL_{m,l}$ are given in Equation III.9.32:

$$\text{Net input: } f_{m,l}^{IV} = \max_k(\alpha_k^{III})$$

$$\text{Net output: } o_{m,l}^{IV} = f_{m,l}^{IV} \tag{III.9.32}$$

where o_k^{III} = output of the rule node R_k whose consequence is $OL_{m,l}$.

9.7.1.6 Output Linguistic Layer

The neurons in the output linguistic layer are called output linguistic nodes. The output linguistic node O_m represents the output linguistic value of the output y_m. The net input and output of the output linguistic node O_m are given in Equation III.9.33:

$$\text{Net input: } f_m^V = \begin{cases} \sum_{l=1}^{L(m)} (v_{m,l}^{IV} \times (\gamma_{m,l}^{IV} - \beta_{m,l}^{IV}) \times o_{m,l}^{IV}) & \text{if } \gamma_{m,l}^{IV} > \beta_{m,l}^{IV} \\ \sum_{i=1}^{L(m)} (v_{m,l}^{IV} \times o_{m,l}^{IV}) & \text{if } \gamma_{m,l}^{IV} = \beta_{m,l}^{IV} \end{cases}$$

$$\text{Net output: } o_m^V = \begin{cases} \dfrac{f_m^V}{\sum_{l=1}^{L(m)} (v_{m,l}^{IV} \times (\gamma_{m,l}^{IV} - \beta_{m,l}^{IV}))} & \text{if } \gamma_{m,l}^{IV} > \beta_{m,l}^{IV} \\ \dfrac{f_m^V}{\sum_{l=1}^{L(m)} (v_{m,l}^{IV})} & \text{if } \gamma_{m,l}^{IV} = \beta_{m,l}^{IV} \end{cases} \tag{III.9.33}$$

where $v_{m,l}^{IV}$ = the centroid of the output-label node $OL_{m,l}$, and $\beta_{m,l}^{IV} - \gamma_{m,l}^{IV}$ = the width of the membership function for output-label node $OL_{m,l}$.

An illustrative example of the inference process with two inputs and a single output system is given in Figure III.9.22. The two fuzzy if-then rules used are given by Equation III.9.34:

$$\text{If } x_i \text{ is } X_{1,1} \text{ and } x_2 \text{ is } X_{2,1} \text{ then } y \text{ is } Y_{1,1}$$

$$\text{If } x_1 \text{ is } X_{1,2} \text{ and } x_2 \text{ is } X_{2,2} \text{ then } y \text{ is } Y_{1,2} \tag{III.9.34}$$

When the values of x_1 and x_2 are presented to the singleton POPFNN-CRI, they are respectively fuzzified into fuzzy singletons $x_1 = X_1'$ and $x_2 = X_2'$ by the input linguistic layer (Figure III.9.22(a)). The condition layer then determines the degrees of compatibility $c_1(X_1')c_1(X_2')c_2(X_1')c_2(X_2')$ of the given facts and the antecedents of the rules. Next, the following rule-base layer determines the firing strengths r_1 and r_2 based on the total compatibility of the facts with the antecedents of each rule (Figure III.9.22(b)). The consequence layer then determines the firing strengths $s_{1,1}$ and $s_{1,2}$ for the consequences $Y_{1,1}'$ and $Y_{1,2}'$, respectively, from r_1 and r_2 (Figure III.9.22(c)). Finally, the output linguistic layer integrates the consequences $Y_{1,1}'$ and $Y_{1,2}'$ into Y_1' (Figure III.9.22(d)). The output layer also performs the defuzzification of Y_1' to yield the real output value y_1.

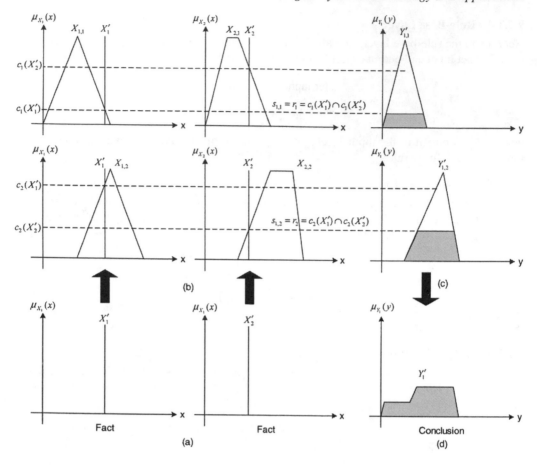

FIGURE III.9.22   An example of inference using POPFNN-CRI(S).

9.8   Learning Process of POPFNN-CRI(S)

The learning process of POPFNN-CRI(S) consists of only two phases: the fuzzy membership learning and the POP learning. Similar to the POPFNN-TVR architecture,[27] a self-organizing type of learning algorithm is employed in the first phase to determine the membership functions of the condition and consequence layers. The difference between the learning process of POPFNN-CRI(S) and POPFNN-TVR is that the former only requires two phases of learning, while the latter requires an additional supervised learning to adjust the membership functions. The former uses two novel membership-learning algorithms, known as pseudo fuzzy Kohonen partition (PFKP) and fuzzy Kohonen partition (FKP)[32] to determine the membership functions of the condition and consequence layers. A detailed description of the two-phase learning process is presented in the following section.

9.8.1   Fuzzy Membership Learning Algorithms

Fuzzy membership information is stored in the condition and consequence layers of the POPFNN-CRI(S). These membership functions can be identified and stored in the POPFNN-CRI(S) from relevant information contained in the training data. The fuzzy Kohonen partition (FKP) and pseudo fuzzy Kohonen partition (PFKP) algorithms[32] can be employed in POPFNN-CRI(S) to train its condition and consequence layers to derive the membership functions of the input and output variables. The difference between FKP and PFKP is that the latter produces pseudo fuzzy partitions while the former only produces fuzzy partitions. The former is a supervised learning algorithm, while the latter is unsupervised. Both the FKP and PFKP algorithms are described as follows.

9.8.1.1 Fuzzy Kohonen Partition Algorithm

Step 1: Define c as the number of classes, $\lambda \leq \frac{1}{\Omega}$ as the learning constant, η as the learning width, and a small positive number ε as a stopping criterion; where $\Omega =$ number of data vectors in a cluster, $n =$ total number of data vectors.

Step 2: Initialize the training iteration $T = 0$ and the weights $v_i^{(0)}$ with Equation III.9.35:

$$v_i^{(0)} = \min_k(x_k) + \frac{i + \frac{1}{2}}{c}(\max_k(x_k) - \min_k(x_k)) \text{ for } i = 1 \ldots c, k = 1 \ldots n \quad \text{(III.9.35)}$$

Step 3: Initialize $v_i^{(T+1)} = v_i^{(T)}$ for $i = 1 \ldots c$

Step 4: For $k = 1 \ldots n$:

    a. *Determine the ith cluster that the data x_k belongs to from the training data.*

    b. *Update weights v_i of the ith cluster with Equation III.9.36:*

$$v_i^{(T+1)} = v_i^{(T)} + \lambda(x_k - v_i^{(T)}) \quad \text{(III.9.36)}$$

Step 5: Compute $e^{(T+1)}$ using Equation III.9.37:

$$e^{(T+1)} = \sum_{k=1}^{n} |x_k - v_i^{(T+1)}| \quad \text{(III.9.37)}$$

Step 6: Compare $e^{(T+1)}$ and $e^{(T)}$ where $e^{(0)} = 0$, using Equation III.9.38:

$$de^{(T+1)} = e^{(T+1)} - e^{(T)} \quad \text{(III.9.38)}$$

Step 7: If $de^{(T+1)} \leq \varepsilon$, stop; otherwise repeat steps 3–7 for $T = T + 1$.

Step 8: Initialize $\alpha_i = \beta_i = \delta_i = \gamma_i = \varphi_i = v_i^{(T+1)}$ for $i = 1 \ldots c$

Step 9: For $k = 1 \ldots n$

    a. *Determine the ith cluster that the data x_k belongs to from the training data.*

    b. *Update pseudo weights φ_i of the ith cluster using Equation III.9.39):*

$$\varphi_i = \varphi_i + \eta(x_k - \varphi_i) \quad \text{(III.9.39)}$$

    c. *Update the four points of the trapezoidal fuzzy number $(T_r FN)$ with Equation III.9.40:*

$$\begin{aligned}
\alpha_i &= \min(\alpha_i, x_k) \\
\beta_i &= \min(\beta_i, \varphi_i) \\
\delta_i &= \max(\delta_i, \varphi_i) \\
\gamma_i &= \max(\gamma_i, x_k)
\end{aligned} \quad \text{(III.9.40)}$$

9.8.1.2 Pseudo Fuzzy Kohonen Partition Algorithm

Step 1: *Define c as the number of classes, $\lambda \leq \frac{1}{\Omega}$ as the learning constant, η as the learning width, and a small positive number ε as a stopping criterion where $\Omega =$ number of data vectors in a cluster, $n =$ total number of data vectors.*

Step 2: *Initialize the training iteration $T = 0$ and the weights $v_i^{(0)}$ with Equation III.9.41:*

$$v_i^{(0)} = \min_k(x_k) + \frac{i + \frac{1}{2}}{c}(\max_k(x_k) - \min_k(x_k)) \text{ for } i = 1 \ldots c, k = 1 \ldots n \quad \text{(III.9.41)}$$

Step 3: *Initialize $v_i^{(T+1)} = v_i^{(T)}$ for $i = 1 \ldots c$*

Step 4: *For $k = 1 \ldots n$:*

  a. *Find the winner using Equation III.9.42:*

$$|x_k - v_i^{(T+1)}| = \min_j(|x_k - v_j^{(T+1)}|) \text{ for } j = 1 \ldots c \qquad \text{(III.9.42)}$$

  b. *Update weights of the winner i with Equation III.9.43:*

$$v_i^{(T+1)} = v_i^{(T)} + \lambda(x_k - v_i^{(T+1)}) \qquad \text{(III.9.43)}$$

Step 5: *Compute $e^{(T+1)}$, where x_k belongs to the ith cluster, using Equation III.9.44:*

$$e^{(T+1)} = \sum_{k=1}^{n} |x_k - v_i^{(T+1)}| \qquad \text{(III.9.44)}$$

Step 6: *Compare $e^{(T+1)}$ and $e^{(T)}$ where $e^{(0)} = 0$, using Equation III.9.45:*

$$de^{(T+1)} = e^{(T+1)} - e^{(T)} \qquad \text{(III.9.45)}$$

Step 7: *If $de^{(T)} \leq \varepsilon$, stop; otherwise repeat steps 3–7 for next $T = T + 1$.*

Step 8: *Initialize $\alpha_i = \beta_i = \delta_i = \gamma_i = \varphi_i = v_i^{(T+1)}$ for $i = 1 \ldots c$*

Step 9: *For $k = 1 \ldots n$*

  a. *Find the winner using Equation III.9.46:*

$$|x_k - \varphi_i| = \min_j(|x_k - \varphi_j|) \text{ for } j = 1 \ldots c \qquad \text{(III.9.46)}$$

  b. *Update pseudo weights of the winner i with Equation III.9.47:*

$$\varphi_i = \varphi_i + \eta(x_k - \varphi_i) \qquad \text{(III.9.47)}$$

  c. *Update the four points of the trapezoidal fuzzy number (T_rFN) with Equation III.9.48:*

$$\alpha_i = \begin{cases} \min(\alpha_i, x_k) & \text{for } i = 1 \\ \delta_{i-1} & \text{for } i > 1 \end{cases}$$

$$\beta_i = \min(\beta_i, \varphi_i)$$

$$\delta_i = \max(\delta_i, \varphi_i) \qquad \text{(III.9.48)}$$

$$\gamma_i = \begin{cases} \max(\gamma_i, x_k) & \text{for } i = c \\ \beta_{i+1} & \text{for } i < c \end{cases}$$

9.8.2   Pseudo Outer-Product Learning Algorithm

After the membership functions of the condition and consequence layers have been identified, the pseudo outer product (POP) learning algorithm[19] is employed to identify the fuzzy rules. The POP learning algorithm is a simple one-pass learning algorithm. The algorithm is easy to comprehend, as it coincides with intuitive ways of identifying relevant rules.

In POPFNN–CRI(S), each node in the condition and consequence layers represents a linguistic label once the membership functions have been identified. Under the POP learning algorithm, the set of training data $\{\mathbf{X^p}, \mathbf{Y^p}\}$, where $\mathbf{X^p}$ is the input vector and $\mathbf{Y^p}$ is the output vector, is simultaneously fed into both the input linguistic and output linguistic layers. The membership values of each input-label node o^{II} are then determined. These values are subsequently used to compute the firing strength f^{III} of the rule nodes in the rule-base layer. Similarly, the membership values of each output-label node are determined by feeding the output value back from the output layer to the consequence layer. The weights of the consequence layer linking the rule-base layer are then determined using Equation III.9.49:

$$w_{k,m,l} = \sum_{p=1}^{n} f_k^{III}(\mathbf{X^p}) \times \mu_{m,l}(y_m^p) \qquad \text{(III.9.49)}$$

where $w_{k,m,l}$ = weight of the link between the kth rule node and the ith linguistic label for the mth output, $\mu_{m,l}(y_m^p)$ = firing strength of kth rule node when presented with input vector $\mathbf{X^p}$, and $f_k^{III}(\mathbf{X^p})$ = membership value of the mth output of $\mathbf{Y^p}$ with the fuzzy subset $Y_{m,l}$ that semantically represents the lth linguistic label of the mth output.

The weights in Equation III.9.49 are initially set to zero. After performing POP learning, these weights represent the strength of the fuzzy rules having the corresponding output-label nodes as their consequences. Among the links between a rule node and the output-label nodes, the link with the highest weight is chosen and the rest are deleted. The links with zero weights to all output-label nodes are also deleted. The remaining rule nodes after this link selection process subsequently represent the rules used in the POPFNN-CRI(S).

The functions performed by each layer in the POPFNN-CRI(S) correspond strictly to the inference steps in the composition rule of inference method using standard t-norm and fuzzy relation. This gives the proposed POPFNN-CRI(S) a strong theoretical foundation. The learning process of the proposed POPFNN-CRI(S) consists of only two phases. In the first phase, two novel fuzzy membership identification algorithms called the fuzzy Kohonen partition (FKP) and pseudo Kohonen partition (PFKP)[32] are used to identify the fuzzy membership functions of the input and output linguistic labels. In the second phase, a novel one-pass rule-identification algorithm called POP learning[19] is used to identify the fuzzy rules. The proposed two-phase learning algorithm can effectively construct membership functions and identify the fuzzy rules without the need of any supervised training to tune the membership functions.

9.8.3   Experimental Results

Experiments were conducted to study the performance of the POPFNN-CRI(S). In the experiment the input signal-to-noise ratio were varied from 0 to 13 dB. The Pearson product moment correlation coefficient (R^2) was used to compare two columns of data. The relationship between the two datasets was expressed from zero to one, zero being the least correlated and one being the actual match. A correlation of 0.7–0.9 is considered good, while that above 0.9 is considered excellent.

The timing required for learning and recalling processes was measured in term of seconds. Even if the R^2 result is excellent, the system may not have any usefulness if the recalling process takes a long time before it is available. On the other hand, the time taken for the learning process is of least

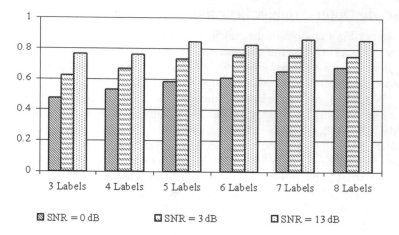

FIGURE III.9.23   Performance plot of the POPFNN-CRI(S) using different number of linguistic labels and the input SNR.

importance, since POPFNN-CRI(S) only needs to train once and can be repeatedly used until the plant characteristic changes.

9.8.4   Learning and Recalling Results

Figure III.9.23 shows the result when different linguistic labels and noise levels were used. From the graph, the performance of the POPFNN-CRI(S) improved with increasing SNR and also increasing the number of input labels. There were no significant improvements made between 7 linguistic labels and 8 linguistic labels used. In fact, the output correlation between 3 linguistic labels and 7 linguistic labels shows an increase in the output correlation of 15%. Notice also (Figure III.9.23) an increase in the SNR of 13 dB improves the correlation output by 20%.

Linguistic labels define the number of fuzzy areas that will be classified. Thus, increasing the linguistic labels will increase the number of fuzzy areas to be classified and more rules will be generated. In the case of the POPFNN-CRI(S), there will be more rules to compare and this results in higher accuracy. However, there is a limit to this, as shown from the results in Figure III.9.23. The output correlation between using 7 and 8 linguistic labels produces no significant change. This may be due to the maximum fuzzification area that can be defined by POPFNN-CRI(S) based on the data provided.

9.8.5   Timing for Learning and Recalling Process

Figure III.9.24 shows the recall timing results using the POPFNN-CRI(S) algorithm. Notice that when 8 linguistic labels were used, the recalling process took 554 sec. This is more than 9 min, which is far too long for most signal processing to be carried out on-line. The results were based on a Pentium II processor running at a speed of 300 MHz. In fact, if the number of linguistic labels were reduced to 5 (Figure III.9.24), the recall timing decreases tremendously, to less than 50 sec. Thus, the accuracy required for the system determines the number of linguistic labels required; but as the number of labels increases, both the learning and recall timing will also increase exponentially.

In the experiment it is also found that changes to the learning constant, pseudo learning constant, and maximum total error have little effect on the training and recall timing when the number of linguistic labels used is small. A data size of one million samples was used to train the POPFNN-CRI(S) network and later recall to test the output correlation. A learning constant = 0.02, pseudo learning constant = 0.01, maximum error = 0.0005, and linguistic labels = 3 were used to carried

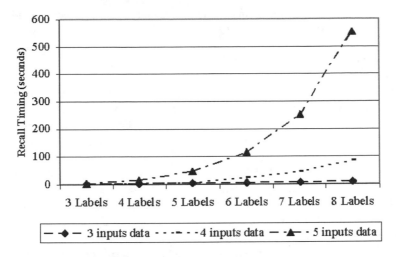

FIGURE III.9.24 POPFNN-CRI(S) timing results.

out the experiment. The POPFNN-CRI(S) took 24 min, 30 sec to learn and 36 min, 8 sec to recall. For 4 linguistic labels, it took 31 min to learn and 43 min, 55 sec to recall. This is again based on the similar system setup using a Pentium II processor at a clock speed of 300 MHz.

9.9 Conclusion

This chapter examined the different techniques used in noise modeling. Classical techniques applying both the time domain and the frequency domain methods have been discussed in great detail. Both the LMS and the ASE methods require a known estimate of the noise signal, either in the time domain or as an energy spectrum, respectively. To date there is no known algorithm that can perform accurate noise modeling in a short time frame to enable on-line cancellation of a corrupted speech.

This leads to the investigation using associative memory map and novel fuzzy neural network approaches. Both approaches demonstrated greater improvement over the existing technique. In addition, they allow the decision process to be impeccable and be made meaningful. The experimental results have clearly demonstrated that the novel AI technique is as good as or sometimes better and easier to implement in practice.

References

1. Succi, G.P., The interior acoustic field of an automobile cabin, *J. Acoust. Soc. Am.*, vol. 81(6) pp. 1688, 1987.
2. Houdt, J.J., Geoman, Th., Breugel, P.A., and Springborn, M., Influence of road surface on traffic noise: a comparison of mobile and stationary measuring techniques, *Noise Control Engineering Journal*, 41(3), 365, 1993
3. Olson, H., Electronic control of noise, vibration, and reverberation, *J. Acoust. Soc. Am.* 28, 966, 1956.
4. Messerschmitt, D., Hedberg, D., Cole, C., Haoui, A., and Winship, P., Digital voice echo canceler with a TMS320C20, in *DSP Applications*, K.-S. Lin, Ed., Prentice-Hall, 1987.
5. Harrison, W.A., Lim, J.S., and Singer, K., (1986). A new application of adaptive noise cancellation, *IEEE Trans. Acoust., Speech and Signal Processing*, 34(1), (21–27).

6. Miller, T.E. and Barish, J., Optimizing sound for listening in the presence of road noise, *Int. Conf. Signal Processing Appl. Tech.*, 1, 97, 1993.

7. Abdul, W., Tan, E.C., and Abut, H., Noise suppression on the interior acoustic field of the vehicular chamber, *Proc. 5th Int. Conf. Automation, Robotics and Computer Vision (ICARCV '98)*, 1703, 1998.

8. Kraft, L.G. and Campagna, D.P., A Comparison between CMAC neural network control and two traditional adaptive control systems, *IEEE Control Systems Magazine*, 36, 1990.

9. Lim, J.S., Evaluation of a correlation subtraction method for enhancing speech degraded by additive white noise, *IEEE Trans. Acoustic, Speech and Signal Processing*, 26, 471, 1978.

10. Berouti, M., Schwartz, R., and Makhoul, J., Enhancement of speech corrupted by acoustic noise, *Proc. IEEE Int. Conf. Acoustic, Speech and Signal Processing*, 208, 1979.

11. Curtis, R.A. and Niederjohn, R.J., An investigation of several frequency domain processing method for enhancing the intelligibility of speech in wideband random noise, *IEEE Proc. Int. Conf. Acoustics, Speech and Signal Processing*, 606, 1978.

12. Preuss, R.D., A frequency domain noise canceling preprocessor for narrow band speech communications systems, *IEEE Int. Conf. Acoustic, Speech and Signal processing*, 212, 1979.

13. Ephraim, Y. and Malah, D., Speech enhancement using minimum mean square error short-time spectral amplitude estimator, *IEEE Trans. Acoustics Speech, and Signal Processing*, ASSP-32, 6, 1109, 1984.

14. Beaugeant, C., Turbin, V., Scalart, P., and Gilloire, A., New optimal filtering approaches for hands-free telecommunication terminals, *Signal Processing*, 64(1), 33, 1998.

15. Pascal, S. and Benamar, A., A system for speech enhancement in the context of hands-free radiotelephony with combined noise reduction and acoustic echo cancellation, *Speech Communication*, 20, 203, 1996.

16. Jeannès, R., Le, B., Faucon, G., and Ayad, B., How to improve acoustic echo and noise cancelling using a single talk detector, *Speech Communication*, 20, 191, 1996.

17. Abdul, W., Tan, E.C., and Abut, H., Robust speech enhancement using amplitude spectral estimator, *Proc. IEEE ICASSP2000 Silver Anniversary*, vol. VI, 3558, 2000.

18. Zurada, J.M., *Introduction to Artificial Neural Systems*, Info Access Distribution Pte. Ltd., Singapore, 1992.

19. Quek, C. and Ng, P.W., Realisation of neural network controllers in integrated process supervision, *Int. J. Artificial Intelligence in Engineering*, 10(2), 135, 1996.

20. Abdul, W., Quek, H.C., and Tan, E. C., ASE-CMAC for speech enhancement in a vehicular environment, *Proc. SPIE 2000 – Appl. Sci. Neural Networks, Fuzzy Systems, and Evolutionary Computation III, 2000, USA*, 4120, 179, 2000.

21. Nauck, D. and Kruse, R., A neuro-fuzzy method to learn fuzzy classification rules from data, *Fuzzy Sets and Systems*, 89(3), 277, 1997.

22. Kasabov, N.K., Learning fuzzy rules and approximate reasoning in fuzzy neural networks and hybrid systems, *Fuzzy Sets and Systems*, 82(2), 135, 1996.

23. Lin, C.T., A neural fuzzy control system with structure and parameter learning, *Fuzzy Sets and Systems*, 70(2–3), 183, 1995.

24. Shann, J.J. and Fu, H.C., A fuzzy neural network for rule acquiring on fuzzy control systems, *Fuzzy Sets and Systems*, 71(3), 345, 1995.

25. Keller, J.M., Yager, R.R., and Tahani, H., Neural network implementation of fuzzy logic, *Fuzzy Sets and Systems*, 45, 1, 1992.

26. Kaufmann, A. and Gupta Madan, M., *Introduction to Fuzzy Arithmetic: Theory and Applications*, Van Nostrand Reinhold Company Inc., 1985.

27. Quek, H.C. and Zhou, R.W., POPFNN: a pseudo outer-product based fuzzy neural network, *Neural Networks*, 9(9), 1569, 1996.

28. Yager, R.R., Modeling and formulating fuzzy knowledge bases using neural network, *Neural Networks*, 7(8), 1273, 1994.

29. Buckley, J.J. and Hayashi, Y., Neural nets for fuzzy systems, *Fuzzy Sets and Systems*, 71(3), 265, 1995.

30. Zadeh, L.A., *A Theory of Approximate Reasoning, Fuzzy Sets and Applications: Selected Papers by L.A. Zadeh,* John Wiley & Sons, 1979.

31. Esteva, F., Garcia-Calves, P., and Godo, L., Relating and extending semantical approaches to possibilistic reasoning, *Int. J. Approximate Reasoning*, 10, 311, 1994.

32. Quek, H.C. and Ang, K.K., Determination of fuzzy membership function using fuzzy Kohonen partition and pseudo fuzzy Kohonen partition algorithms, submitted for journal review (undergoing revision), 1998.

Index

A

A, *see* Artifacts
Abnormal LS (ALS), **III**-105
Acoustic energy, environmental reflections of, **III**-16
Acoustic environments, impulse response in, **III**-27
Acoustic model, **III**-224, **III**-230
Acoustic resonances, frequency-dependent sustaining
    effect of, **III**-3
Acoustic transfer functions, **III**-15
Activation function, **III**-21, **III**-220
Adaptive Network-based FIS (ANFIS), **III**-126
Adaptive noise cancellation (ANC), **III**-12,
    III-32, **III**-110
    classical, **III**-15
    filter coefficient vector, **III**-112
Adaptive processing, multi-microphone sub-band, **III**-16
Adjunction property, **III**-188
Advanced neural networks, **III**-242
Affine filters, **III**-64, **III**-67
AI, *see* Artificial intelligence
Albert
    image, **III**-88, **III**-91
    optimized filters, **III**-91
Algebraic subspace, **III**-33
Alpha-stable distribution, **III**-135
ALS, *see* Abnormal LS
ANC, *see* Adaptive noise cancellation
ANC-FOS, **III**-110
    algorithm, **III**-111, **III**-113, **III**-115
    application of on real data, **III**-113
ANC-WT, **III**-115
    application of on real data, **III**-120
    efficiency of, **III**-121
ANFIS, *see* Adaptive Network-based FIS
Animation, **III**-237
ANN, *see* Artificial neural network
Anti-Hebbian learning, **III**-27, **III**-30
    based on feedback network, **III**-31
    maximum likelihood estimation, **III**-34
    memory and, **III**-40
    second-order temporal, **III**-33
    temporal Infomax learning compared with temporal
        linear, **III**-37

Anti-Hebbian model, temporal linear, **III**-29
AR, *see* Autoregressive
Artifacts (A), **III**-147
Artificial intelligence (AI), **III**-2, **III**-237
Artificial neural network (ANN), **III**-2,
    III-158, **III**-215, **III**-238
    application of to speech recognition, **III**-220
    biased, **III**-165
    as classifier, **III**-163
    generalization ability, **III**-164
    learning in, **III**-19, **III**-20
    operations modes in, **III**-19
    output vectors of, **III**-168
    paradigm, signal processing methods
        based on, **III**-26
    performance, **III**-167
    presumption of creating, **III**-18
    speech recognition systems, **III**-27
    training, **III**-164, **III**-166, **III**-167
    use of with speech recognition, **III**-215
Artificial neural networks and speech recognition,
    future of, **III**-215 to **III**-236
    ANNs combined with HMMs, **III**-225 to **III**-231
        advances in large-vocabulary continuous speech
            recognition, **III**-229 to **III**-231
        context-dependent models, **III**-226 to **III**-229
        generating probabilities used in underlying HMM
            models, **III**-226
        implementation of components
            of HMMs, **III**-225
    ANN in robust speech recognition systems, **III**-231
        to **III**-232
    artificial neural networks, **III**-217 to **III**-222
        concepts, **III**-218 to **III**-220
        ANN applied to speech recognition,
            III-220 to **III**-222
    HMMs for speech recognition,
        III-223 to **III**-225
        acoustic model, **III**-224
        language model, **III**-224
        mathematical foundation, **III**-223
        weaknesses of HMMs, **III**-225
    speech recognition, **III**-216
        to **III**-217

G

H

I

T

U